Proteins of the Cerebrospinal Fluid: Analysis and Interpretation in the Diagnosis and Treatment of Neurological Disease

Proteins of the Cerebrospinal Fluid: Analysis and Interpretation in the Diagnosis and Treatment of Neurological Disease

Second Edition

Edward J Thompson PhD MD DSc FRCPath FRCP
Institute of Neurology
The National Hospital for Neurology & Neurosurgery
University College London Hospitals
Queen Square
London WC1N 3BG
UK
e.thompson@ion.ucl.ac.uk

ELSEVIER
ACADEMIC
PRESS

AMSTERDAM • BOSTON • HEIDELBERG • LONDON
NEW YORK • OXFORD • PARIS • SAN DIEGO
SAN FRANCISCO • SINGAPORE • SYDNEY • TOKYO

This book is printed on acid-free paper

First published as *The CSF Proteins: a Biochemical Approach* 1988
This second edition published 2005

Copyright © 2005, Elsevier Ltd. All rights reserved

No part of this publication may be reproduced, stored in a retrieval system, or transmitted in any form or by any means electronic, mechanical, photocopying, recording or otherwise, without the prior written permission of the publisher

Permissions may be sought directly from Elsevier's Science & Technology Rights Department in Oxford, UK: phone: (+44) 1865 843830, fax: (+44) 1865 853333, e-mail: permissions@elsevier.co.uk. You may also complete your request on-line via the Elsevier homepage (http://www.elsevier.com), by selecting 'Customer Support' and then 'Obtaining Permissions'

Elsevier Academic Press
525 B Street, Suite 1900, San Diego, California 92101-4495, USA
http://www.elsevier.com

Elsevier Academic Press
84 Theobald's Road, London WC1X 8RR, UK
http://www.elsevier.com

British Library Cataloguing in Publication Data
A catalogue record for this book is available from the British Library

Library of Congress Cataloging in Publication Data
A catalog record for this book is available from the Library of Congress

ISBN   0-12-369369-1

Printed and bound in Great Britain
05 06 07 08 09 9 8 7 6 5 4 3 2 1

Contents

Foreword ix
Acknowledgment xiii
Introduction xv
Abbreviations xix

1 Brief historical review of CSF proteins 1
Introduction 1
Investigators of CSF proteins 3

Part 1: Normal CSF proteins, an overview 7

2 Functional glossary 9
Introduction 9
Immunoglobulin synthesis 10
Barrier terms 10

3 The roster of CSF proteins 13
Introduction 13
Amounts 14
Functions 15

4 Differences between proteins in CSF and serum 33
Introduction 33
Quantitative 34
Qualitative 38

5   Different blood–CSF barriers 43

Introduction 43
Different sources of CSF proteins 43
Conclusion 63

6   Qualitative versus quantitative analysis 65

Introduction 65
Conclusion 79

Part 2: Normal CSF proteins, detailed discussion   81

7   Methodologies and their limitations 83

Introduction 83
General methods 83
Further examples of the discrepancy problem 87
Radiolabeled proteins 99

8   Fitting the data to curves 105

Introduction 105
Mathematics of the multiple sources of CSF proteins 105
Conclusion 124

9   Normal blood–CSF barrier values 127

Introduction 127
Expressions of values for normal data 127
Degrees of biologic variability 128
Critique of CSF immunoglobulin levels 137

Part 3: Individual diseases, and the future   139

10   CNS immunoglobulin synthesis 141

Introduction 141
Efficient diagnosis 142
Differential diagnosis 144
Specific antigens 146
Technology 150

Serum source of CSF antibodies 161
Conclusion 166

11 Non-immunoglobulin proteins 167

Introduction 168
Serum proteins modified within the CNS 168
General problems of any protein index or quotient 181
Conclusion 183

12 Monitoring therapy 185

Multiple sclerosis 185
Neurosyphilis 187
Tuberculous meningitis 188
CNS leukemia 190
Hydrocephalus 190
Herpes simplex encephalitis 190

13 Prognostic protein levels 193

Introduction 193
Multiple sclerosis 193
Optic neuritis 196
Meningitis 197
Guillain–Barré syndrome 197
Head injury 197
Other diseases with possibly treatable and/or infectious origins 197

14 Future growth areas 201

Introduction 201
Clinical 202
Biochemical 202
Technology 203
Conclusion 206

Part 4: Appendices   207

15 Appendices of methods 209

Methodology notes 209
Appendix 1: Separation 209

Appendix 2: Estimation 214
Appendix 3: Specific methods 215

16 Cross-index of references 241

References 251

Index 325

Foreword

In this book, Professor Thompson, an MD and DSc (neurochemistry), has continued to focus on cerebrospinal fluid (CSF) proteins by organizing information and knowledge on the nature of CSF proteins in health and disease, as well as the best methods to detect them; in addition, he has integrated his decades of experience and discoveries. This makes for an encyclopedia type of book for laboratory directors, experimentalists, lecturers, writers and newcomers.

An important message of the book is his defense of his method to identify oligoclonal IgG bands, the most sensitive and specific test to support the clinical/MRI (head and spinal cord) diagnosis of multiple sclerosis (MS). It is my opinion that his method to detect oligoclonal IgG bands will replace the silver technique we standardized and many clinical laboratories use today. This new technique has been approved by the Federal Drug Administration (FDA) (USA) thanks to Mark Freedman, Ottawa, Canada, and marketed by Helena Laboratories Company (www.helena.com). It is a consensus that oligoclonal IgG bands are a gold standard test 'unique' to MS patients, so clinic laboratories will adopt Professor Thompson's technique as recommended by the FDA.

Of great importance is an MS CSF consensus meeting organized by Professor Thompson at the National Hospital for Neurology, Queen Square in London, United Kingdom, in May 2002. The invitees were experienced clinical CSF laboratory directors with extensive experience with MS CSF, other neurological diseases and normals. The names of invitees (alphabetically) were:

Florian Deisenhammer MD
Mark S. Freedman MD, FRCPC
Gavin Giovannoni MD, PhD
Guy Grimsley PhD
Geoffrey Keir PhD
Sten Ohman PhD
Michael K. Racke MD
Hans-Otto Reiber PhD
Finn Sellebjerg DMSci

Mohammad Sharief MD, PhD
Christian Sindic MD, PhD
Edward J Thompson MD, DSc
Wallace W Tourtellotte MD, PhD

The meeting was sponsored by the USA Consortium of Multiple Sclerosis Centers (www.mscare.org).

The purpose of the meeting was to serve as an extension to the recently proposed McDonald et al. (2001) Taskforce Report, Recommended Diagnostic Criteria for Multiple Sclerosis: Guidelines from the International Panel on the Diagnosis of Multiple Sclerosis [730]. McDonald et al. redefined MS diagnosis criteria utilizing symptoms and signs, MRI and a CSF profile indicative of MS. At the Consensus meeting it was agreed that the CSF part of McDonald's proposal needed a more detailed methodology and explanation for clinical CSF laboratories. The transactions of the meeting will be published in *Archives of Neurology* (2005). Further, the evidence-based information and knowledge for a CSF profile indicative of MS which will be presented in the *Archives of Neurology* article is presented in more detail in this book by Professor Thompson. Hence, this book on CSF proteins is a timely book.

I will give a brief opinion on some of the information in this text. Having witnessed the chronological assembly of progress (1942 to 1981), as did Professor Thompson, it can be seen that progress was driven by methods.

Regarding the chapters on the roster of CSF proteins and normal blood–CSF protein values, both Ig and non-Ig are current.

Differences in CSF versus serum proteins and the different blood–CSF barriers provide important information because assessment of CSF proteins can only be understood in light of the nature of individual proteins and their barriers, as well as synthesis of proteins inside the blood–brain barrier which sink into the CSF.

Professor Thompson discusses fitting data to curves, which has been championed by Professor Reiber of the University of Göttingen, Germany.

Of great interest to me is the issue of quantitative versus qualitative analysis. Professor Thompson identifies oligoclonal IgG bands by inspection of gels comparing oligoclonal IgG bands with controls to obtain the most specificity and sensitivity; we do the same with the silver technique. In my opinion, this type of quantitation satisfies Lord Kelvin's axiom:

> When you cannot measure it, when you cannot express it in numbers, you have scarcely, in your thoughts, advanced to the stage of science whatever the matter may be.

The chapter on central nervous system synthesis of immunoglobulins in particular, IgG, is an issue I have experimented on for decades. It is the basis for the presence of oligoclonal IgG bands in MS CSF. This is a very important chapter, and this chapter alone makes the book worthy of inclusion in all medical, neurological, and neuroscience libraries, as well as in the offices of clinical laboratory directors.

Monitoring therapy via CSF in neurosyphilis, meningitis, meningo-encephalitis is routine. However, in MS my research group has found only one treatment which remarkably reduces intrathecal IgG synthesis; namely, cortico-steroids.

In the Appendices Professor Thompson has detailed methods used in his CSF laboratory in a cookbook fashion; this will be well received.

There is clearly still much to be done on the nature of CSF proteins in health and disease.

Written by an experienced and disciplined neurochemist, this book is fair and balanced. It will be a classic when tested by time; it will be used as the basis to obtain competitive research funding; and it will be cited in lectures and in publications. It is an honor to be invited to write the Foreword to this book.

Wallace W. Tourtellotte MD, PhD

Acknowledgement

I remain indebted to all of my students, who continue to stretch my breadth with their youthful enthusiasm and plumb my depths with their thirst for knowledge. I must mention Drs Geoffrey Keir, Nick Davies, Rosalyn Davies, Michael Johnson, Rodney Walker and Peter Kaufmann. Among my colleagues are all those physicians and surgeons at the National, in particular Dr Peter Rudge and Professor Ian McDonald. For expert help with IT, I thank Tom Adams. I am also grateful to Dr Johannes Menzel for being such an enlightened and kind editor.

In the wider world I am indebted to many others with similar interests, including Professors Wallace Tourtellotte (who continues to stimulate, based on his vast experience) and Hans Link. Remembered with gratitude and affection are the late Bert Morton and Professors Klaus Felgenhauer and Armand Lowenthal. Finally I thank my faithful personal assistant, Miss Jan Alsop, who has helped and guided my efforts since the last edition and thus has made the task easier.

Introduction

Most previous books on CSF have been written by celebrated physiologists or clinicians. Few have been attempted by biochemists. Since a large body of information has been generated more recently on larger-sized molecules (proteins) rather than smaller molecules (ions, sugars, etc., well studied by physiologists), it was felt that a neurochemical pathologist should broach the subject of the CSF proteins. The rationale was simply that a critical rather than a catalogue approach was not yet available. Obviously individual books will have critical portions, but their primary aim has often been to be as up to date as possible. All books (including this updated monograph) are soon out of date, but the present rationale has also been to take a longer view of the subject in question. Time is the best referee. It now seems that 16 years after the earlier edition, there is much more ground to be covered. Once again there seems to have been a recent flurry of books about spinal fluid, with one of the most illuminating being that by a dear colleague, the late Klaus Felgenhauer, which was initially published in German. It was planned to translate this into English, but unfortunately due to illness he was unable to complete this task [279]. The other book which has struck a chord, although not primarily to do with spinal fluid at all, is that of Richard Johnson, a neurovirologist from Johns Hopkins [506], whose approach to his topic seemed somehow very comfortable and sympathetic to my own although we have never compared notes on such matters. In addition I share his particular slant, namely towards the molecular side, which is perhaps an easy explanation since we are both interested in the brain; however his own infrastructure relies more on virology whereas mine is based more on biochemistry. As it happens, he also had an interval of 16 years before publication of his second edition.

Following a brief historical review in Chapter 1, the monograph is divided into four mini-books:

Part 1. A mini-textbook (Chapters 2 to 6), which deals with the normal CSF proteins in a general fashion without many references being specifically discussed

Part 2. A mini-reference book (Chapters 7 to 9), which deals with normal CSF proteins but in a specific fashion, giving critical details from individual papers

Part 3. A mini-clinical book (Chapters 10 to 14), which deals with individual diseases, and looks to the future

Part 4. A mini-lab book (Appendices), which can be used at the bench, either giving some examples as a 'cookbook' or as a reference to abnormal CSF values.

The main thrust of the monograph, in terms of new material, is to put forward two hypotheses, based mainly (but not exclusively) on our data from the analysis of some 100 000 CSF samples:

1. There are multiple, discrete sources (or barriers) for the CSF proteins, rather than a single barrier which is responsible for the filtration of all CSF proteins (Chapter 5)
2. A logarithmic model gives a better fit to the quantitative analysis of IgG data (as well as alpha-2-macroglobulin) than a simple linear model fitted to the same data (Chapter 8).

There are also two practical advances which have emerged from our work over the past decade:

1. The introduction of Eastern blotting, which involves fractionation of the antibodies using their endogenous charge (isoelectric focusing) before binding to a homogenous antigen, as opposed to the more traditional Western blotting whereby the antigen is fractionated according to molecular weight and then bound to a homogenous source of antibody
2. The US Food & Drug Administration (FDA) has recently approved our technique as the 'gold standard' for the demonstration of IgG by immunofixation following isoelectric focusing.

The reader may detect a degree of repetition with this chosen outlay. It is quite intentional, using the heuristic concept that it is better to state something simply the first time, then in later context it may not only be reinforced but also embellished with more fine details. Those who are already experts in the field will hopefully forgive me, once they realize my purpose. For those who are called on to lecture about various CSF topics, much of the data is given in summary tables, which might be used as slides.

For best use of this book, the reader is particularly encouraged to focus primary attention on the tables since much of the data is not repeated in the text except to reinforce certain points. For orientation, I have given a brief summary at the beginning of each chapter.

In addition, there is a cross-index of references. This is for the expert who wishes to go directly to the seminal papers. It also saves several hundred pages of redundant text in which one simply reiterates what has already been stated in these summaries of individual papers. The catalogue of over 1300 cross-indexed references is necessary, lest one be thought unscholarly. This cross-indexed database is available on disc for Reference Manager, and can also be supplied in a delimited format.

Rather than clutter up the text with perhaps an excessive number of references (which can impede the flow of ideas), the cross-index of references (Chapter 16) should be used as a first look, and the second look should be the text index. Do not assume that the few references given in the text (as opposed to the cross-index of references) represent the entire scope of the particular subject. To reiterate: the vast majority of references are given primarily in the cross-index of references. You will miss the major benefits of the computer-aided search if you do not always consult this cross-index of references for your topic of particular interest.

Abbreviations

AlAT	Alpha-1-antitrypsin
A2M	Alpha-2-macroglobulin
Aby	Antibody
Alb	Albumin
An	Antigen
C'3	Complement protein 3 (C'1...9)
CPK	Creatine kinase
CRP	C-Reactive protein
CV	Coefficient of variation (% standard deviation/mean)
EDTA	Ethylene diamine tetra-acetic ccid
EID	Electro immunodiffusion (Laurell 'Rockets')
ELISA	Enzyme-linked immuno-sorbent assay
FDP	Fibrin degradation products
G5	Gamma 5 zone (or gamma 1...4)
Gc	Group components protein (carry vitamin D)
GFAP	Glial fibrillary acidic protein
Hb	Hemoglobin
HLA	Human lymphocyte antigen
Hp	Haptoglobin
HRP	Horseradish peroxidase
IEF	Isoelectric focusing
Ig	Immunoglobulin (IgG, A, M, D, E)
MAG	Myelin-associated glycoprotein
MOG	Major oligodendrocyte glycoprotein
NANA	N-Acetyl neuraminic acid (sialic acid)
N/C	Nitrocellulose paper
N-CAM	Neural cell adhesion molecule
NEPH	Nephelometer (light scattering)
NPH	Normal pressure hydrocephalus

Oro	Orosomucoid (alpha-1-acid glycoprotein)
PACIA	Particle-counting immuno-assay
PAGE	Poly-acrylamide gel electrophoresis
pI	Isoelectric point
PLP	Proteolipid protein
Prealb	Prealbumin (transthyretin)
PV	Population variance
RIA	Radio immuno-assay
#	Number of patients

CHAPTER

1

Brief historical review of CSF proteins

Introduction
Investigators of CSF proteins

Introduction

CSF is somewhat like a Cinderella fluid, but continuing interest is seen by the publication of several books on the subject [202, 279, 282, 903]. The brief review given in this chapter is not intended to be exclusive but rather a short catalogue of some of the present facts concerning CSF which we take for granted, as well as some implications to further our understanding of this subject. While some may take umbrage at not being included, I have opted mainly for the 'communicators', who in some cases were better at promulgating particular ideas than were the 'originators'.

By a strange quirk of fate, I now find myself in the same institution (UCLH) as the late Dr Essex-Wynter, who was a medical registrar at the Middlesex Hospital over a hundred years ago. When Quincke wrote his classical paper in September 1891 [894], he referred to the earlier paper of Essex-Wynter published in May 1891 [258] saying that he (Quincke) followed the technique of Essex-Wynter. Although people typically think of Quincke as measuring total protein and specific gravity, in fact Essex-Wynter measured not only 'albumen' and specific gravity but also sugar, chloride, and pH (alkaline). It is therefore appropriate to show a picture (Figure 1.1) of Essex-Wynter as the first man to measure proteins in human CSF.

FIGURE 1.1 Dr W Essex Wynter, the first to measure CSF proteins (albumen) from patients.

Investigators of CSF proteins

Although one can always return to the past and reawaken some of the earliest interests in the 'water of the brain', a clear start to modern biochemical analysis began in 1942 [515]. Elvin Kabat, a biochemist working at Columbia Presbyterian Hospital in New York City, showed, using the Tiselius electrophoresis apparatus, that CSF differed from serum in that prealbumin (which he termed protein 'X') was specific to spinal fluid. For these analyses, he required up to 70 ml of CSF – i.e. approximately half of a patient's total fluid volume (about 125 ml)! Hence he used mainly pooled CSF samples. The next major step, again taken by Kabat [514], was the quantitative immunochemical precipitation of albumin and IgG, thereby reducing the volume of CSF required to about 2 ml. This heralded the beginning of routine CSF analysis, and not only allowed large numbers of patients to be studied but also led to practical aids for the diagnosis of individual patients. Another early investigator of CSF was Scheid, who also performed electrophoresis in 1944 (see [268]).

Although many previous workers had performed electrophoresis on a micro-scale using paper [46, 180], it was Armand Lowenthal, a neurologist working in the Institute Bunge in Berchem-Antwerp, Belgium, who wrote the classic book *Agar Gel Electrophoresis in Neurology* [671]. This plus his seminal paper [677] focused attention on the CSF in MS, showing that the presence of bands in the gamma region was an important clue to the pathogenesis of the disease. Yet again, a technique of qualitative analysis would lead to practical diagnosis. The careful work of Denise Karcher (with Lowenthal) showed that the individual bands in the gamma region from patients with subacute sclerosing panencephalitis in fact represented individual idiotypes of IgG [534, 1104].

Over a similar course of time another neurologist, Wallace Tourtellotte, working in Ann Arbor, Michigan, was amassing data to show that plaques of MS brain contained higher amounts of IgG than could be found in the corresponding CSF compared with parallel serum [1175, 1176], thus leading via a different route to the same seminal notion of local synthesis of IgG within the brain. On a more practical level, but nonetheless of fundamental relevance to clinical practice, there are still too few physicians who realize the value of Tourtellotte's little book *Post-lumbar Puncture Headaches* [1170], in which he showed convincingly that the small size of the needle is more important than all of the other machinations which have been performed in vain attempts to alleviate the headache which sometimes follows lumbar puncture. Using the 'needle within a needle' technique [1171] the incidence of headache was not 30 per cent (as with an 18-gauge needle) or even 3 per cent (as with a 21-gauge needle), but only 0.3 per cent with a 25-gauge needle. Some would argue that no pressure reading is obtained, since the CSF is removed by syringe. This is circumvented by prior filling of the monitor with saline to 300 mm, and if the fluid goes down, the pressure is normal. Slightly more worrying is the notion carried by some younger neurologists that to perform a lumbar puncture is rather old-fashioned, especially if they are accustomed to MRI or PET scans. Neither the most powerful magnet nor the most oblique X-ray will visualize the bacterium or the virus, and hence the possible diagnosis of infection should always be sought from the spinal fluid as a matter of priority for a potentially curable inflammatory disorder. Returning to practical CSF analysis, Tourtellotte's other classic contribution must be his use of the Laurell rocket technique [1182] whereby both albumin and IgG estimation are performed on the same agar dish (and since they carry opposite charges he arranged for them to migrate in opposite directions), which again allowed him to amass a large number of results from his bank of tissue samples. Last but not least, he was also one of the first to recognize the importance of immunofixation for the study of CSF proteins [127].

Another neurologist, Klaus Felgenhauer, working in Cologne, Germany, popularized the idea of molecular size as a relevant factor during the process of CSF filtration; namely that small molecules in serum enter CSF more readily than larger proteins [271]. Although there was previous work by Rosenthal and Soothill [951], the wider audience was more often drawn to the author of 'Protein size and CSF composition', than of 'An immunochemical study of the proteins in the CSF'. Felgenhauer also wrote a mini-classic (in German) on acrylamide electrophoresis, where he successfully combined quantitative scanning of the gel following the qualitative fractionation of up to 40 CSF proteins. He also showed that high molecular weight haptoglobin oligomers were not present in normal CSF, and this remains one of the most subtle but definite abnormalities to be found in CSF, long before any gross elevation of the level of CSF total protein [268]. His basic discovery of the important role of alpha-2-macroglobulin as a parameter of barrier function is still insufficiently appreciated.

Bodvar Vandvik, a neurologist working in Oslo, Norway showed that more than 90 per cent of the antibody in SSPE CSF was specifically anti-measles, using adsorption to measles-infected Vero cells [1228]. However, in MS the amount was trivial (less than 5 per cent).

Hans Link, a neurologist working in Linkoping, Sweden, showed in his MD thesis that the bands in the gamma region on CSF electrophoresis were in fact IgG [641]. Although there had been previous immunochemical work on brain-specific proteins [220, 455] Link showed, using biochemical analyses, that the beta trace and gamma trace were neither IgG nor its fragments.

Elizabeth Bock, a psychiatrist working in Copenhagen, Denmark, showed that brain-specific antigens could be found in the CSF [71]. She later showed that one of Blake Moore's [772] original brain-specific proteins (14–3–2) was in fact enolase [74], an enzyme in the Emden–Meyerhoff pathway of extra-mitochondrial glycolysis. This now plays a role in Sydenham's chorea.

Magnhild Sandberg-Wollheim, a neurologist/neuro-ophthalmologist working in Lund, Sweden, showed that CSF lymphocytes from patients with MS synthesized radiolabelled proteins which migrated in the gamma region after incubation of the CSF cells in tissue culture [976]. Once again, there had been prior work of similar nature by Cohen and Bannister [161].

Christian Laterre, a neurologist working in Louvain, Belgium, wrote his post-doctoral thesis on an extensive survey of abnormalities of CSF proteins (which had been separated by agar gel electrophoresis) in several diseases, which was rather encyclopedic in its day [616]. He then condensed this into quite a readable article for the practising neurologist [617], giving the differential diagnosis for the presence of oligoclonal bands. It was he who gave common parlance to the term 'oligoclonal'.

Last and by no means least is perhaps the somewhat unusual choice of Carl-Bertil Laurell, a chemical pathologist working in Malmo, Sweden, since he is principally known for his 'rocket' method of quantitative immunochemical precipitation enhanced by an electrophoretic technique [622]. Although he was clearly preceded by Delpech and Lichtblau [219] with their idea of an IgG index, his particular method [335] of dividing albumin in CSF by albumin in serum (rather than Alb/IgG) has considerable intuitive appeal, as we shall discuss later. However, his important caveats regarding the limitations of quantitative immunochemical techniques are still too poorly understood [623]. His wife demonstrated, in an often-neglected paper [839], that under pathological conditions (syphilis) the CSF contained altered molecular forms of IgG, which could therefore give an incorrect answer in quantitative analysis. Thus it has been argued that it is important always to compare 'like with like' and be wary of fragments and/or aggregates.

PART 1

Normal CSF proteins, an overview

CHAPTER

2

Functional glossary

Introduction
Immunoglobulin synthesis
Barrier terms

Introduction

Since this monograph contains composite summary tables of data from many other investigators, a short glossary is given here rather than at the end. This is because much of the effort in compiling these summary tables lies not in simply collecting all of the data, but also in attempting to simplify some of the conflicting terms which have been used by various authors to mean diverse things (e.g. quotient versus ratio). Without wishing to add to the problem of terminology, the ensuing 'definitions' are based on the simple concept that blood–CSF barriers can be thought of as allowing a certain percentage of the serum level of a given protein (taken as 100 per cent) to be 'transferred' into the CSF, hence the idea of 'percentage transfer'. Since the normal level of plasma protein is around 7000 mg/dl and the normal level of CSF protein is around 35 mg/dl, the average percentage transfer is $35 \times 100/7000$, or 0.5 per cent. As we shall see in subsequent chapters, the 'percentage transfer' can also be extended to correct for the proportion which is locally synthesized within the central nervous system. Graphical/mathematical methods for distinguishing between a serum source (leak) and local synthesis in the CNS are given in Chapter 8.

An analogous concept is the 'percentage of total IgG'. This can be particularly helpful in comparing CSF with serum, since no degree of barrier alteration can make the CSF percentage any different from the serum percentage for any individual antigen. Therefore an increased percentage in CSF over serum must represent local synthesis regardless of the barrier state, as will be seen in the final definition below, namely relative specific antibody.

Immunoglobulin synthesis

Electrophoresis of CSF reveals different protein spectra. In addition to the normal pattern, there are three abnormalities of the gamma region (see Figure 2.1):

1. *Monoclonal* (one clone). A single band in the gamma region implies a single clone of plasma cells secreting a homogeneous protein. On isoelectric focusing, the single band can be resolved into many (e.g. three to eight) bands. However, these bands may be artefactual, since many more are seen with IgD, a subtype notoriously prone to fragmentation.
2. *Polyclonal* (many clones). Many thousands of 'bands' merge to yield a population distribution in a bell-shaped curve. This also shows 'bands' on isoelectric focusing which are artefactual, since very similar bands can be seen without any protein being applied, merely by staining the ampholytes in the gel (see Chapter 15, Figure 15.3).
3. *Oligoclonal* (small number of clones). Usually two to five bands are seen in the gamma region, which implies a restricted antigenic response to a few immunodominant antigens that have induced the proliferation of a few plasma cell clones.

Barrier terms

There are six measurements of barrier function which have been used in the literature, and in the discussion below the following abbreviations have been used:

Albc = albumin concentration in CSF.
Albs = albumin concentration in serum.
IgGc = IgG concentration in CSF.
IgGs = IgG concentration in serum.

The six measurements are:

1. *Ratio* – the result of dividing the amount of one protein (e.g. IgG) in CSF by the amount of another (e.g. albumin) also in the CSF. The IgG to Alb ratio is therefore IgGc/Albc.
2. *Quotient* – the result of dividing the amount of a given protein (e.g. albumin) in the CSF by the amount of the same protein again (e.g. albumin) in the serum. The albumin quotient is thus Albc/Albs.
3. *Percentage transfer* – similar to the quotient above, but multiplied by 100. For example, the percentage transfer of albumin is 0.5 per cent, i.e. (Albc/Albs) × 100, hence one albumin molecule of each 200 albumin molecules in serum is transferred into the CSF.

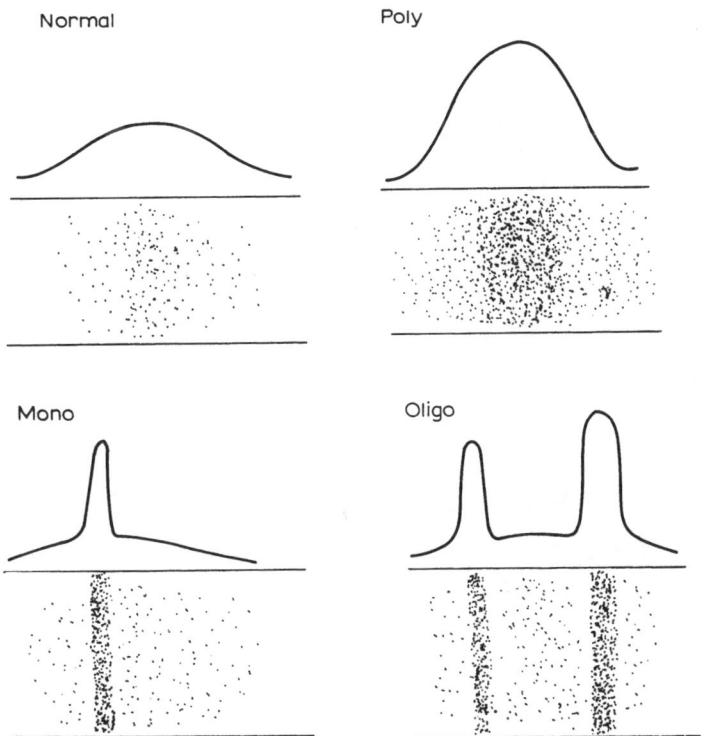

FIGURE 2.1 Clonal distribution of immunoglobulins.

4. *Index* – usually the IgG index, unless another protein is specified (e.g. alpha-2-macroglobulin). It is the ratio of two quotients (normally albumin as denominator). Hence the IgG index is (IgGc/IgGs)/(Albc/Albs).
5. *Relative specific antibody* (RSA) – the quotient of two quotients (usually the fraction of IgG which is antigen-specific). Hence the IgG directed specifically against, for example, the measles antigen (M) is derived as (Mc/IgGc)/(Ms/IgGs).
6. *Selectivity* – the slope of the line which derives from plotting the percentage transfer values of the various proteins against their respective molecular sizes.

CHAPTER

3

The roster of CSF proteins

Introduction
Amounts
Functions
   Chemical groups: molecular properties
     Molecular properties: size
     Molecular properties: charge

Introduction

Up to 20 per cent of CSF proteins may be synthesized within the brain. Smaller proteins from the plasma protein pool enter the CSF more readily than the larger proteins.

This chapter has three main sections:

1. *Amounts of specific proteins.* This section lists the amounts of individual proteins in the order of their decreasing concentrations in the CSF. A useful comparison can also be made of the relative proportions of proteins in CSF and serum. The quotients for the levels in CSF divided by the levels in serum are presented as the percentage transfer. This enables those CSF proteins which are derived from the serum to be distinguished from those which are derived from the brain itself.
2. *Functions of specific proteins.* This section regroups the proteins according to their physiological functions, e.g. carrier proteins versus enzyme inhibitors.
3. *Chemical groups of specific proteins.* The final section then re-groups the proteins according to their differing chemical properties, e.g. acidic versus basic charges.

Amounts

An examination of the rank order of the most abundant CSF proteins provides quite an instructive introduction, and is given in Table 3.1.

Arranging the amounts of CSF proteins reveals an order of magnitude difference between the most abundant protein (albumin) and the second most abundant protein (beta trace or prostaglandin synthase which is derived from the brain). Following these, in decreasing amounts, are: IgG, prealbumin (a choroid-enriched protein, i.e. transthyretin), transferrin, alpha-1-antitrypsin, apo-E-lipoprotein, and gamma trace (also a brain protein, i.e. cystatin C). The rank order of serum proteins shows quite a different pattern, with IgG being the second most abundant protein. After albumin is IgG (only about three-fold less), followed by: beta lipoprotein, alpha lipoprotein, fibrinogen, transferrin, alpha-1-antitrypsin, and alpha-2-macroglobulin. Examination of the rank order of the proteins in CSF and serum reveals a fundamental difference: the CSF principally contains three brain-derived proteins (beta trace, gamma trace and prealbumin), which replace the three proteins of much larger molecular size found in serum (alpha-2-macroglobulin, fibrinogen and beta lipoprotein). One can therefore discern two principles governing CSF composition:

1. A relatively selective exclusion of the larger molecular weight proteins during the formation of the CSF from the plasma.
2. Local release of other brain proteins into a much smaller CSF pool (125 ml). These eventually become diluted in the plasma (3000 ml), i.e. twenty-five-fold.

A separate examination of the relative amounts (versus absolute amounts, as discussed above) of the CSF proteins, specific to their presumed primary source (the plasma proteins), leads to the second principle concerning the nature of

TABLE 3.1 Comparative rank order of the major proteins in CSF versus serum, presented in order of decreasing amounts

Ranking	CSF	Serum
1	Albumin	Albumin
2	Beta trace (ten-fold less)	IgG (three-fold less)
3	IgG	Beta lipoprotein
4	Prealbumin	Alpha lipoprotein
5	Transferrin	Fibrinogen
6	Alpha-1-antitrypsin	Transferrin
7	Apo-E-lipoprotein	Alpha-1-antitrypsin
8	Gamma trace	Alpha-2-macroglobulin

TABLE 3.2 CSF proteins with the highest percentage transfer from the serum

Protein (source)	% Transfer	Mol. weight (kD)	Charge
Beta-2-microglobulin (brain)	88	12	Neutral
Lysozyme (polymorphs)	7	14	Very basic
Prealbumin (choroid)	6	61	Very acidic
Albumin (liver)	0.5	69	Neutral
For comparison: lowest percentage transfer (large molecular size)			
Beta lipoprotein	0.01	2200	Neutral
IgM	0.02	800	Neutral
Fibrinogen	0.02	340	Neutral

the CSF protein mixture (see Table 3.2); namely that the proteins are locally synthesized (by the brain, its vessels and its cellular coverings) rather than by simple filtration from the serum. As a general rule, it appears that CSF protein levels greater than 1 per cent (the so-called 'percentage transfer') of the serum levels are associated with local synthesis.

It is clear that three proteins are present in extraordinarily high amounts when compared to their corresponding relative levels in the serum:

1. Beta-2-microglobulin (11.5 kD) shows by far the highest level of percentage transfer, with the concentration in CSF essentially equivalent to that in serum
2. Lysozyme (14 kD) has a percentage transfer one order of magnitude less
3. Prealbumin (61 kD) has a similar percentage transfer to lysozyme.

These proteins will be considered in further detail later in this chapter. The point here is that the simplistic notion that the CSF is merely an ultrafiltrate of plasma is clearly false. A more complete list of the CSF proteins is given in Table 3.3. In the next chapter, these proteins are regrouped in order of percentage transfer (Table 4.4). Details of lower molecular weight substances, such as peptides, are given elsewhere [284].

Functions

The proteins can be grouped according to their specific physiologic functions (Table 3.4). Since the majority of the functions are similar in both CSF and serum, the reader is referred to the standard texts [15, 68, 891, 892]. We shall focus mainly on the more specialized CSF functions, but a very brief general sketch is presented here with the appropriate details listed in the traditional reference sources for the plasma proteins.

3. THE ROSTER OF CSF PROTEINS

TABLE 3.3 The roster of the CSF proteins

Protein	mg/l	% Transfer	% Total
1. Albumin	200	0.5	67
2. Beta trace	26	3400	9
3. IgG	22	0.2	7
4. Prealbumin	17	6	6
5. Transferrin (Tau about 1/3)	14	0.6	5
6. Alpha-1-antitrypsin	8	0.4	3
7. Apo-E-lipoprotein	8	6	3
8. Apo-A-lipoprotein	6	0.4	2
9. Gamma trace	6	500	2
10. Orosomucoid	3.6	0.6	<1
11. Hemopexin	3.0	0.3	
12. Group components	2.5		
13. Haptoglobin	2.1	0.1	
14. Antichymotrypsin	2.1		
15. Alpha-2-macroglobulin	2.0	0.2	
16. Alpha-2 Hermann Schultz	1.7	0.3	
17. Complement C'3	1.5	0.3	
18. Complement C'9	1.5	2	
19. Fibrinogen degradation products	1.4	50	
20. IgA	1.3	0.1	
21. Beta-2-microglobulin	1.0	88	
22. Ceruloplasmin	1.0	0.5	
23. Complement C'4	1.0	0.3	
24. Lysozyme	1.0	7	
25. Beta-2-glycoprotein I	1.0	0.5	
26. Fibrinogen	0.65	0.02	
27. D-2 antigen (N-CAM)	0.64	40	
28. Beta lipoprotein	0.59	0.01	
29. Zinc alpha-2-glycoprotein	0.27		
30. Plasminogen	0.25	0.2	
31. IgM	0.15	0.02	
32. Aldolase C4	0.085		
33. Alpha-1-microglobulin	0.035	0.1	
34. Enolase (gamma)	0.010	100	
35. Lactoferrin	0.0066	2	
36. Glial fibrillary acidic protein	0.0033		
37. Ferritin	0.0023	2	
38. S-100	0.0020	2000	
39. Creatine kinase-BB	0.0009		
40. Eosinophil cationic protein	0.0009	5	
41. Myelin basic protein	0.0006		
Peptides (for reference)			
Beta endorphin	0.000105	[344]	
BL1, SS, Ga, CCK	0.000030	[900]	
Calcitonin	0.000030	[49]	
Vasopressin	0.0000005	[912]	

TABLE 3.4 Physiologic functions

1. Carry ('non'-specific)
   - Albumin (non-esterified fatty acids)
   - Alpha lipoprotein (cholesterol, phospholipids)
   - Beta lipoprotein (triglycerides)
2. Carry (specific)
   - Red blood cells (recycled)
        Transferrin (Fe)
        Hemopexin (porphobilinogen)
        Haptoglobin (hemoglobin)
   - Hormone
        Prealbumin (thyroxin)
   - Vitamins
        Retinol-binding protein (vitamin A)
        Group components (vitamin D)
3. Enzyme (inhibitors)
   - Alpha-2-macroglobulin
   - Alpha-1-antitrypsin
   - Alpha-1-antichymotrypsin
4. Enzyme (active)
   - Ceruloplasmin
   - Lysozyme
   - Aldolase
   - Enolase
   - Creatine kinase
5. Enzyme (inactive)
   - Plasminogen
   - Complement C'3
   - Complement C'4
6. Structural subunits
   - Fibrinogen
   - Glial fibrillary acidic protein
   - Myelin basic protein
7. Protein–protein binding
   - Homogeneous
        Haptoglobin di, tri, tetra, penta, heptameric
        Albumin di, tri, meric (*in vitro* artefacts)
   - Heterogeneous
        Haptoglobin–Hemoglobin
        Alpha-1-antitrypsin–IgA
        Prealbumin–retinol binding protein (less in CSF)
              (but no binding in CSF or urine)
        Thrombin–enzyme
        Alpha-2-macroglobulin–enzyme

(continued)

TABLE 3.4 (Continued)

- IgM–IgG (rheumatoid factor)
- Antigen–antibody
- B2 microglobulin–HLA (surface marker)
- Serum amyloid A–AI Apoprotein (high density)

8. Acute phase
   - Increase
     Rapid (6–24 hours):
        By 100-fold
           C-reactive protein (200 mg/l serum)
           Serum amyloid A (2000 mg/l serum)
        By 3-fold
           Orosomucoid
           Alpha-1-antichymotrypsin
           Alpha-1-antitrypsin
           Haptoglobin
           Fibrinogen
     Slower (2–3 days):
        By 0.5-fold
           Ceruloplasmin
           Complement C'3
   - Decrease (associated with increases noted above)
        Albumin
        Transferrin
        Alpha-2 Hermann-Schultz
        Prealbumin
        Fibronectin
        Alpha lipoprotein

9. Brain
   - Beta trace (prostaglandin synthase)
   - Gamma trace (anti-elastase) or cystatin C
   - Glial fibrillary acidic protein (astrocytes)
   - Myelin basic protein (oligodendrocytes)
   - Enolase gamma (neurons) and alpha (glia)
   - S-100 (glia)
   - Aldolase C4 (glia?)
   - Creatine kinase–BB (glia?)
   - Ferritin (glia?)
   - B2 microglobulin (glia?)
   - Prealbumin (choroid plexus?)

10. Brain (modified by)
    - Tau, i.e. hydrolyzed beta-2 transferrin (minus sialic acid)
    - Fibrinogen degradation products (meningeal)
        Fragment D 83 kD
        Fragment E 41 kD
    - Retinol-binding protein (not bound to prealbumin)

TABLE 3.4 (Continued)

11. CSF (cells)
    - Complement C'1q (macrophages)
    - IgG, A, M (lymphocytes)
    - Lysozyme (polymorphs)
    - Lactoferrin (polymorphs)
    - Eosinophilic cationic protein (polymorphs)
12. Homologies (functional significance?)
    - Immunoglobulins with beta-2-microglobulin (domain)
    - Haptoglobin, trypsin, chymotrypsin, plasmin, clotting Factor X, complement (serine proteases)
    - C-reactive protein with serum amyloid protein (pentraxins)
    - Orosomucoid with carcinoembryonic antigen
    - Alpha-2-macroglobulin with complement C'3 and C'4
    - Alpha-1-antitrypsin with antithrombin III
    - Albumin with alpha fetoprotein

Non-specific carriage is mainly the domain of albumin, which can readily transport non-esterified fatty acids, bile pigments, hormones, drugs, etc.

The transport of other fats is performed by the two major lipoproteins, which are synthesized in two different tissues:

1. Apo-A-lipoprotein is made principally in the liver and carries mainly cholesterol and phospholipids (to a lesser extent the non-esterified fatty acids)
2. Apo-B-lipoprotein is made principally in the gut and carries mainly triglycerides (i.e. the esterified fatty acids).

Specific carriers include those proteins dealing with breakdown products of the relatively short-lived red cell: transferrin to carry iron, hemopexin to carry the heme porphyrin, and haptoglobin to carry the protein moiety of hemoglobin. Other specific carriers include prealbumin (transthyretin) for thyroxin (T4), retinol-binding protein for vitamin A, and group components for vitamin D.

Enzyme inhibitors serve to keep any lytic actions on a local level (rather than allowing the entire body to become caught up in the destructive actions), and are thus necessary at only one circumscribed place at any given time. They are present in very large amounts when one considers their total CSF level is about 13 500 mg/l, and they could be counted as the second most abundant protein group, even before IgG (which is present at 12 500 mg/l). The most abundant enzyme inhibitor within the group is cystatin C (gamma trace), followed by alpha-2-macroglobulin (also the largest in molecular size) which blocks trypsin, plasmin and thrombin as well as the clotting factor X. Alpha-1-antitrypsin primarily blocks trypsin, but also chymotrypsin, plasmin, thrombin and elastase. Alpha-1-antichymotrypsin essentially blocks chymotrypsin only.

Enzymes that are already circulating in an active form include ceruloplasmin, which scavenges free radicals from polymorphs and xanthine oxidase in rapidly-growing tissues (e.g. fetus and, to a lesser degree, tumors). Lysozyme hydrolyzes muramic acid, which is a constituent of bacterial cell walls. Aldolase cleaves fructose 1,6-diphosphate to produce two three-carbon fragments in the Emden–Meyerhof pathway of extra-mitochondrial glycolysis. Enolase hydrolyzes water from 2-phosphoglycerate to produce phosphoenol pyruvate, also in the Emden–Meyerhof pathway. Creatine kinase transfers a high-energy phosphate bond to ATP for eventual release.

Enzymes that are present in an inactive form (proenzymes or zymogens) include plasminogen which, when activated by plasminogen-activator, lyses clotted fibrin to allow blood to flow once again. Complement C'3, when acted on by either the classical pathway (C'2 and C'4) or the alternative pathway (factor B, D and antigen–antibody complexes), will then hydrolyze C'5, which subsequently inserts the hydrophobic probe into the membrane along with C'6, C'7, C'8 and C'9. Complement C'4, when acted on by C'1qrs, will combine with C'2 to hydrolyze C'3 in active form (see above).

Certain proteins act as subunits for building a larger structure, and include fibrinogen, which can first be polymerized and then cross-linked to form a blood clot. It is worth noting in passing that CSF would not be collected in EDTA (versus plasma) unless there were a wish to estimate fibrinogen levels. Glial fibrillary acidic protein can be polymerized by astrocytes in large quantities (to become one-third of the cellular protein), forming scar tissue 'plaques' as in multiple sclerosis. Myelin basic protein is a hydrophobic insulator that can be synthesized by the oligodendrocyte and/or Schwann cell as part of the sheath which is wrapped around the axon, and again may constitute as much as one-third of myelin protein.

Among the many acute phase proteins that change during the course of an insult, the most useful for CSF purposes is orosomucoid (alpha-1-acid glycoprotein). It may also be synthesized by macrophages within the thecal confines in meningeal diseases (Chapter 11).

Several other proteins may be synthesized by specific brain cells or CSF cells, or even the cells of the surface linings, e.g. prealbumin by choroidal epithelial cells, or possibly tau protein (beta transferrin) by ependymal cells.

Chemical groups: molecular properties

The two main biochemical properties of any protein are its molecular size and its net charge. These two attributes allow almost any given protein to be uniquely distinguished from any other. For example, all the proteins of a single bacterial

TABLE 3.5 Proteins with a broad range of size or charge

Polydisperse in size		
Protein	Size range	
	Larger	Smaller
1. Lipoprotein A	Pre-gamma	Transferrin
2. Beta trace	Pre-gamma	Beta
3. IgG	Alpha-2-macroglobulin	Pre-gamma
4. IgA	Mid-gamma	Pre-gamma

Polydisperse in charge		
Protein	pI range	
	Basic	Acidic
1. IgG	Gamma 5	Beta
2. Beta trace	Beta	Alpha 2
3. IgA	Gamma 2	Beta
4. Lipoprotein A	Alpha 1	Albumin
5. Gamma trace	Gamma 1	Gamma 2
6. Gamma 'acidic' ?Apo E	Gamma 1	Gamma 2

species can be separated by two-dimensional electrophoresis (measurement of size and charge). Essentially the same can be said of individual tissues [480], including the plasma proteins [23].

Although two-dimensional gels provide a plethora of information about denatured (unfolded) proteins, most of the information given in this section deals with the parameters of size and charge which are found under physiologic conditions, but size is by far the more relevant for the selection of CSF proteins. Charge appears to be of greater relevance for urinary proteins, but once again molecular size has the more important role in both fluids.

Under physiologic conditions, several proteins are found not as discrete, tight bands on electrophoresis but tend to extend over a larger range of sizes and/or charge. This molecular characteristic has allowed them to be termed 'polydisperse', and these proteins can be listed in decreasing order of their tendency to be heterogeneous under normal conditions. Since they are polydisperse, they do not have a single value, but a broad range of values for either size or charge (Table 3.5).

Molecular properties: size

Size plays a dominant role in the transfer of CSF proteins from the plasma. Felgenhauer [271] showed an approximately log-linear relationship between the

3. THE ROSTER OF CSF PROTEINS

CSF levels of the ten most abundant serum proteins and their molecular size. The spectrum for different sizes of proteins in Ångstroms is shown in the Table 3.6, and includes three broad ranges: very small (less than 20 Å), intermediate (32–70 Å) and very large (greater than 95 Å).

The idea that many CSF proteins were derived from serum on the basis of molecular size was most convincingly argued by Felgenhauer [271]. Figure 3.1

TABLE 3.6 Increasing molecular size (versus molecular weight) of CSF proteins (the molecular size is given in Angstrom units, Å)

	Size (Å)	Mol. weight (kD)
Apo-C-lipoprotein II, III	13.0	9
Gamma trace	14.0	11
Beta-2-microglobulin	15.0	12
Lysozyme	19.0	14
Apo-A-lipoprotein II	20.0	17
Retinol-binding protein	21.0	21
Kappa, lambda light chains	23.0	22
Beta trace	24.0	31
Apo-E-lipoprotein	25.0	34
Hemoglobin	31.3	67
Alpha-1-antitrypsin	32.8	61
Prealbumin	32.8	61
Group components	33.6	62
Alpha-2 Hermann Schultz	34.1	64
Antichymotrypsin	34.2	64
Hemopexin	36.0	68
Albumin	35.6	69
Transferrin	36.7	81
Orosomucoid	40.6	44
Haptoglobin 1-1	41.8	80
Plasminogen	42.7	143
Ceruloplasmin	44.2	152
Aldolase	48.1	165
C'3 (post conversion)	53.3	150
IgG	53.4	155
Haptoglobin 2-2	54.2	155
IgA	57.0	168
Haptoglobin 1-1 (with Hb)	62.6	150
C'3 (pre-conversion)	69.5	155
Apoferritin	74.0	473
Alpha-2-macroglobulin	96.0	798
Fibrinogen	108.5	340
IgM	121.0	800
Beta lipoprotein	124.0	2200

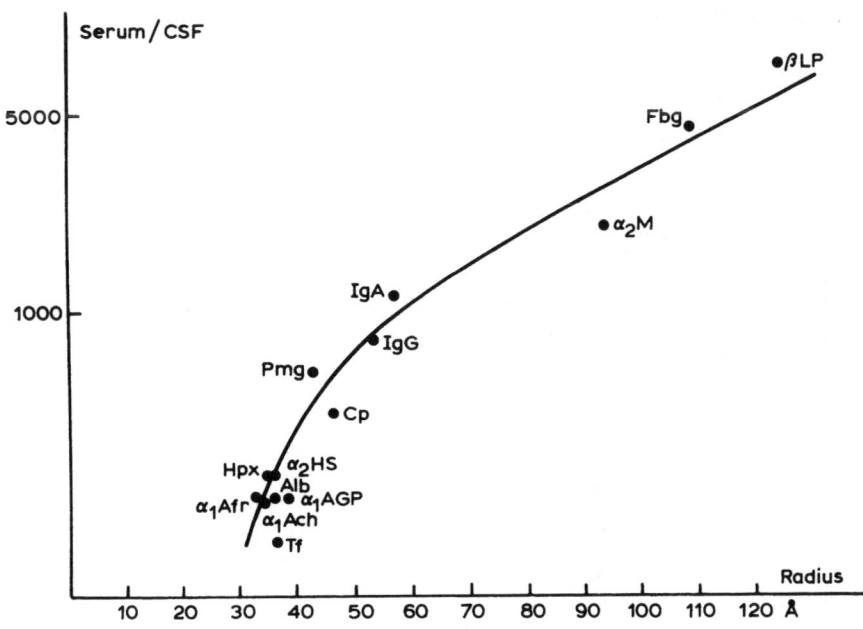

FIGURE 3.1 Original figure from Felgenhauer (1972) [271].

(modified from his paper) shows a curvilinear function with equal weight given to all proteins, rather than a statistical approach using linear regression. In later work Felgenhauer drew a straight line, with 'correction factors' [290] for IgG and IgA since they did not fall on the straight line (Figure 3.2).

His idea seems not to have gained wide acceptance, perhaps because of the inverse presentation of the CSF/serum quotients. However, by inverting of the Y-axis values and expressing them as 'percentage transfer', perhaps the fundamental nature of the concept becomes apparent. Two other notions have also been partially confused with this plot, namely the IgG index (with various statistical presentations as we shall see in Chapter 6) and the various IgG formulae, especially those from Tourtellotte [1168, 1180] and Schuller [1002].

Using our own version of the 'size' plot we can recalculate the original data of Felgenhauer, and this is shown in Figure 3.3.

The important difference is not simply the inversion of the quotients (from serum/CSF to CSF/serum), but also the use of the concept of percentage transfer – i.e. at 100 per cent transfer there is complete equilibrium between CSF and serum (like a Froin's syndrome). At 1 per cent transfer, obviously only 1 molecule in 100 goes from serum across the barrier into CSF, and hence at

3. THE ROSTER OF CSF PROTEINS

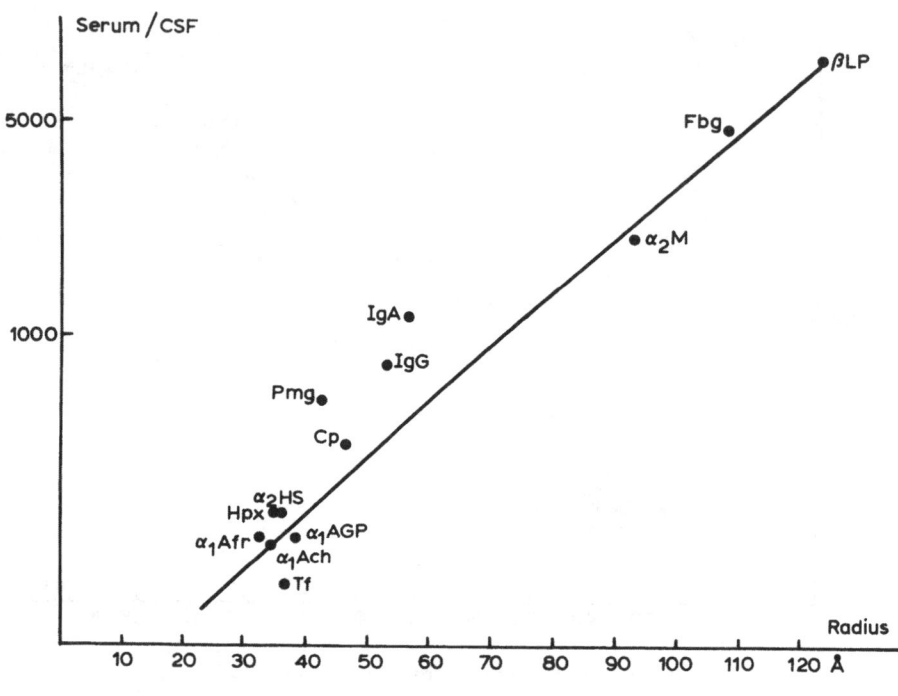

FIGURE 3.2 Data of Felgenhauer (1972) redrawn [271].

0.5 per cent (i.e. the normal albumin level) this becomes 1 in 200. Since IgG is twice as large as albumin, only 0.2 per cent, or 1 molecule in 500, crosses the barrier. The general rule is therefore easily visualized: the larger the molecule (to the right on the X-axis as in Figure 3.3), the less its percentage transfer into the CSF (lower on the Y-axis).

There is one additional point to be clarified: the quotient of CSF/serum amounts can either be a 'calculated' average, in which the average of many individual CSF results is then divided by an average of many individual serum results. Much to be preferred, however, is the use of pairs of CSF and matching serum levels from any given person, which then have an individual quotient calculated, and then finally an average of all the quotients from several individuals is prepared. This average of quotients is more reliable than taking quotients of averages. We can thus see that Felgenhauer's latter data is an average of quotients, as given in Figure 3.4.

Note the slight difference in the data points from Figure 3.3, which is based on Felgenhauer's early data i.e. a quotient of averages. A similar comparison is seen

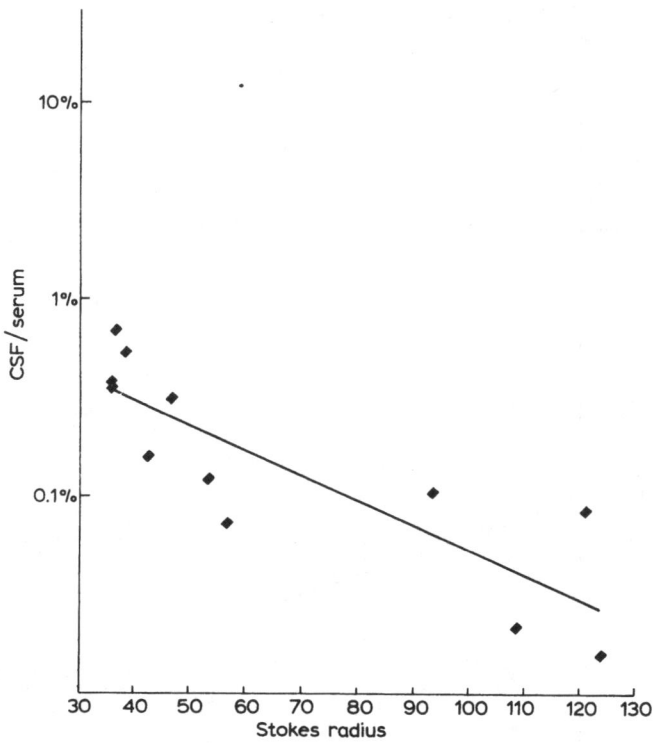

FIGURE 3.3 Averages from data of Felgenhauer (1972) replotted [271].

for Bock's data, using first the older method (see Figure 3.5) and then the improved average of quotients as seen in Figure 3.6.

The data of Tourtellotte (using Behring antibodies) [217] in Figure 3.7 agree with the data of Felgenhauer in Figure 3.3 (also Behring antibodies), and of Reiber (again using Behring antibodies), which is used as an example in a later chapter (Figure 5.3). However there is less agreement with Bock's data in Figure 3.6 (for which she used Dako antibodies). Our own data show yet a different slope in Figure 3.8 (using Seward antibodies). It might be assumed that the differences in antiserum would cancel out, since CSF results are being divided by serum results. However, there are still differences between the immunodominant polyclonal mix of anti-IgG and anti-albumin from Behring (slope 0.010), Dako (0.020) and Seward (0.007). Note also that the IgG and alpha-2-macroglobulin results do not parallel one another using different sources of antiserum. This is a perennial problem with immunoassays, and is discussed with the more complex immunochemical problems in Chapter 7.

3. THE ROSTER OF CSF PROTEINS

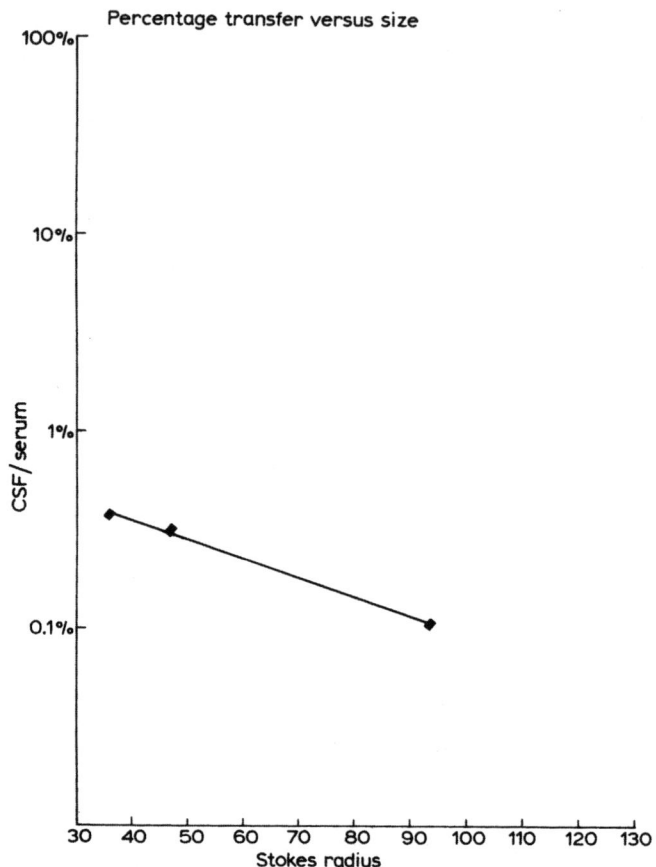

FIGURE 3.4 Strict quotient data of Felgenhauer (1976) replotted [290].

Since these different sources of antibody yield different values for the slope of the line (plotting molecular size against percentage transfer), it follows that to impose a particular pore size on any general model which derives a particular slope, e.g. Rapoport [896], is subject to the same vagaries as the differences in slope. Indeed the degree to which the points actually fall on (or off) the line in a curve (be it exponential or some polynomial) is determined by the reproducibility of the values of percentage transfer for each protein. In other words, the coefficient of variation for individual proteins is another source of potential error [665]. Reiber subsequently changed from a linear to a curved representation, i.e. the hyperbolic formula, which will be discussed in detail in Chapter 8, as will Ohman's exponential model.

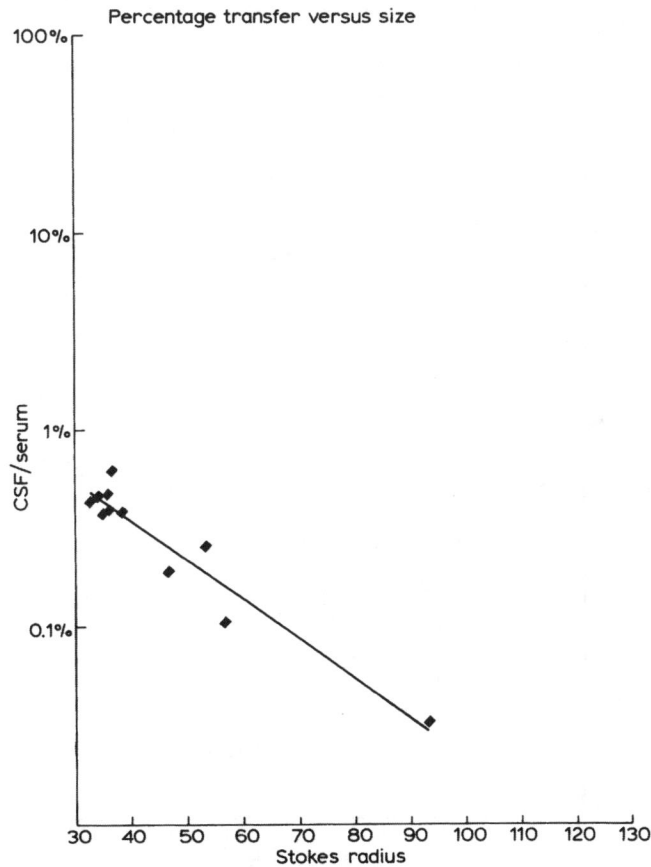

FIGURE 3.5 Averages of Bock (1973) replotted [72].

In any case, there is a consensus that small molecules have a major advantage over the larger molecules in terms of entry into the CSF from the plasma. As noted above (Table 3.2), there are at least three proteins which appear not to follow the size rule in that they are present in disproportionately high amounts in CSF: beta-2-microglobulin, prealbumin and lysozyme. Neither does transferrin (including tau protein or beta-2 transferrin), which is discussed in Chapter 7. IgG and IgA are found in disproportionately low amounts, as can be seen in most of the figures above; these are also discussed in Chapter 7.

In summary, size is the dominant feature in the composition of CSF, although there are some important discrepancies. Detailed discussions of the discrepancy problems are given in Chapters 6 and 7.

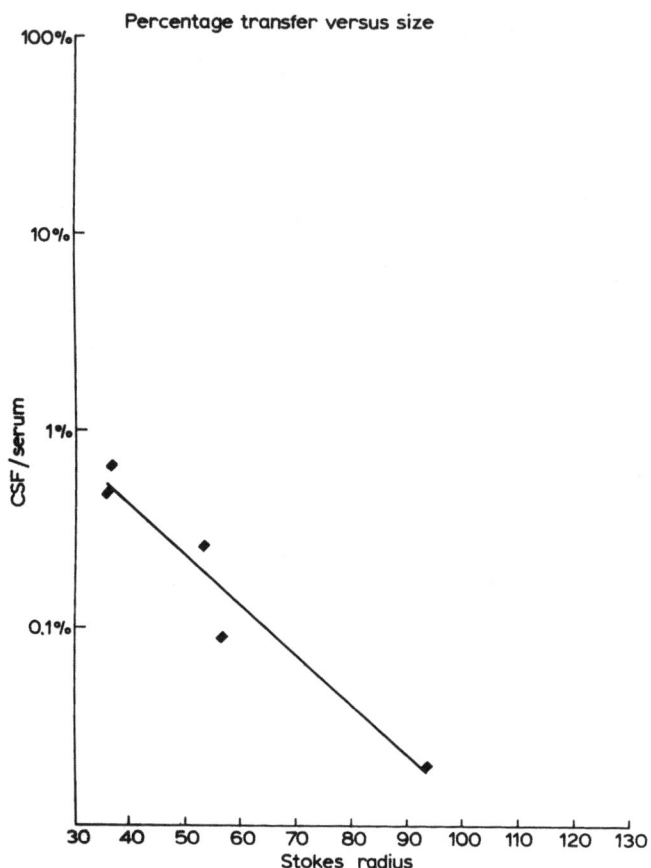

FIGURE 3.6 Strict quotient data of Bock (1978) replotted [72].

Molecular properties: charge

The second most important molecular characteristic is that of net charge. The majority of proteins are acidic because they carry a net negative surface charge at physiological pH, with prealbumin being relatively more acidic under physiologic conditions. The few basically-charged proteins, which carry a positive charge under physiologic conditions are IgG, IgM, fibrinogen, hemoglobin, C-reactive protein, gamma trace, plasminogen and, the most basic of all, lysozyme (pI 10.8). Their pI values are listed in Table 3.7; some are polydisperse.

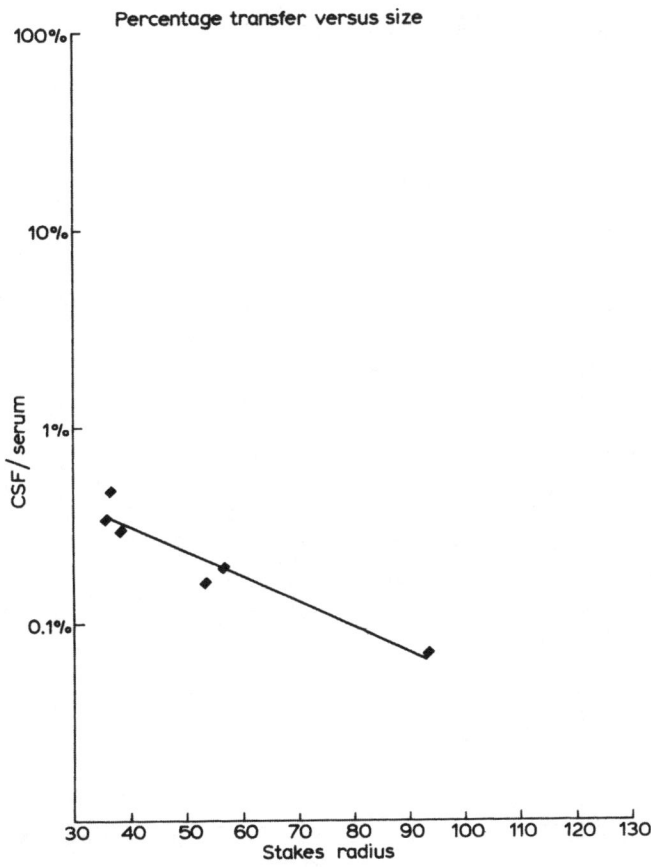

FIGURE 3.7 Strict quotient data of Tourtellotte (1980) replotted [217].

Although it has been shown that producing a more basic charge on albumin induces higher permeability of the blood–brain barrier [400], nevertheless the amount of (basic) IgG is under-represented rather than over-represented in CSF (see the many plots of percentage transfer versus size); hence this point remains to be clarified.

Although proteins are typically found to have sugar residues added when they are being prepared for extracellular use (as opposed to inside the cell for its own housekeeping needs), orosmucoid must be singled out for comment since the sugar content accounts for 45 per cent of the total weight of the protein. Among the various sugars, the most acidic (strongly negatively charged) is sialic acid (N-acetyl neuraminic acid, or NANA), and the proteins which have the largest

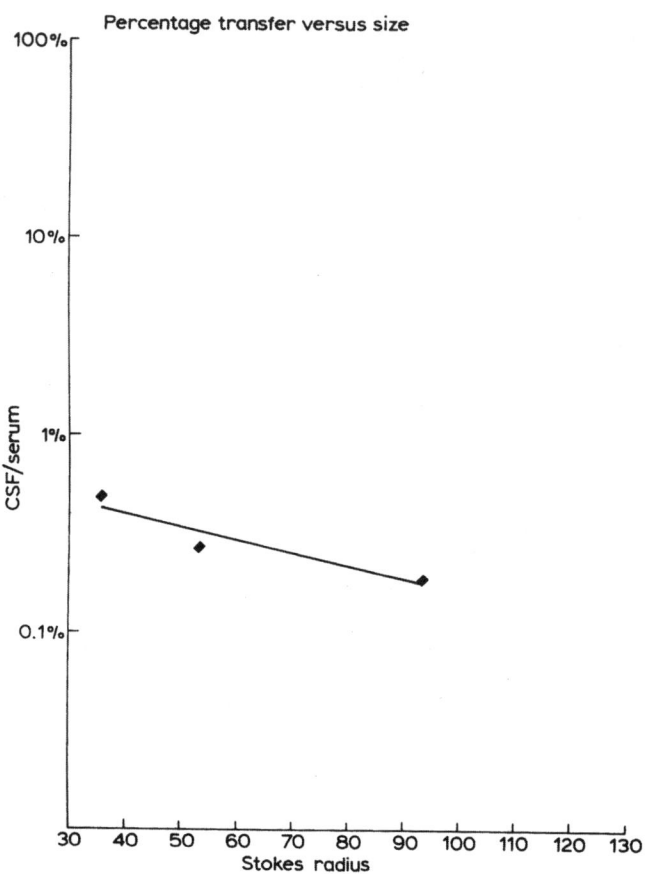

FIGURE 3.8 Strict quotient data from our lab.

TABLE 3.7 Basic proteins: carry net positive charge at pH 7.4

Protein	pI
Lysozyme	10.8
Gamma trace	9.2
IgG	6.0–8.5
IgM	5.0–8.0
Fibrinogen	8.5
Plasminogen	6.3–8.6
C-reactive protein	8.0
Hemoglobin	7.5

amount of NANA are alpha-1-antitrypsin, C′1 esterase inhibitor and (once again) orosomucoid. The former two proteins are also both serine protease enzyme inhibitors.

The other important role which is performed by the acidic sugar NANA is to indicate to the reticulo-endothelial system that the protein is 'normal', since following removal of NANA there is rapid extraction of the protein from the circulation, followed by its complete catabolism.

CHAPTER

4

Differences between proteins in CSF and serum

Introduction
Quantitative
  Higher absolute levels due to obvious synthesis within the CNS
  Higher relative levels due to selective entry into the CSF and/or local synthesis
Qualitative

Introduction

Some CSF proteins are synthesized within the brain while others may have a selective advantage, such as the relatively acidic charge of prealbumin (in addition to local synthesis). Such proteins are readily distinguished from other plasma proteins, due to the physiologic isolation of CSF from other body fluids. Percentage transfer allows quantitative distinction between proteins synthesized or released by the brain and/or CSF cells from other proteins which are mainly filtered from the serum.

This chapter has two main sections:

1. *Quantitative*. This section deals initially with proteins present in much larger *absolute* amounts in CSF than in serum. These are derived from the central nervous system and then subsequently find their way into the systemic circulation. Also discussed are CSF proteins present in much larger *relative* amounts (i.e. percentage of total proteins). Those may have had a selective qualitative advantage in gaining access to the CSF (see next section).
2. *Qualitative*. The second section deals with differences in the physical–chemical composition of CSF proteins from that of their serum counterparts, due to some local modification by the brain parenchyma or choroid itself. This modification could confer a selective advantage for entry into the CSF.

Quantitative

There may be larger amounts of proteins in CSF than in serum either in *absolute* or *relative* terms, and these two options will now be considered separately.

Higher absolute levels due to obvious synthesis within the CNS

The most abundant CSF protein (after albumin) is beta trace, which has an absolute level five times higher than that in the corresponding serum. Beta trace (prostaglandin synthase), which causes sleep [435], may be synthesized by astrocytes and/or oligodendrocytes. It is polydisperse in charge [642], and levels are increased in people over 70 years of age, in some tumor cases and, in some patients, several weeks following a stroke. The level is otherwise remarkably constant in the CSF across a broad range of neurological diseases [652]. On electrophoresis there may be up to four bands within a rather diffuse zone. Removal of sialic acid (the negatively-charged sugar group) will produce a protein which travels just behind the tau protein (beta-2 transferrin, see below). Gamma trace is less abundant in absolute amounts than beta trace, but still comes in the top ten of the most abundant CSF proteins and, like beta trace, is present in absolute amounts five times higher in CSF than in serum. It is a very basic protein (pI approx 9), but with storage it may be come more acidic (pI 8) – unlike beta trace, which becomes more basic. It is thus a common artefact in the gamma region, and on focusing it may be mistaken for an immunoglobulin (as noted by Olsson [835]). It has recently been considered to be a precursor of the hypothalamic peptide ACTH [405] or (more likely) a protease inhibitor [42], specifically cystatin C [939].

Tau protein (beta-2 transferrin) is essentially absent from serum, and may be produced within the CNS by the removal of sialic acid from transferrin. It may also be produced systemically, but the reticulo-endothelial system in the systemic circulation is very efficient at removing it. Conversely, it accumulates in the CSF because of the paucity of reticulo-endothelial elements within the CNS (see below and Chapter 7 for further details.) An alternative scenario is that no neuraminic acid is added to the transferrin which is synthesized by the brain.

A general, semi-quantitative examination of the most abundant CSF (not serum) proteins using the more 'physiologic' approach of two-dimensional separation on agarose (for protein charge) followed by gradient acrylamide (for protein size) reveals that, after albumin, the following are prominent: prealbumin, transferrin, IgG and 'gamma acidic' protein (? ApoE) [272]. Other proteins present in more moderate amounts are also given in Table 4.1.

Several proteins are selectively depleted (probably filtered out due to their large size), and these are given in Table 4.2.

TABLE 4.1 Comparison of protein patterns for CSF and serum

Proteins prominent in CSF	
Major	Albumin
	Prealbumin
	Transferrin
	IgG
	Gamma 'acidic' (? ApoE)
Moderate	Tau
	C'3 complement
	Alpha-2-macroglobulin
	Alpha-2 Hermann Schultz
	Group components
	Alpha-1-antitrypsin
	Orosomucoid

TABLE 4.2 Proteins filtered out of the CSF

Extent removed	
Major	Beta lipoprotein
	Haptoglobin polymers
Moderate	Hemopexin
	Protein 'No. 23'

Three proteins seem, by visual inspection of 1D electropherograms, to be relatively rich in CSF when compared with serum; namely beta trace, prealbumin and tau.

Higher relative levels due to selective entry into the CSF and/or local synthesis

The protein with a level in CSF that most nearly approaches (90 per cent) the serum level is clearly beta-2-microglobulin. This protein forms an integral part of the human lymphocyte antigen (HLA) protein complex, which is also found on most cells and thus may normally be released from the brain to the same degree as from other tissues, leading to essentially the same level in all body fluids. The source may be the microglia, which constitute about one-third of brain cells. The CSF levels show no significant changes in most neurological diseases [1118].

There is an order of magnitude drop from the 90 per cent 'transfer' of beta-2-microglobulin to CSF to the 7 per cent shown by lysozyme. Nevertheless, 7 per cent transfer is very much higher than most of the other CSF proteins. In most cases the amount present in CSF parallels that found in serum [427] and is in accord with the degree of barrier damage as estimated by the so-called 'index' of lysozyme with respect to albumin, i.e.

$$\frac{(\text{lysozyme CSF})/(\text{lysozyme serum})}{(\text{albumin CSF})/(\text{albumin serum})}$$

However, in some cases of meningitis lysozyme is probably synthesized within the CNS (possibly by polymorphs), since the 'index' becomes greater than 1.0. Lysozyme also carries a uniquely strong basic charge, having a pI of 10.8.

Prealbumin, like lysozyme, has a very high CSF percentage transfer (6 per cent). However, it differs from lysozyme in that there appears to be a gradient for prealbumin within the CSF circulation. This means that the highest percentages (relative to total protein) are found in the ventricles and the lowest in cisternal and lumbar fluids. Since the absolute level is the same throughout the neuraxis, there must be an influx of other proteins to account for the well-known increase in the gradient of CSF total protein from the ventricles, proceeding along into the lumbar sac. In various neurological diseases, when the total protein rises as a result of damage to various barriers, the CSF levels of lysozyme tend to go up in proportion with the degree of barrier damage [427] while the prealbumin stays relatively static [656, 1274]. As noted above, there may even be local synthesis of relatively large amounts of lysozyme by local, inflamed polymorphs [283, 427]. Nevertheless, there is of course also a serum source of prealbumin which has been estimated using an analogous index [1274]. Prealbumin, unlike lysozyme carries a very acidic charge.

Several other proteins are probably synthesized within the confines of the theca or may have some selective advantage (e.g. very acidic charge) when it comes to entering CSF from the plasma protein pool. Such proteins, with levels of percentage transfer greater than 1 per cent, are mentioned in Table 4.3.

Polymorphs, although typically absent from CSF (except in pus), produce two other proteins: eosinophil cationic protein (5 per cent transfer) and lactoferrin (2 per cent transfer).

Macrophages typically constitute one-third of normal CSF cells and can synthesize complement components (C'1q), therefore it would not be surprising if they could also synthesize C'9. Not only does C'9 have a higher percentage transfer (2 per cent), but the C'9 index is also 3.2, i.e. much higher than 1.0, and this in spite of a molecular weight of 79 kD (versus the denominator protein albumin at 69 kD) [774].

TABLE 4.3 Other cell-derived proteins with high percentage transfer (> 1 per cent but < 6 per cent)

Protein	% Transfer	Source
Ferritin	2	Microglial
Lactoferrin	2	Polymorphs
Complement C'9	2	Macrophages
Eosinophil cationic protein	5	Polymorphs

There are several protein populations, or subfamilies, whose relative proportions have been altered by passage from serum into CSF, i.e. selection in favor of smaller molecular size. Haptoglobin oligomers have been the best studied, and although the monomers may normally enter the CSF, the genetic type Hp 1-1 (41.8 Å) is smaller than type Hp 2-2 (54.2 Å). The occurrence of polymers in CSF is always a pathological sign [268]. IgM monomers are present in significant amounts in CSF [271], but at much lower levels in corresponding serum samples. The ratio of monomeric to dimeric IgA is also tipped in favor of the monomer when measured in the CSF [1061]. Since there are varieties of alpha-2-macroglobulin molecule [41], it would be expected that there would be more of the smaller variety in CSF. There are prominent 'tails' [271] in the distribution of IgG on the basis of molecular size, due to smaller 'fragments' as well as larger 'aggregates' (? immune complexes).

In conclusion, when regarding the quantitative amounts of proteins in CSF by correlating CSF concentrations with serum concentrations, the notion of percentage transfer (PT) is very useful in that it suggests different sources of proteins (see Table 4.4):

1. If the PT is approximately 500 per cent, i.e. the protein level is five-fold higher in CSF than in serum, then the brain is the principle source of synthesis (e.g. gamma trace or cystatin C)
2. If the PT is approximately 100 per cent, the brain is a source approximately equivalent to other tissues (e.g. beta-2-microglobulin)
3. If the PT is 2–7 per cent, there is probably a minor proportion of cells within the brain and/or theca which synthesize the protein compared with similar cells outside the CNS (e.g. polymorphs synthesizing lysozyme)
4. If the PT is < 1 per cent, proteins are probably derived mainly from the plasma pool.

However, careful attention should be paid to the molecular size, to see if this protein behaves in accordance with the principles of filtration (small molecules pass into CSF more readily than larger molecules). A CSF protein with behaviour

TABLE 4.4 Proteins regrouped in order of 'percentage transfer'

500 per cent (approx.) Primarily intrathecal synthesis	Beta trace Gamma trace Tau protein (beta transferrin) Myelin basic protein Glial fibrillary acidic protein S100
100 per cent (approx.) Intrathecal synthesis similar to systemic levels	Beta-2-microglobulin Enolase (gamma) D-2 antigen (N-CAM protein) Fibrinogen degradation products
10–1 per cent Partial intrathecal synthesis	Prealbumin (transthyretin) Lysozyme Eosinophil cationic protein Complement C'9 Lactoferrin Ferritin Apo-E-lipoprotein
< 1 per cent Mainly filtration from plasma	Rest of proteins (from Table 3.3)

at odds with the size rule is probably synthesized locally (e.g. beta-2 transferrin or tau protein).

Qualitative

There is a number of CSF proteins which differ qualitatively from their serum counterparts. For example, ostensibly (by immunochemical analysis) there is more transferrin relative to albumin in CSF than serum. This is due in part to the inclusion of desialated transferrin (tau protein) in CSF measurements. However, since removal of the sialic acid would result in fewer antigenic sites per molecule, the data must be interpreted with some caution (see Chapter 7).

There are various other as yet only partially-characterized CSF proteins which are relatively more prominent in CSF [280] than in parallel serum (Table 4.5), and these are best considered by their physical–chemical characteristics.

TABLE 4.5 Proteins which show extremes of size or charge in CSF compared to serum

Protein size	Small: five proteins (e.g. beta-2-glycoprotein III; 35 kD)
	Large: four proteins (e.g. complement C'lq; 410 kD)
Protein charge	Acidic: 'pre'-albumin
	Basic: gamma trace

First, on the basis of molecular size, there are five proteins, including beta-glycoprotein III, which are smaller than prealbumin and four, including complement protein Clq, which are larger than alpha-2-macroglobulin.

Secondly, on the basis of molecular charge at least one additional protein is more acidic than prealbumin and one, gamma trace (or cystatin C), is more basic than IgG. Further work is required to ascertain the functions of these proteins and the reasons for their relative abundance in CSF.

Thirdly, several protein congeners have been slightly modified by the CSF environment relative to their counterparts in serum – e.g. the prealbumin in CSF shows an extra 'spur' on electro-immunodiffusion, thus revealing an altered antigenic state. There is a 'double' peak on the Laurell rockets (electroimmunodiffusion) which also demonstrates prealbumin's heterogeneous nature. The net charge is also more acidic in CSF than in serum [621]. However, these differences may reflect the higher relative amounts of prealbumin to albumin in CSF than in serum [265]. Transferrin loses sialic acid to become tau protein. This may be due to its uptake from the serum by the brain, and its subsequent release minus the acidic sugar ([197]; see also Chapter 5). There does not appear to be a similar change in orosomucoid since the pI is unchanged [1040, 1041]. Alpha-2-macroglobulin also has a

TABLE 4.6 CSF Proteins which differ from their serum counterparts in serum

Protein	Variation
Prealbumin	Abnormal 'spur'
Tau	Sialic acid removed
IgG	Fragments and aggregates
IgM	More monomers
Haptoglobin	Fewer polymers
Alpha-1-antitrypsin	Different pI pattern
Alpha-2-macroglobulin	Different pI pattern
Retinol binding globulin	Not bound to prealbumin
Fibrinogen	Without EDTA, becomes fragments?

TABLE 4.7 Paraneoplastic antibodies and their properties [474, 1257]

Antibody	Alternate	Sex	Tumor	Clinical	Protein	MW (kDa)
Hu	ANNA1	M > F	SCLC	Neuropathy/encephalitis (OCB+)	RNA bind	40/35
Yo	PCA1	F ≫ M	Br/Ov	Ataxia	DNA bind	58/34
Ri/NOV2	ANNA2	F	Breast	Opsoclonus-myoclonus-ataxia	RNA bind	80/55
CRMP5	CV2	F > M	SCLC	Ataxia/dementia/cranial nerve/chorea	Normal development	66
Ta/Ma2		M ≫ F	Testis	Encephalitis	?	40
Recoverin		M = F	SCLC	Visual loss	PhotoR	23
ANNA3		M = F	SCLC	Ataxia/encephalitis/neuropathy	?	170
PCA2		M = F	Lung	Ataxia/encephalitis/autonomic	?	280
Amphiphysin		F	Breast	Stiff	SV endocytosis	128/125
Non-malignant						
GAD		M = F	Diabetes	Stiff (OCB+)	GAD	67/65
PANDAS		M = F	Strep	Chorea/psychiatric	? enolase	45
MAG		M = F	Para	Neuropathy	MAG	30

Abbreviations: SV = synaptic vesicles; SCLC = small cell lung cancer; Br = breast; Ov = ovary; Para = paraprotein

different pI in CSF, being more cathodic. Likewise, alpha-1-antitrypsin is different between serum versus CSF [1111]. Native C'3 (69.5 Å) is cleaved to produce C'3a (53.3 Å). Fibrinogen is initially consumed to form a clot, and then hydrolyzed in many different pathological syndromes to yield fibrin degradation products (FDP). Retinol-binding globulin is not bound to prealbumin [881], it may be speculated that this is due either to a much lower affinity for prealbumin after having delivered vitamin A, or to having been taken up by the brain (as in the case of transferrin) and released in a slightly altered form. Table 4.6 provides a short summary of the major changes in proteins after leaving the serum for CSF.

Finally we should consider serum 'only' tests (as opposed to including parallel CSF) performed to diagnose various neurological diseases. There is a whole series of anti-ganglioside antibodies which are measured in patients with peripheral neuropathy. Although Pestronk at Washington University is quite keen to measure these antibodies (www.neuro.wustl.edu), P. K. Thomas has been less enthusiastic (Personal communication, 2003).

There has also been much controversy over the years concerning the pros and cons of using a paraneoplastic classification based primarily on either a cellular or a molecular approach. The cellular approach involves histological characterization of staining patterns produced by the different antibodies, whereas the molecular approach involves Western blotting against antigens of different molecular weights. Historically these antigens were Hu and Yo, but the family now includes Ri and several others, and continues to grow as experience in this area increases. Our hospital's preference is to perform both cellular and molecular tests. Some of the recent molecular data are set out in Table 4.7.

CHAPTER

5

Different blood–CSF barriers

Introduction
Different sources of CSF proteins
Conclusion

Introduction

Normal CSF is derived mostly from the choroid plexi and also from interstitial fluid. Under various pathological conditions, significant sources can include: the meninges, the dorsal root ganglia, the CSF cells and brain cells. CSF from each source (normal as well as pathologic) has a typical protein pattern.

Different barriers produce specific fluids with individual mixtures of different proportions of proteins. As the CSF flows along the neuraxis, the different fluids will intermingle to give rise to differences in both the quantitative and qualitative nature of the protein mixture.

Different sources of CSF proteins

The individual blood–CSF barriers (or sources) produce characteristic fluids, each with different proteins (differentiated mainly by their size). These discrete fluids then mix according to the unusual flow dynamics of the CSF. Circulation of the blood through the brain is unlike that in any other organ of the body. This was dramatically demonstrated by intravenous injection of the blue dye (which is immediately bound to albumin), when the entire animal became blue except for the brain which remained white, hence the concept of the blood–brain barrier was born [90]. Several books have since been written on this topic [91]. The circulation of CSF is almost static relative to the circulation of blood, i.e. CSF may be renewed four times per day (i.e. approximately every 6 hours), but the heart typically beats once per second. The volume of CSF circulated is much less than the volume of blood moved, since only approximately 0.3 ml/min of CSF is formed, whereas the heart has a stroke volume of about 50 ml/s. As a proportion of the total volume moved per unit time, therefore, the CSF circulates over 500 times more slowly than blood. CSF flow is even slower in the lumbar sac (being a 'dead end') than over the rest of its course, during which it flows up and over the cortex to be reabsorbed in the arachnoid villi. As we shall see, this relatively static CSF circulation makes for some rather striking differences between the CSF proteins and the typical plasma proteins.

Aside from the dramatic difference in speed, the circulation is quite different for the 'dead end' down to the spinal cord compared with the path of less resistance up over the cortex. Proteins released into the CSF by the brain (through the white matter) may be carried down into the lumbar sac [225], but any that are released through the cerebral cortex are swept into the arachnoid villi and back into the venous blood. In the spinal cord, the white matter is mainly on the surface, with the grey matter buried inside (analogous to the ventricles). We would thus expect to see proteins in the lumbar CSF derived more from the oligodendrocytes, and perhaps the astrocytes or microglia, than from the neurons.

Barriers to the egress of plasma proteins are found in other tissues, notably the kidney as depicted in Figure 5.1. Here the most important portion of the barrier is the basement membrane, since after removal of all the cells using detergents (Triton), the remaining extracellular skeleton still performs the macromolecular sieving action to the same degree as the intact kidney [94].

Nevertheless, there are rather striking differences between various tissues in their ability to filter proteins, as seen in Figure 5.2. Indeed there may also be changes in different pathological states within the same tissue such that the filtration process is quite different depending on the specific pathological circumstances. The two parameters which have been used to characterize these different tissues and pathological processes are: (1) the degree of selectivity (or slope) in relation to molecular size; and (2) the absolute level of albumin transfer (or 'Y' intercept).

These two values, rather like the parameters for a straight line equation, allow the distinctive characterization of each barrier function. Using Reiber's data [902] (see Figure 5.3), the slope of the line is determined by dividing the 'Y' increment (0.4–0.1 per cent) by the 'X' increment (90–60 per cent). Since the 'Y' value is a dimensionless fractional number and the 'X' value is expressed in Angstroms, the value of the slope is reciprocal Angstroms (1/Å). On the logarithmic scale, the most useful value for the 'Y' intercept would be the value of percentage transfer for albumin (0.4 per cent), rather than extrapolation to zero Angstroms. Another alternative has been the calculation of the percentage transfer of the total protein, treated operationally as an equivalent protein of a particular radius as read from

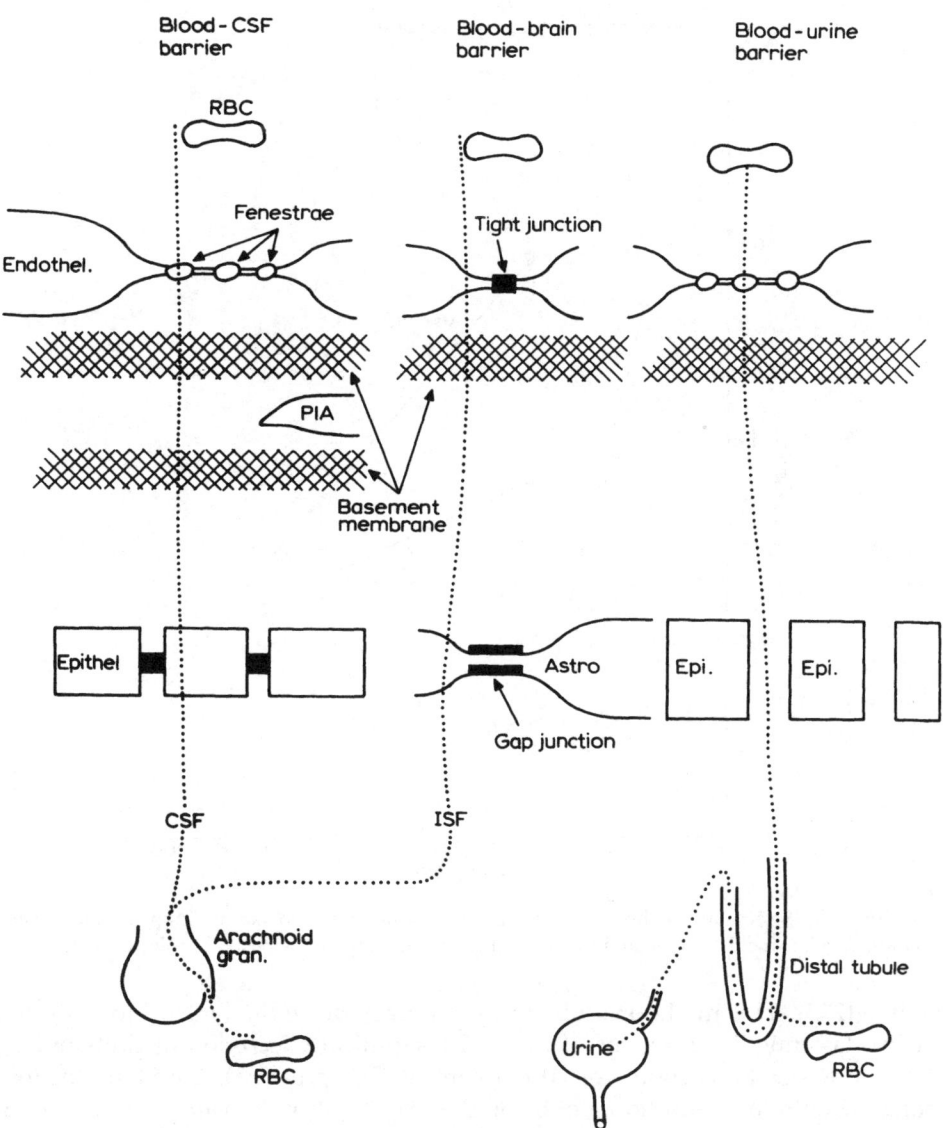

FIGURE 5.1 Anatomy of three different barriers.

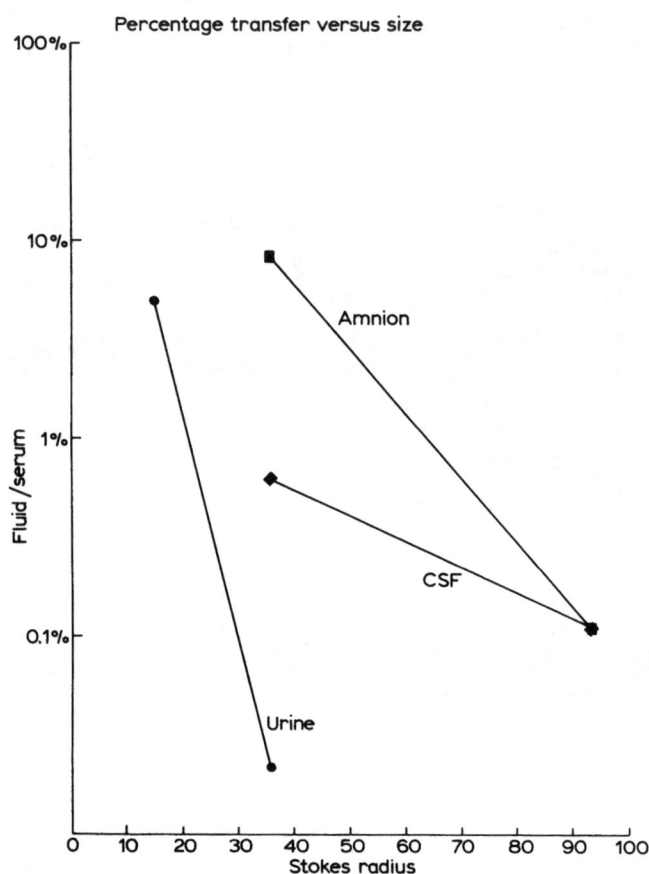

FIGURE 5.2 Selectivity of three different barriers: amniotic fluid has the highest total protein (albumin) level, urine the lowest and CSF is least discriminatory on the basis of molecular size.

the line [273]. This method can be rather indirect, since the level of total protein is often 'assumed'. Moreover, if there is a significant secretion of proteins (e.g. local synthesis can account for 20 per cent of CSF proteins), the value for total protein should, by definition, not be on the line. Most of the values published are for lumbar CSF, whereas ventricular fluid has much more prealbumin and much less albumin and IgG, resulting in a different Y intercept, i.e. a lower value of total protein (mainly albumin).

Although it is widely believed that the 'blood–brain barrier' keeps most proteins out of the brain, it is also possible to look on this and indeed other barriers (e.g. the 'blood–CSF barriers') as being sources of protein-containing fluids, with different sources (different barriers) giving rise to different types of

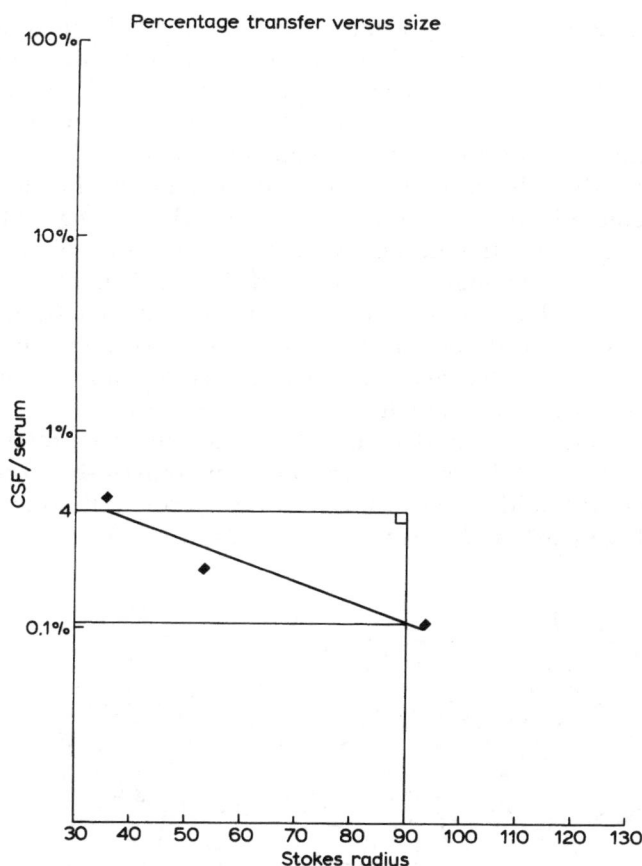

FIGURE 5.3 Parameters of the selectivity line: slope and possible 'Y' intercepts.

protein-containing fluids, and varying in composition according to the different sources (barriers). Perhaps the best known example is the difference between ventricular fluid and lumbar fluid. While it is widely appreciated that the level of total protein is approximately 300 per cent higher in lumbar than ventricular fluid, it is not always realized that this does not simply involve an increase in the concentration of all proteins by the relative removal of water. For example, one protein which remains unchanged in absolute concentration in both fluids is prealbumin. Historically, prealbumin was noted in the first electrophoretic study of CSF [515] as a 'new' protein, i.e. thought not to be present in serum. However, it soon became clear that this was not a brain-specific protein (i.e. derived primarily from the brain parenchymal tissue), but rather was also derived from the serum. There is nevertheless a particular mechanism (local synthesis in the choroid plexus) responsible for prealbumin being selectively included in the CSF. The important point is that the relative amounts of prealbumin are clearly different in ventricular and lumbar CSF, i.e. prealbumin constitutes a much higher proportion of the proteins in the ventricles than in the lumbar sac. However, by immunochemical quantification, the absolute level was found to be the same in both fluids, and thus the relative decrease of prealbumin in the lumbar fluid reflects the absolute increase of other proteins, i.e. the mixing with the interstitial fluid. Note again that the increase of other proteins is absolute, such that the total protein is 300 per cent higher in lumbar than in ventricular fluid. This increase primarily represents albumin, since this is the major protein in the CSF. Although prealbumin decreases relatively, there is a relative 6-fold increase in IgG [1271]. It can therefore be concluded that choroidal fluid is rich in prealbumin, whereas interstitial fluid is rich in albumin, IgG and other proteins.

Therefore, not only does this demonstrate that the kinds of protein patterns are different, but there must be at least two different sources (barriers) to produce these rather different fluids. These fluids have been termed choroidal (ventricular) and interstitial (lumbar) fluids, respectively (see Figure 5.4).

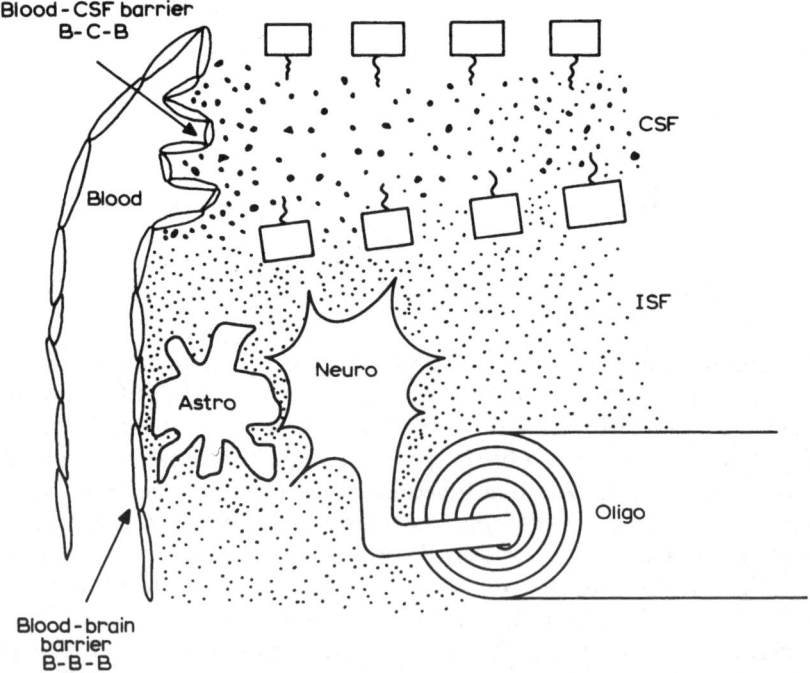

FIGURE 5.4 Two normal brain barriers which proteins transgress to reach the CSF.

Interstitial fluid is estimated to account for up to one-third of the total CSF, and appears to be analogous to brain 'lymph'. The few lymphatic channels in the brain are principally located near the major blood vessels [886]. There is compelling evidence from electrical impedance studies for the existence of the interstitial fluid (distinct from ventricular CSF and the CSF passing over the cortex). Although electron microscopists had previously assumed that there was no interstitial space, more recent techniques for tissue fixation have concurred with the electrical studies of the physiologists that different fluids have quite different protein compositions.

Various minor sources of CSF proteins, e.g. dorsal root ganglia and ventral roots can become major sources under pathological conditions. The passage of plasma through these particular areas occurs relatively freely, i.e. the endothelial barriers are quite leaky [838] and in Guillain–Barré syndrome, the barriers can become even more damaged, such that the level of total protein in lumbar fluid can be extremely high, while being quite normal in the ventricles of the same patients (Figure 5.5).

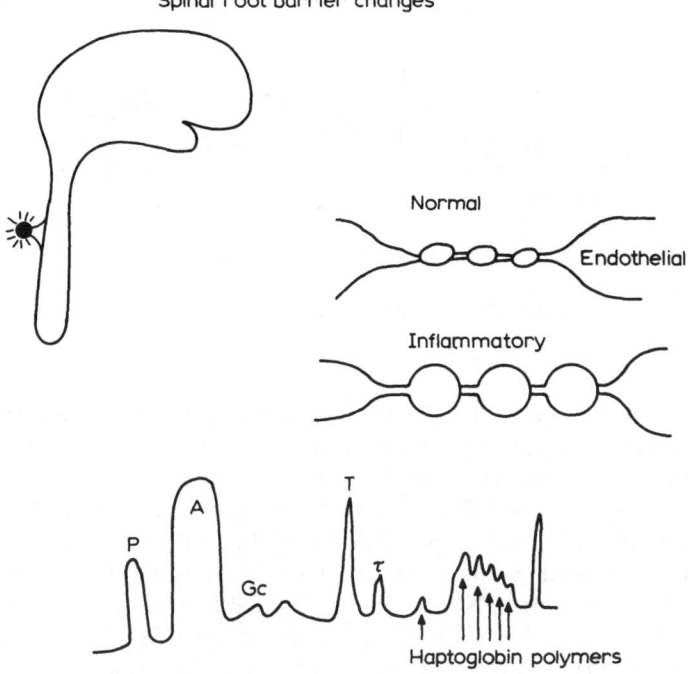

FIGURE 5.5 Normal barriers which can become abnormal: spinal roots.

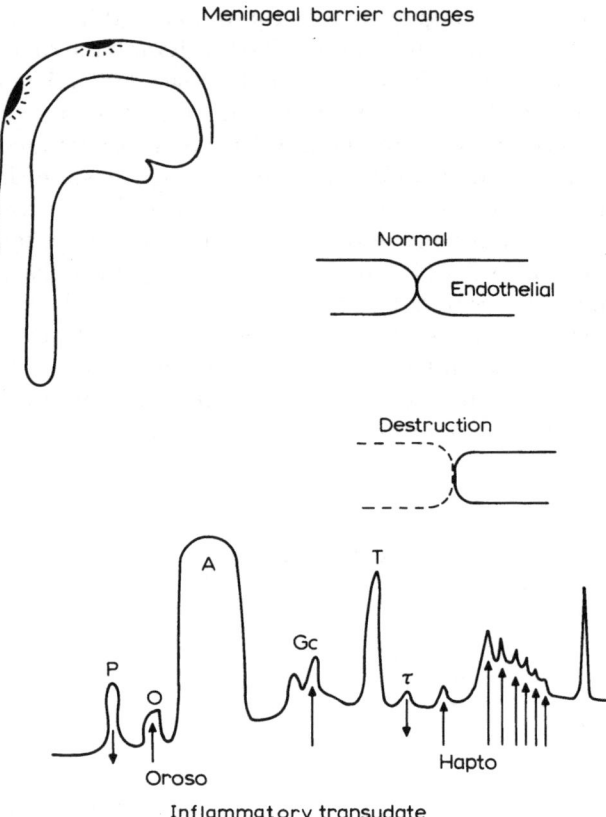

FIGURE 5.6 Abnormal barriers: meninges.

The meninges, another normally minor source, become prominent in disease. They may become very leaky when inflamed as a result of bacterial or viral infection or even secondary carcinoma, as depicted in Figure 5.6.

Under various pathological conditions (described in detail in Chapters 10 and 11) there are eight other protein patterns (see Table 5.1).

Once again, because of the relatively slow flow of CSF, proteins which are released locally within the CNS do not readily diffuse. Given the small pool size (120 ml CSF of compared with 3000 ml of serum), and since there is no filtration either by size (as in the kidney: low molecular weight fragments) or by charge (as in the reticulo-endothelial system where proteins which have no neuraminic acid are selectively removed from the systemic circulation), any 'local' CNS changes are easily reflected in the relatively protected or isolated environment of the CSF.

TABLE 5.1 Abnormal protein patterns on polyacrylamide gels

Monclonal (one band) IgG
Oligoclonal (2–9 bands) IgG
Polyclonal (diffuse cathodic gamma) IgG
High molecular weight transudate of haptoglobin polymers
Frank transudate of all proteins
Inflammatory transudate of acute phase proteins
Elevated tau protein
Elevated prealbumin

The best example is that of oligoclonal bands seen on electrophoresis. In this case, a restricted number of plasma cells within the CNS release their clonal products, which are in excess of the low polyclonal background and are therefore clearly discernible. Typically, serum from the same patient does not show these bands because they are effectively 'diluted out' by the much higher level of IgG ($\times 400$) as well as the larger volume ($\times 25$), the net protein 'dilution', i.e. per gram IgG of specific clones, being over 10 000-fold, as we shall see later.

When plasma cells synthesize immunoglobulins, disproportionately more light chains than heavy chains are produced, leading to excess amounts of 'free' light chains (i.e. not bound to heavy chains). In the serum these are readily filtered out by the kidney and thus are typically found in myeloma urine (i.e. as Bence–Jones proteins). Another protein which is typically found only in CSF is tau protein, which is a desialated (i.e. sialic acid or N-acetylneuraminic acid has been removed) version of transferrin. When Fe^{2+} (from transferrin) is delivered to cells [197], the acidic lysosomal environment may also cleave the sialic acid and the systemic reticuloendothelial cells which are not present in white matter (i.e. ventricles and most of the spinal cord) may remove the recycled protein (minus the sugar residues). This hypothesis is depicted in Figure 5.7.

There are three different types of classification each reflecting a different vantage point:

1. *Anatomy* – barriers and cells as sources (Table 5.2);
2. *Pathology* – mechanisms of pathophysiology (Table 5.3);
3. *Chemistry* – profiles of protein patterns (Table 5.4).

One of the fundamental concepts of protein barriers is that they are relative rather than absolute. The classical work of Felgenhauer [271] has shown that the single most important variable for proteins is their molecular size, specifically their Stokes–Einstein radius as tumbling molecules in solution [270]. There is a natural extension to this hypothesis: no single protein has been described as being completely excluded from the brain by an absolute barrier. As mentioned elsewhere,

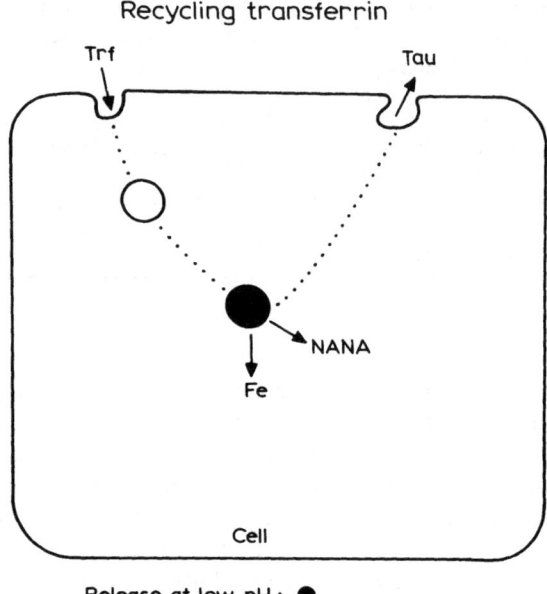

FIGURE 5.7 The cellular cycle of transferrin with possible loss of the acidic sugar NANA.

it is not an uncommon fallacy to think that IgG does not enter the brain (1 in 400 molecules normally passes from blood into the brain). There is also a variation on this fallacy whereby the presence of a given 'brain-specific' protein is measured in the cerebrospinal fluid, but unfortunately the investigators neglect to measure the amount of the same protein present in the serum. If they find the same level of serum immunoreactivity, then it is almost a contradiction in terms to think that what is being measured is brain-specific; at the least this is a serious devaluation of the definition of the term 'brain-specific'. One striking example is that of neuron-specific enolase (NSE), found in CSF and serum in the same amounts, i.e. 0.01 mg/l [905]. Clearly, there is an epitope of systemic origin in serum which is being produced in equal amounts to the enolase produced by the brain.

The blood–CSF barrier can simply be thought of as a concentration gradient whereby successively higher levels of a protein in serum will be passively reflected in similar levels in CSF, as typified by Q_{alb} (i.e. the albumin quotient = the concentration of albumin in CSF divided by the concentration of albumin in serum).

Mother Nature has conveniently provided markers of different molecular sizes in the serum, which have allowed us to study normal, 'physiological' barrier

TABLE 5.2 Anatomical basis for classification: different sources of CSF proteins

Barriers as sources
  Normal (filtrate)
    Blood–CSF barrier produces choroidal fluid (about 2/3 of CSF)
    Blood–brain barrier produces interstitial fluid (about 1/3 of CSF)
  Pathological (transudate)
    Blood–CSF barrier: cholera toxin
    Blood–brain barrier (via CSF): endothelial damage
    Blood–CSF 'barrier' (isolated from choroidal fluid); Froin's
  Minor sources which become major sources
    Meninges: meningitis or secondary carcinoma
    Spinal roots: Guillain–Barré syndrome
    Cauda equina: local pressure
Cells as sources
  Normal
    Macrophages: complement C'1q; C-reactive protein
    Lymphocytes: IgG
    Polymorphs: lactoferrin; lysozyme; eosinophil cationic protein
  Pathological
    Macrophages: bacteria
    Lymphocytes: plasma cells
    Polymorphs: pus
  Minor sources which become major sources
    Oligodendrocytes: myelin basic protein
    Astrocytes: glial fibrillary acid protein
    Neurons: 14-3-2 (enolase, gamma type)
  Tumors
    Lymphocytes: paraprotein; beta-2-microglobulin

function *in vivo*. These haptoglobin polymers come in three convenient sizes, the smallest being Type 1-1 (see Figure 5.8A), the largest is Type 2-2 and the intermediate is a hybrid, Type 2-1 [132]. We examined CSF donated by normal Swedish volunteers of increasing ages, who had been carefully examined to exclude any detectable neurological abnormality. Since it has been well documented for many years that the level of CSF total protein increases with age, it was therefore interesting to be able to demonstrate a statistically significant difference between the normal passage of the intermediate-sized haptoglobin, Type 2-1, compared with the larger Type 2-2, the latter normally being retarded by the barrier up to the age of 45 years, but thereafter passing more readily through the normal (i.e. non-diseased) blood–CSF barriers.

A rather different approach to the 'absolute' permeability of barriers was previously fostered by electron microscopic studies of brain capillaries with horseradish peroxidase (HRP). The molecular weight of HRP (80 kD) is rather similar to that of albumin (69 kD), but nevertheless, quite dramatic pictures of black, heavily

TABLE 5.3 Pathological basis for classification: mechanisms of pathophysiology

1. Decreased flow of CSF
    Decreased production: (alcoholic) increased beta-2-tau (? secondary to liver failure)
    Normal production
        Obstruction: Absent (Jakob–Creutzfeldt) increased tau (secondary to degeneration
                                                                with no compression)
        Obstruction: Present
            Diffuse site (communicating hydrocephalus)
            Single site
                Intermittent: (prolapsed disc) increased tau
                        (compression produces degeneration)
                Complete
                    Above site: (non-communicating hydrocephalus)
                        Increased prealbumin (choroidal)
                    Below site: (Froin's) decreased prealbumin (interstitial fluid)
2. Degenerative processes (see above) increased tau
3. Transudate (serum leak)
    Non-inflammatory (or minimal inflammation)
        Minimal: high molecular weight (local inflammation)
        Maximal: frank (group components > tau)
    Inflammatory
        Mainly acute: increased orosomucoid
        Mainly chronic: diffuse gamma 4-5 (polyclonal)
4. Immunogenic (oligoclonal)
    Primarily inside CNS: gamma 4 gamma 5 (serum negative)
    Primarily outside CNS: gamma 4 gamma 5 (serum positive)
5. Paraprotein (monoclonal)
    Primarily inside CNS: (serum negative) CNS lymphoma
    Primarily outside CNS: (serum positive) myeloma or benign?
    Primarily outside but secondarily burn through: (myeloma) i.e. CSF > serum, therefore
                                                                intrathecal therapy
6. Hemic ('no' barrier)
    Traumatic tap: decreasing number of red cells in tubes 1–3
    Stroke: no change of red cells in tubes 1–3
    Monocytic: C-reactive protein

stained enzyme products from the HRP enzyme were shown on the luminal (blood) side of the endothelial cells, as opposed to the virtual absence of any black staining within the parenchyma of the brain [899], see Figure 5.9. This gave rise to a rather simplistic or 'absolute' quality regarding the blood–brain barrier, as had been introduced many years earlier, when the concept of the blood–brain barrier was first born [90] using trypan blue dye bound to albumin. The HRP 'absolute' argument was even more persuasive, given the simultaneous demonstration of the tight junctions between endothelial cells which again showed, in effect, the damming up of HRP enzyme to prevent the entry of this protein into the brain.

TABLE 5.4 Chemical basis for classification: pathological protein patterns

Pathological type	Cell source	Protein pattern
1. Transudate (high molecular weight)	'Endothelial min'	Increased pregamma
2. Transudate (frank)	'Endothelial max'	Group components > tau
3. Transudate (inflammatory)	'Meningeal'	Increased orosomucoid
4. Degeneration (+ compression)	'Parenchymal'	Increased tau
5. Obstruction (above site)	'Choroidal'	Increased prealbumin
6. Obstruction (below site)	'Interstitial'	Decreased prealbumin
7. Local synthesis (CNS)	'Immunogenic'	Increased gamma 5
8. Traumatic (previous bleed)	'Hemic'	Increased gamma 2
9. Plasmatic (vs. primary local)	'Paraprotein'	Increased monoclonal
10. Complement (+ conversion)	'Monocytic'	Increased gamma 1

This point can be further confirmed from the results of the original experiments on the blood–brain barrier, whereby the blue dye (although this is a low molecular weight substance, it is tightly bound to albumin) was injected into animals, and yet essentially all the tissues of the animal turned blue, except the brain, which remained white.

These 'black blood (i.e. stopped by tight junctions) versus white brain' pictures of HRP staining easily encourage the notion of an absolute barrier with the total exclusion of the HRP protein, in spite of its molecular size (which again is essentially the same as albumin, the most abundant protein within the CSF). However, what we are really seeing is not an absolute difference, but a relative difference, which on balance equals two orders of magnitude (i.e. the ratio for albumin of 1 to 200, CSF to serum). It could therefore be a kind of 'optical illusion'. The principle remains that, however visually dramatic, there is no absolute barrier for any protein, whatever its Stokes–Einstein radius.

It is also worth considering the other end of the spectrum of barriers, namely a complete absence of any barrier on the basis of molecular size for the classical model of the superior sagittal sinus. Specifically, the rate at which red cells at one extreme of size and individual water molecules at the other, pass from the CSF back into the venous blood is exactly the same despite the wide spectrum of molecules which have previously crossed over from the endothelium into the interstitial fluid itself. There is once again no barrier to the movement of any proteins, regardless of their molecular size, restricting the migration of extracellular/interstitial fluid between the various cells found in the brain parenchyma. The fluid eventually moves into the ventricles, and here also there is no hint of any barriers to impede the flow of proteins between the ependymal cells lining the ventricular surfaces of the 'inside' of the brain. This was originally shown by injection of the blue dye into the ventricles of animals and the entire brain quickly became blue.

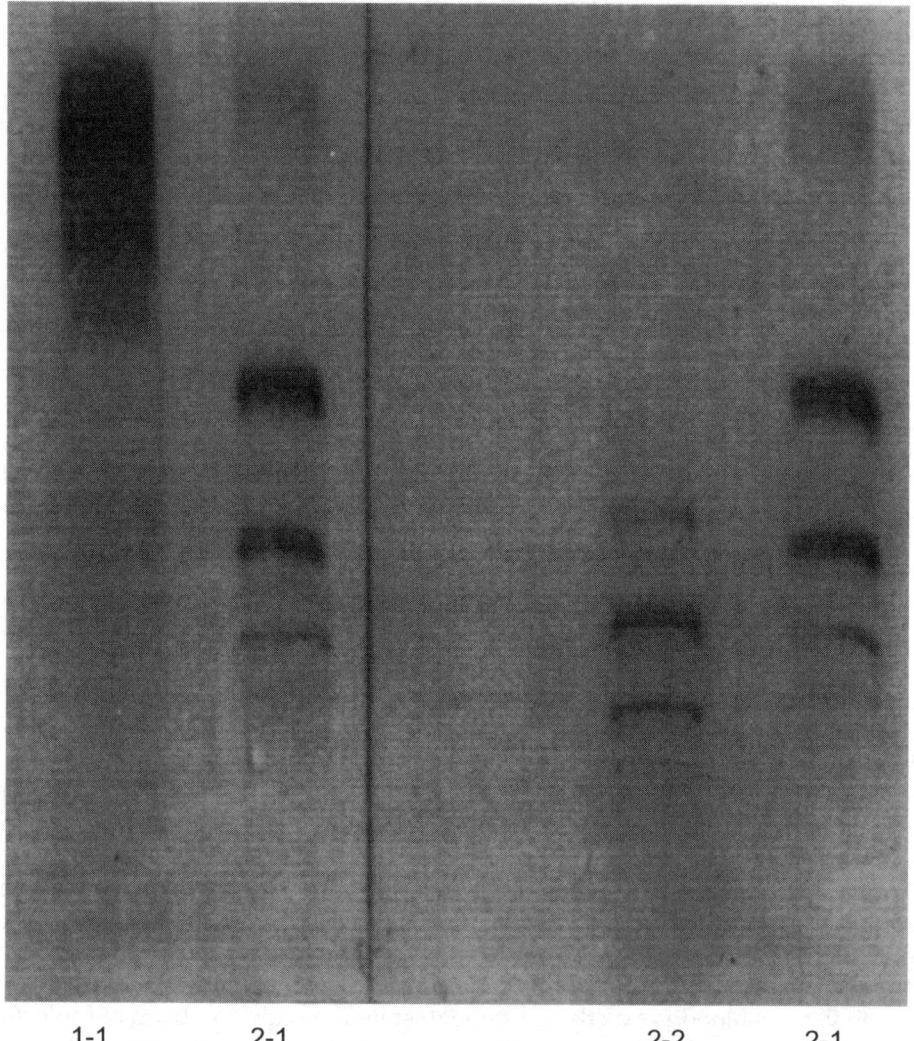

FIGURE 5.8 Three types of haptoglobins of increasing size: 1-1, 2-1 and 2-2.

It is also worth noting that on the 'outside' of the brain, namely the cortex, the Virchow–Robin spaces plunge into the cortex following the penetration of the major arteries, and the cellular constituents within these spaces [255] are effectively a continuum of the cortical CSF. There are however, differences between the

"Black" blood

Endothelial cell

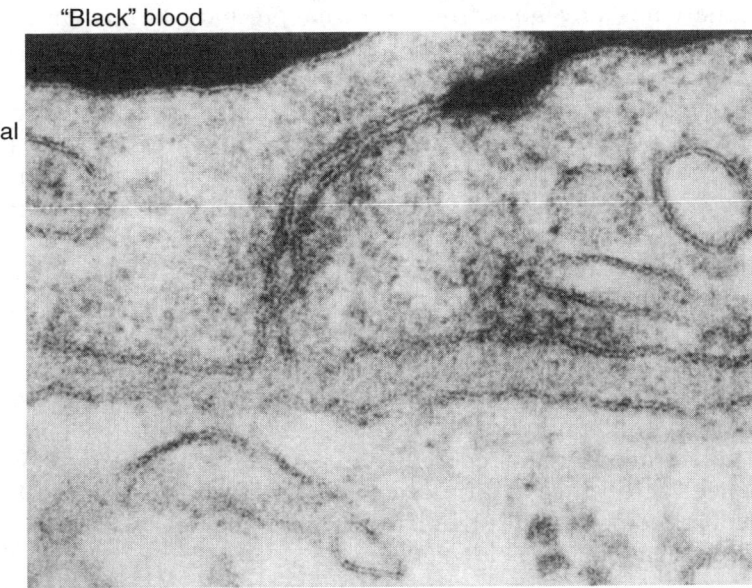

"White" brain

FIGURE 5.9 Typical HRP stain of 'black' blood versus 'white' brain.

migrating cellular constituents found in the CSF and those cells migrating between the static parenchymal cells. Specifically, the CD4/CD8 lymphocyte ratio is 2.4 in the CSF versus 0.4 in the CNS (i.e. six times more CD4 lymphocytes in CSF), which reinforces the differences between the ratios found in the blood and CSF. The ratios of the two cell types are also dependent upon the different cytokine mixtures found within the brain parenchyma itself.

Perhaps the most dramatic difference between the blood and the spinal fluid is the fact that the most common white cell in the blood is the polymorph, whereas this cell is essentially absent in normal spinal fluid (although there may be the odd polymorph following a slightly traumatic tap). Lymphocytes move across the endothelium of the brain via processes which have been described in considerable detail utilizing Inter-Cellular Adhesion Molecules (ICAM). Plasma cells, like polymorphs, are exquisitely rare in normal spinal fluid.

In summary, there are four reasons why the blood–CSF barriers will induce differing relative/absolute amounts of individual proteins, as one follows the fluid along its tortuous route(s):

1. The principal bulk flow (2/3) from the choroid plexi through the ventricles down to the foramen magnum, and then eventually up/over to the superior

sagittal sinus for venous return, does not preclude the movements which could be considered 'lateral', not only along the main route but also typically down into the cul-de-sac covering the spinal cord. The residual flow (1/3) derives from the interstitial/extracellular fluid, which originates further upstream from the capillaries of the blood–brain barrier, in addition to the local proteins/fluid released by static parenchymal cells of the brain. It is worth remembering that 'pure' water is also synthesized locally in substantial amounts (about 10 per cent) from the actions of the Krebs cycle on glucose.

2. The second principal difference among the several blood–CSF barriers is that they are quite distinctive in various locations. The most easily visible example of these different loci is seen through a more careful examination of the classical experiments, whereby an animal is injected intravenously with Evan's Blue dye and although the brain is essentially white, nevertheless one can see ready diffusion of albumin bound dye into the dorsal roots [838]. This is also seen in the so-called circumventricular area, where again there is normally a much higher degree of easy diffusion of the dye from blood into CSF. This major point reinforces the idea that barriers are dramatically different at various loci within the brain (see below).

3. The next major difference is time-related, namely the circadian rhythm, in which there is a 600 per cent increase in the rate of production of CSF, peaking at about 12:00 midnight and falling to its lowest level at 12:00 noon [1204].

4. Last, but not least, there are dramatic differences among various pathophysiological mechanisms, the most dramatic example of this being seen in meningitis. In a normal person the degree of fluid movement from the blood vessels of the meninges into the CSF is minimal at best, whereas bacteria can spread along the inside of the dura causing a purulent meningitis. This reaction thereby allows a massive influx of serum albumin and other proteins through the vessels of the meninges, and thus directly into the CSF.

The five different loci shown in the diagram (Figure 5.10) [1139], illustrate the main sources of egress from the blood or other tissues by which proteins enter the CSF, depending upon the normal/pathological circumstances of the individual patient (which are also age-related) [338].

It is also worth calculating the differences in concentration between CSF and serum in a mathematical consideration of the protein gradients as one moves from the brain back into the blood (again bearing in mind that there is basically no size barrier to this movement). If we choose as an example the IgG molecule, we know that the total volume of the spinal fluid is of the order of 120 ml, whereas the total volume of blood is approximately 3 L. There is therefore an initial factor of a twenty-five-fold difference between the two fluids. We also know that the absolute concentration of IgG between the two fluids is of the order of 1 to 400. Hence the final product when these two factors are cross-multiplied is essentially

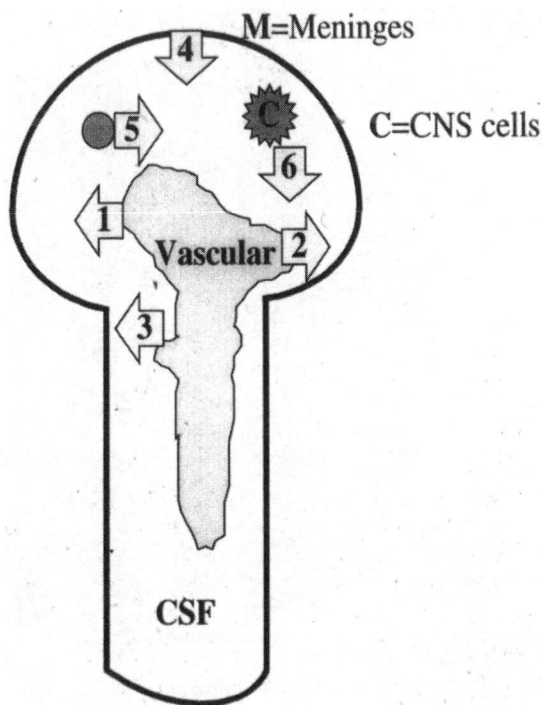

FIGURE 5.10 Cartoon of the 5 different sources of CSF in normal versus abnormal conditions.

four orders of magnitude (10 000) (see Figure 5.11). Trying to find any unique IgG molecule derived from an individual plasma cell synthesizing its antibody within the brain (and then releasing its clonal products into the vast ocean of the blood to be diluted 10 000-fold by its neighbouring IgG molecules) is therefore worse than looking for the proverbial needle in a haystack. Experiments to demonstrate this have been published in the form of IgG titrations [1317] (see also Chapter 10, Figure 10.10), but the main point in the case of multiple sclerosis, is that protein bands in the blood are not just coming from the brain, but rather there must be a systemic source for most of these antibodies. This remains one of the (as yet) unexplained paradoxes of this disease.

Results of studies performed on rats [586] indicated that there was so much intrathecal synthesis of IgG that the normal ratio of CSF to serum was almost completely reversed. There was a relatively large amount of antibody present in the brain of the rats, not withstanding all of the 'contiguous' (cervical) lymph nodes, yet essentially no IgG was found in the serum, see Figure 5.12. The antigen was instilled directly into the brain via an indwelling catheter. This does not cause

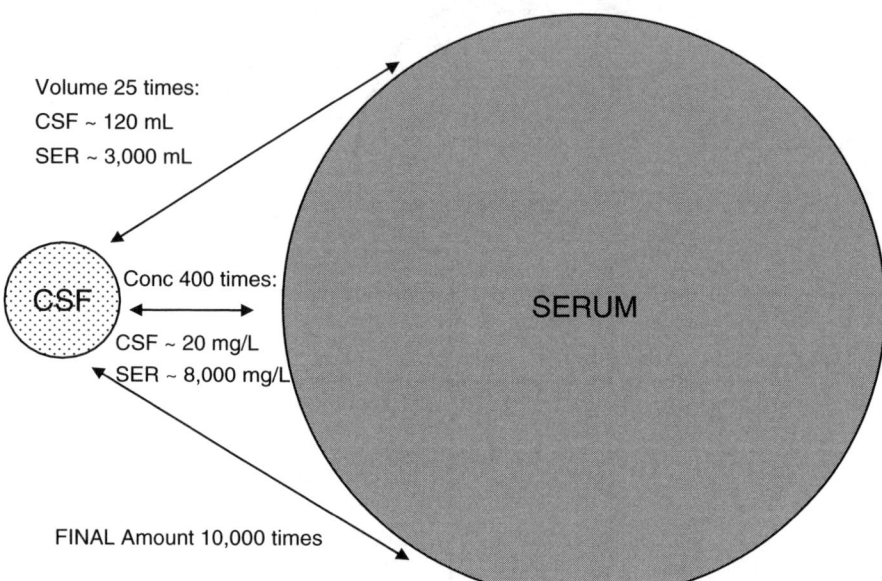

FIGURE 5.11 The idea of a 10 000-fold dilution of IgG in CSF versus serum.

the animal any distress, since we know that otherwise normal children can lead an active life in spite of having chronic hydrocephalus and an indwelling ventriculo-peritoneal shunt.

Since all serum proteins pass into CSF down a gradient (depending on their molecular size), the diagnosis of myeloma can therefore also be made via lumbar puncture. This is obviously not the method of choice to make this particular diagnosis, but we must always be attuned to this possible diagnosis by recognising the characteristic 'harmonic' pattern which occurs when a paraprotein is revealed by isoelectric focusing. Given a systemic source of IgG as the principal component of CSF IgG, the original work of Kahn et al. [516] showed that there was a much higher incidence of so-called 'benign' paraproteins associated with various neurological diseases than was found with other tissue- or organ-specific diseases.

The interstitial fluid which emerges in the region of the spinal cord is strikingly different from that which flows down from the brain, in that the albumin/total protein level is about 300 per cent greater. How much of this is due to decreased flow on the one hand versus higher permeability from the large number of dorsal roots (i.e. 62) has yet to be determined. It is well known that when patients are

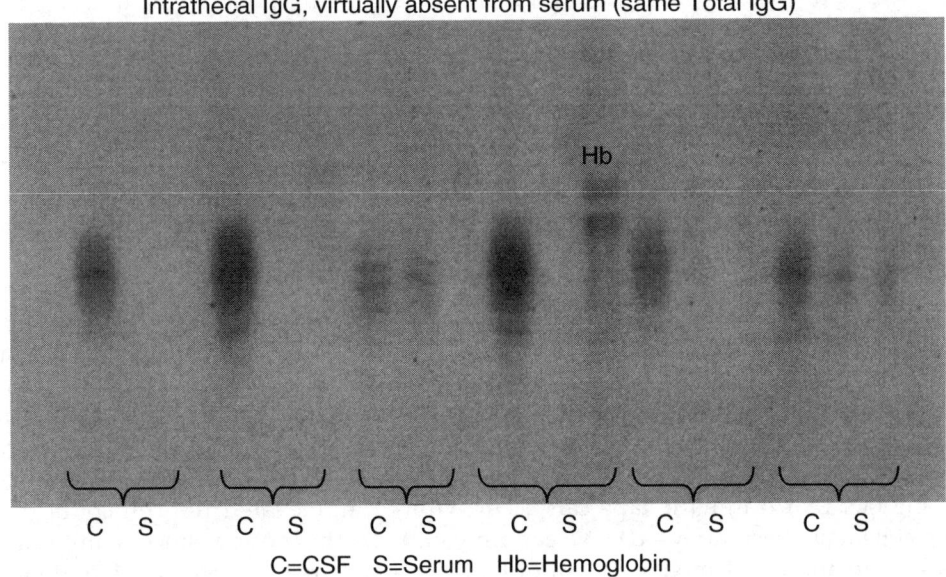

FIGURE 5.12 Parallel, similar amounts CSF versus serum (1:300 dilution) showing 'only' CSF synthesis of antibody against human albumin (in rats). Note the endogenous peroxidase activity of hemoglobin from the lysed red blood cells, contaminating the serum.

examined by fluoroscopy there are quite dramatic movements of the spinal fluid from the cul-de-sac up and down (i.e. to and from the foramen magnum during coughing or other movements of the torso), which basically amount to several Valsalva maneuvers. It is certainly clear from the older literature that patients with prolonged immobility in bed, due to coma or other causes, could have quite striking elevations in total protein [181]. This can definitely occur in catatonic schizophrenia or other psychoses characterized by severe withdrawal and a relative lack of body movements. The limited movement of the torso over a prolonged period will result in a lack of 'refluxing' of the fluid around the spinal cord. The classical example of limited movement of fluid is Froin's syndrome, where there is a complete block to the flow of fluid below the level of the occlusion. This can consequently lead to the levels of total protein becoming so high that they approach (although they still never reach) levels found in the corresponding serum. What is therefore quite interesting, is that the filtration effect still takes place, the barrier has never convincingly been shown to be at a level of 45° in any case (namely the slope = 1.0, where there is complete equilibration between CSF and serum). Felgenhauer reported this in 1978, [989] when he studied alpha-2 macroglobulin and albumin in gross abnormalities of the blood–CSF barrier. Although there were hints that the line connecting the two protein quotients in the occasional patient seemed to have an altered slope, the overall impression was that the slope remained essentially the same 'by eye', in spite of dramatic alterations seen during the course of, for example, Guillain–Barré syndrome. We know now, from the more careful work of Reiber, that not only does the slope change with elevated levels of barrier dysfunction, but also that the most dramatic changes are found in the higher molecular weight proteins, as the original Felgenhauer formula would have predicted, namely that small molecules enter the CSF more easily.

The converse point about large molecular size is made by the work of Ohman in which he stresses the dramatic importance of leakage of even small volumes of blood following a traumatic tap. In the specific case of IgM there will be a relatively striking increase relative to its large molecular size [307]. When 0.1 µl of blood is added to 1 ml CSF, it will more than double the concentration of CSF IgM. Thus it is important to raise the caveat when finding elevated CSF red cell counts, since this can cause confusion as to the real levels of IgM in CSF. Of even greater concern is the fact that traumatic taps can occur without an elevated red cell count (i.e. the cells have become lysed). We certainly endorse this notion, since many years of examination and interpretation of banding patterns from polyacrylamide gel electrophoresis have convinced us that the presence of hemoglobin/haptoglobin complexes (unequivocal markers of red cell damage), can frequently be found in patients who have only had quite a trivial number (or indeed an ostensible absence) of red cells when CSF was examined under the hemocytometer.

Flow was studied by Rapoport [729] in seven elderly people and seven controls (average age 29 versus 77) and a 100 per cent decrease of flow was found (0.41 versus 0.19 ml/min), but the increased total protein was only 25 per cent (0.47 versus 0.59 g/l). However, it was disappointing that there was no correlation of flow rate with CSF total protein. Therfore, as Rapoport comments 'the elevation of CSF protein concentration with ageing is not attributable solely to reduced CSF production or bulk flow'. Reiber's notion that flow is the main parameter of CSF albumin [904] should thus reflect the 600 per cent change over 12 hours, but no one has ever reported any 'molecular' consequence of this diurnal variation, e.g. no change in total protein with time of day, namely the highest flow is at midnight, and the lowest at midday [1204] (see Figure 5.13). There are also dramatic differences between flow rates in blood and spinal fluid: think of it as 'refreshing', i.e. new CSF being produced every 6 hours, whereas the blood is pumped through the brain in only about 6 seconds, or 6 heartbeats.

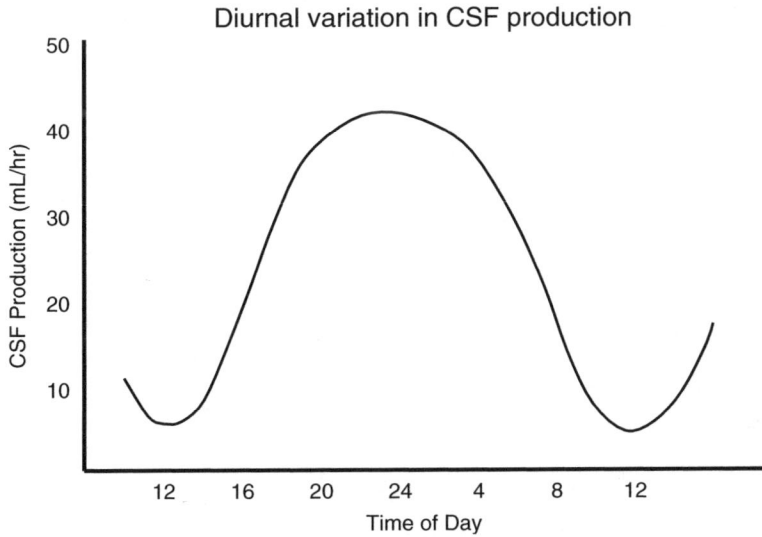

FIGURE 5.13 The six-fold diurnal variation in the rate of production of CSF.

Conclusion

It is clear that the entry of proteins into the CSF from serum is not a simple, monolithic notion which can entirely be explained by the rate of flow [904], although this parameter was mentioned initially in the Felgenhauer formula of 1976 [290]. The actual flow rate is much more variable, e.g. the 600 per cent diurnal variation, but we must also account for the 300 per cent regional differences in protein concentration, between the brain and the spinal cord. In addition, although the rate of flow has been plotted against age by Rapoport [729], this again does not simply relate to the elevation of CSF total protein with age, which has been well described for many years, nor is it directly coupled to the decreased rate of flow, i.e. inversely correlated with the increased level of total protein. Further work is required to illuminate the multifactorial nature of these interactions rather than relying on a singular, homogeneous view of CSF production.

CHAPTER

6

Qualitative versus quantitative analysis

Introduction
  Pros and cons
  Discrepancy between qualitative versus quantitative amounts
Conclusion

Introduction

The glib assertion that 'numbers are more objective' is misleading if one ignores the diverse qualities that proteins clearly possess. It is much more informative first to discern their separate qualities and then generate the numeric estimates.

There are two sections:

1. *Pros and Cons.* Qualitative analysis of CSF proteins is often preferable to quantitative analysis (see also in Chapter 10), but one can relatively easily do both; specifically, quantification following qualitative analysis.
2. *Discrepancy.* There remain several discrepancies between qualitative versus quantitative analysis and these are introduced in this chapter and discussed in more detail in Chapter 7.

Pros and cons

One of the central issues in the studies of CSF proteins is how best to assess the immunoglobulins that are produced by a subset of plasma cells localized to the CNS. This has implications not simply for diagnosis (e.g. local synthesis of IgG

TABLE 6.1 Pros and cons of qualitative versus quantitative methods

Contra	Pro
1. Qualitative methods	
Arguments against qualitative	*Arguments in favor of qualitative*
Less sensitive	but only 1 µl required
Less speed	but only 4 h required
Less reproducible	but CV of less than 5 per cent
Less objective	but scanning densitometer
Less reagent cost-effective	but less than $0.05/test
Fewer large numbers	but maximum 24 per day
Less equipment cost	but 'have it' (electrophoresis)
Less easy interpretation	but computer program available
Less commercially available	but agarose now standardized
Less quality control	but commercial positive controls
Less labor intensive	but other tests done 'in between'
Less automated	true, but greater information
2. Quantitative methods	
Argument against quantitative	*Argument in favor of quantitative*
Fragments and/or aggregates give false stoichiometry	Answer is an objective number (but scanning densitometry pro qualitative)

in multiple sclerosis), but also for therapy and even prognosis as we shall see in later chapters. At this stage it is useful to consider an artificial distinction between quantitative versus qualitative analysis. In a word the best answer, if faced with the decision concerning which is preferred, is 'both'. Let us consider separately the pros and cons of the two methods (see Table 6.1).

And how do we 'test' a test? How does one decide which laboratory method is most useful for a particular diagnosis? This is done firstly by dividing the patients into two groups: with (positive) and without (negative) the disease in question [1239]. Secondly, one reads the laboratory results (blind to the diagnosis, of course) – positive and negative for the test in question (see Table 6.2).

The sensitivity of the test is $A \times 100$ divided by $(A + C)$, i.e. the percentage of the subtotal who have disease $(A + C)$, who show positive (correct) results.

The specificity of the test is $D \times 100$ divided by $(D + B)$, i.e. percentage of the subtotal who do not have the disease $(D + B)$, who show negative (correct) results.

The efficiency of the test is $A + D \times 100$ divided by $(A + D + C + B)$, i.e. the percentage of the total group who are correctly diagnosed (this value estimates the sum of false negative and false positive results).

In general terms, with a disease like MS it is better to have a test which, if it 'must' give false results, gives false negative results. Usually false negative results do not

TABLE 6.2 Basis for comparison of sensitivity specificity and efficiency for a given test for a given disease

		Disease +	Disease −
Test	+	A	B
	−	C	D

A The number of patients who have the disease correctly diagnosed by a positive test.
B The number of patients with a false positive test who do not have the disease.
C The number of patients with a false negative test who do indeed have the disease.
D The number of patients who have been correctly excluded from having the disease by the negative result of the test.

prevent the test from being repeated at a later date, by which time the disease may have progressed further and thus the result is more likely to be positive. A serious problem with false positive results is that too many of these may induce clinicians to stop requesting the test. The same can also be said, although to a lesser degree, for false negative results.

In Chapter 10, we shall see that qualitative tests for MS are generally much better than quantitative tests; however, one should realize that the combination is even better, i.e. not simply electrophoresis plus immunochemical analysis, but also densitometry on the electrophoresis, the latter being enhanced by the immunochemical specificity. For example, we previously measured albumin and IgG in serum and CSF, and followed this by PAGE (polyacrylamide gel electrophoresis). We then stained the gel with Coomassie blue and scanned it to produce peak height ratios. We also performed the immunochemical estimation of kappa and lambda light chains, complement C'3 and IgA, and scanned each of them to produce peak height ratios. Not only were the ratios calculated by a computer, but the decision as to what constituted a peak (or a valley) and the measurement of its height, width and area (including all the normal values as well as the appropriate comments for abnormal values) were all fully programmed (see Appendix, Chapter 15, Figure 15.7).

Different types of information are derived from qualitative and quantitative tests and these are set out in Table 6.3.

Our own recommendations for laboratory practice are given in order of preference in Table 6.4.

There are four fundamental reasons for preferring qualitative analysis to quantitative analysis, but none bears any relation to whether the quantitative formula is linear or nonlinear. The latter is preferred, but again this is a separate issue, especially in MS patients where it makes little practical difference.

The first and the most dramatic difference is the degree of discrimination achieved by qualitative analysis, namely, a single band can readily be discriminated

TABLE 6.3 Qualitative versus quantitative tests for IgG

1. Quantitative: amount of IgG
   (A) Serum level: serum increase produces increase in CSF
   (B) Barrier 'setting': more leakage produces increase in CSF
   (C) Local synthesis: distinct from either of the above
2. Qualitative: kind of IgG (more 'important' than above, but again 'both' performed, i.e. quantitative analysis of clones)
   (A) Polyclonal: normal
   (B) Monoclonal: usually serum source
   (C) Oligoclonal: usually CSF source, rule out serum source

TABLE 6.4 Recommendations for detecting local synthesis in the order of increasing sophistication

NOT preferred, but if you do albumin and IgG on CSF and serum: plot data per log–log (Figure 8.5), rather than 'fixed' index cut-off value

Simple test
  Agarose electrophoresis + nitrocellulose immunofixation for IgG (Fc)

Complete test
  IEF + nitrocellulose immunofixation for IgG (Fc) and nitrocellulose immunofixation, with additional kappa and lambda in 'difficult' samples as a 'tie-breaker'

from the polyclonal background, although it may constitute only 0.5 per cent of the total IgG population. However with quantitative analysis, the IgG may need to be as high as 25 per cent of the total protein to provide adequate distinction, since the normal serum IgG level in the same patient can be in the range of 15–20 per cent of the total protein.

The second issue, almost as striking as the first, is the difference between the individual patient and the total normal population. In this case, for any given method the so-called population variance is typically much wider than any individual method's coefficient of variation. In addition, individual laboratories are invariably measuring two proteins, typically albumin and IgG (each in CSF and serum), giving a total of four determinations, with further sample manipulation to make two rather large dilutions of serum, typically albumin 1 in 200 and IgG 1 in 400, while each dilution will carry its own inherent coefficient of variation, depending upon the nature of the pipetting techniques.

The third reason to prefer qualitative to quantitative analysis is less obvious (and thus much more difficult to detect in quantitative analysis), namely that a systemic IgG response can clearly be detected on qualitative analysis as the so-called 'mirror' pattern. Although this is perhaps of less significance in the diagnosis of multiple sclerosis (nevertheless, it raises important questions concerning the pathophysiology of this disease), it is much more relevant in the case of brain infection, e.g. herpes zoster encephalitis, where one typically sees a rather strong systemic IgG response, and in addition, an intrathecal response in the form of extra bands in the CSF which are absent from the corresponding serum.

The fourth reason to use qualitative analysis is its superiority in the determination of antigen-specific IgG. The so-called 'MRZ' test has been used to measure the quantity of IgG which is bound to measles, rubella and varicella zoster. However, the similar technique of separation on the basis of isoelectric focusing, which has routinely been applied to the total IgG population, can then be followed by antigen-specific binding to a nitrocellulose membrane that was previously coated with the antigen in question. This technique has been shown to be more sensitive for detecting antibodies against these three as well as other antigens [780–782, 1064]. This reinforces the fundamental belief accepted internationally in the case of total IgG, namely that qualitative rather than quantitative is the 'gold standard'. It is unfortunate that this distinction was missed by the International Consensus of Neurologists [730], which suggested (without any evidence) that the detection of local synthesis of IgG in CSF could be either by isoelectric focusing or by calculating the IgG index. Certainly the latter is not sufficient, as clearly stated by the previous European Consensus [25] and reinforced by the more recent International Consensus [316].

In order to check for local synthesis of antibodies against any intrathecal antigen, e.g. herpes simplex, measles etc., we initially screen parallel CSF and serum (the latter diluted to 1 in 400, as we routinely do for isoelectric focusing). For pathological controls we then use the six most common antigens provoking immune reactions in the brain: herpes simplex, measles, mumps, rubella, CMV, and toxoplasma. If the patient has a non-specific response against several different antigens, as is usually seen in multiple sclerosis, it is not likely to be of diagnostic significance for the individual antigens. However, in patients with herpes encephalitis, SSPE or, indeed zoster encephalitis and other infections, there will typically be a single response against the relevant antigen. The response tends to be stronger in CSF than in parallel serum, since similar amounts of total IgG have been applied, based upon the prior routine isoelectric focusing of the serum and CSF (where they have already been assessed 'by eye' for the same amount of total IgG staining following the appropriate dilutions of serum and CSF). In the screening phase (dot blots), if one of the antigens clearly demonstrates a dominant antigenic response, compared with all the other pathologic controls, we then continue with further analysis using isoelectric focusing, but this time using immunoblotting against the specific antigen in question, rather than simply staining all the IgG molecules (as had been done initially in the routine isoelectric focusing). The patterns seen on isoelectric focusing, following IgG binding to the antigen in question, fall into the five classical types (see Chapter 15, Figure 15.2). Once again the interpretation is greatly facilitated using a qualitative approach. This is rather different from the quantitative approach, which is based on a cut-off line which may be derived from the mean slope for IgG or the interpolated line also called 'QLim' (the Quotient Limit or regression line of QIgG versus QAlb), which involves a number of additional assumptions [911]. Qualitative testing can then be enhanced by two additional techniques:

1. The use of IgM rather than IgG to determine whether the individual response is recent and/or ongoing. It is well documented that IgM is not only the first antibody to appear, but also the first to disappear upon removal of the antigenic stimulus. The IgG response may, however, persist for many years, depending on the antigen in question.
2. We can also probe the affinity of the Ab/Ag bond using increasing concentrations of sodium thiocyanate, as has been demonstrated using the ELISA technique [686], and more recently with immunoblotting. Here the same principles of biological significance apply in terms of pathophysiology, where a low affinity bond is non-specific in terms of its pathological relevance [221, 820].

It is also worth noting that although Sindic used the term 'affinity' blotting, he was in fact using immunoblotting against either IgG or free light chains (kappa or lambda) and that technique did not test the affinity per se, at least as we and others have done, by using chaotropic ions such as sodium thiocyanate [1052].

Another general point of interest is that the diagnosis of viral disease of the brain has benefited more from studies of antibody than of antigen (PCR, but not to disparage this test for Herpes). This was found in a large national study from Finland where 3231 patients were examined and 270/3173 (8.5 per cent) were positive by CSF antibody (IgG or IgM) versus only 133/2590 (5.1 per cent) who were found by antigen [592].

The first European Consensus paper was unanimous in its agreement that qualitative analysis was to be preferred to quantitative analysis in the detection of abnormalities of IgG [25]. This has been confirmed in the second International Consensus by Freedman et al. [316]. Relatively few studies have begun with a careful reading of the clinical details of the discharge summaries followed by the comparison of qualitative and quantitative methods. McLean studied over 1000 patients, Souverijn had about 1000 patients and Ohman had nearly 600 patients. The most comprehensive study was that of Souverijn [1081], who published his data in great detail so that any interested party could check and/or recalculate as the need arose. His massive Table 2 has been summarized in Table 6.5, and gives the sensitivities and specificities for all the formulae concerned. Saying 'never' is never an option in clinical medicine, but very few patients with clinically definite MS are negative for IgG on isoelectric focusing with immunofixation (i.e. in the

INTRODUCTION

TABLE 6.5 Comparison of qualitative and quantitative methods of detection of abnormalities in IgG. Modified from data of Souverijn [1081]

Formula	Reference	Cut-off level	TP #	FN #	TN #	FP #	TP − FP as %	Signal #	Noise #
ln IgG+1	[683]	0.60	162	59	669	109	59.3	0	0
Reiber	[906]	0.00	162	59	665	113	58.8	0	+4
Ohman	[829]	1.25	161	60	664	114	58.2	−1	+5
IgG index	[219]	0.70	157	64	677	101	58.0	−5	−8
CSF IgG/Alb	N/A	0.15	194	27	438	340	56.2	+32	+231
Tourtellotte	[1180]	2.50	160	61	618	160	51.0	−2	+51
Schuller	[1002]	15.00	155	66	609	169	48.4	−7	+60

Abbreviations: TP = true positives; FP = false positives; TN = true negatives; FN = false negatives

range of 1–3 per cent), and yet these particular patients also have elevated quantitative IgG values. Although there have been widely accepted differences between the nonlinear (preferred) and the linear representations, the point has been made by several authors that on an average the majority of MS patients do not develop major changes in their blood–CSF barrier function. Here, the values in Table 6.5 also make the same point that the values for the differences between individual quantitative formulations are of the order of 3 per cent (5/157) (e.g. numbers of patients incorrectly classified: Index = 3, Ohman = 4, Reiber = 4 or LnIgG = 0). The most important column, labeled 'Cut-Off Level', i.e. COL by Souverijn, was the subtraction from the True Positives of the False Positives (TP − FP, shown as a fraction of 1.0). The rows otherwise list the numbers of patients for the best value of the COL. The final two columns show the numbers of patients who either detracted from the signal (TP: 162 patients was the highest number), who are shown as a minus value, or on the other hand added to the noise (FN: 59 patients was the lowest number), who are shown as a positive value. Although it is possible to measure the total area under the curve using an ROC analysis, the only way to use this information in studying individual patients on a day-to-day basis, is to decide on the best cut-off level (COL).

In our study of 1007 patients [733], only four out of a total of 79 patients with clinically definite MS were negative for oligoclonal bands. Three had IgG index values < 0.65 and thus were unequivocally normal, however the fourth patient had a value of 1.83, which would have been grossly abnormal according to any of the other calculations. We have, in retrospect, found an unfortunate mathematical error: the CSF albumin was 239 mg/l and the serum albumin was 4129 mg/l, giving an albumin quotient of 0.058. The CSF IgG was 4.4 mg/l and serum IgG was 415 mg/l, hence the IgG quotient was 0.011. Thus the index should have been 0.183 (normal) not 1.83 (abnormal) as listed in this paper.

The following pros and cons of the three studies in question:

1. Dealing first with the paper by Souverijn et al.: the total number of patients was of the order of 1000, but they used concentrated CSF, which is associated with a number of difficulties. They also used Coomassie staining rather than IgG-specific immunostaining (which is now the 'gold standard'). However, on the positive side, its strongest recommendation is that they used not only all the formulae concerned, but they left for posterity all the specific details in the very large table which can be revisited at any time in the future.

   There is an additional, minor, criticism that Ohman wrote in a letter to the journal [828], followed by the appropriate response from Souverijn [1082], which was not felt to make any substantial difference to the overall conclusions. In brief, there was a misunderstanding, but 'COL' was the important point, i.e. True Positives minus False Positives (TP − FP).

2. The second paper by McLean simply did not take into account the calculations of the Reiber formula, nor indeed of the Ohman formula, since these were not as yet so widely accepted in the wider world of CSF analysis. Fortunately, however, we still have the original data from these 1007 patients (the error notwithstanding), thus we can now say that in this particular study there were no patients who were positive by any quantitative analysis, yet negative by qualitative analysis.

3. The third paper by Ohman (558 patients) unfortunately simply dismissed any of the calculations from the log index, and did not bother to show the numbers in a table, or plot the details in a ROC curve as was done by Souverijn, although the latter had demonstrated that there were indeed differences and one should not therefore assume that the results would be the same as for the index. On the positive side, Ohman showed that only 4 per cent of MS patients were negative on focusing.

Discrepancy between qualitative versus quantitative amounts

By way of introduction, the discrepancy problem (shown not infrequently by CSF samples) relates to the striking differences between the amounts of IgG, which seem to be quite at variance using qualitative versus quantitative methods. In specific terms, this derives from the widely recommended idea that one should use the same amounts of IgG (e.g. 40 ng) not only in parallel CSF and serum, but also between specimens from different patients when samples are applied to the same plate following isoelectric focusing and then immunofixation. Using this qualitative analysis, the discrepancy is not only clear by eye, but can also be quantified by scanning densitometry, in order to express the results of the area under the curve in optical density units.

INTRODUCTION

The discrepancy is revealed despite the expectation that there will be the same amount of IgG in all the samples, e.g. 40 ng (based on the prior quantitative analysis, since different volumes will be pipetted in inverse relation to the measured IgG concentration). When scanned quantitatively, there should be comparable optical densities for all the samples on the plate (qualitative analysis 'by eye' followed by quantitative analysis by densitometry). However (despite the original hope for equal levels of staining), in practice the IgG concentrations obtained either by visual inspection or by densitometry do not show uniform amounts in up to two out of ten samples. This is due to a number of factors which reflect the individual patient's oligoclonal IgG pattern (see further in this chapter).

This discrepancy issue is well understood by those with considerable experience (not to say exasperation) with silver staining of IEF, since individual specimens (which can vary from 1–2 out of 10 per plate) will need to be run again on the following day. This is because either an increased or a decreased volume must be applied on the repeat run, due to the discrepancy between the initial 'quantitative' analysis (from a standard curve) and the final 'qualitative' interpretation ('by eye'). Interpretation 'by eye' can be quite difficult when the amounts of IgG are either too high or too low to be compared fairly with other samples on the same plate.

Laboratories generally use either silver staining of all proteins or immunofixation of IgG in particular when analysing samples for oligoclonal bands. Immunofixation is preferable, but currently too many laboratories still use the silver staining technique.

For silver staining, before running the samples on isoelectric focusing, a known set of IgG standards is run (e.g. by nephelometry) to obtain a calibration curve. This will allow the measurement of the amounts of IgG in each of the patient's CSF and serum samples. The appropriate volumes of each sample (selected according to their individual IgG concentrations) are then applied to a plate (see Figure 6.1) and the IgG content can also be subsequently measured using scanning densitometry to verify what is 'obvious' by eye (see Figure 6.2), namely that the first two CSF samples have a ten-fold difference in optical density. There are other wide variations among the individual tracks, and thereby this vividly demonstrates the 'discrepancy problem', i.e. between the quantitative amounts versus the qualitative amounts of IgG. To make the diagnosis by any qualitative technique, there must be the same amount of IgG 'by eye'. If there is an inadequate visualization, then the dangerous 'iceberg' situation may prevail, i.e. one cannot appreciate the true banding pattern, hence the sample must be re-run with a higher CSF volume to properly see the underlying IgG pattern. Conversely, if there is overstaining, bands may become 'blurred' into a pseudo-polyclonal pattern and thus lower volume of CSF must be re-run to visualize the true IgG pattern. In this case, the 'iceberg' is too high in the water, as compared to the previous example, where it is too low in the water.

6. QUALITATIVE VERSUS QUANTITATIVE ANALYSIS

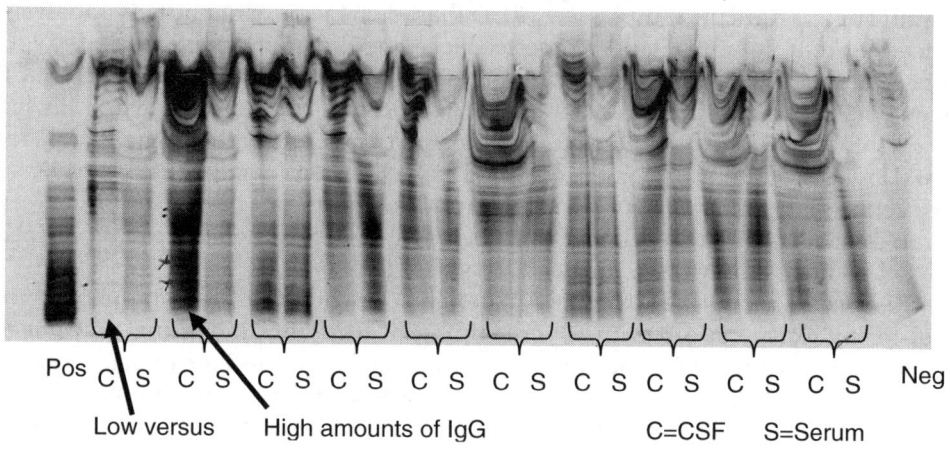

FIGURE 6.1 Silver stain of all proteins, explained in text, but note the dramatic difference in amounts of IgG between adjacent tracks (especially track 1 versus track 3).

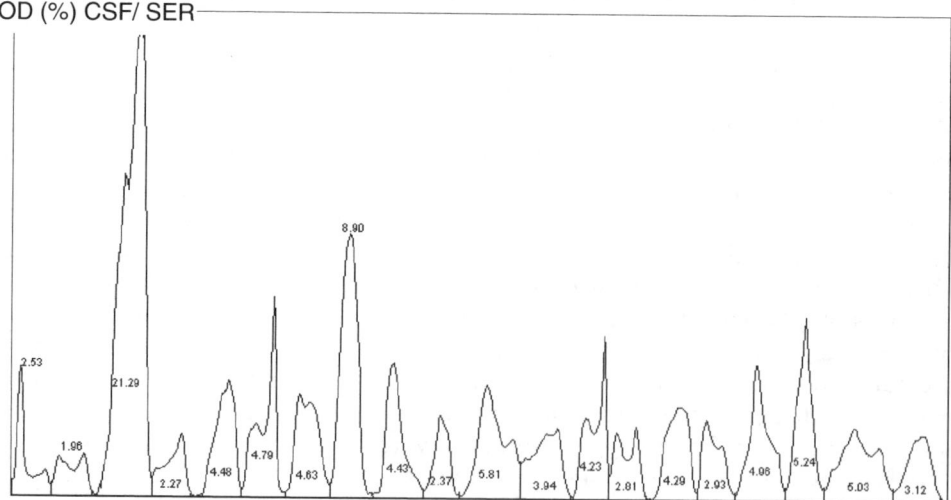

FIGURE 6.2 Densitometer scan of the IgG region of the silver stain of the previous Figure 6.1. Note the ten fold difference in amounts, which were all 'measured' to be 50 nanograms.

Immunofixation with ethylaminocarbazole staining removes the need for the initial quantitative determinations because, unlike the silver method, ethylaminocarbazole can accommodate a much wider range of IgG concentrations, so it is unusual to have to re-run samples (1 in 50, compared with 1 in 5 for silver). Hence the simplest solution to the discrepancy problem is not to attempt the initial calibration of IgG. This also saves the cost of reagents and technician time, and avoids any conflicts between initial nephelometry and final densitometry.

There are various factors that influence this discrepancy problem, and each derives from the individual patient's oligoclonal pattern:

1. *Molecular aggregates.* These can either arise from aggregation of antibody molecules together or with various antigens, depending upon the affinity of their binding to the exposed epitopes. These problems are considered further in Chapter 7.
2. *Three-dimensional configuration, i.e. exposure of epitopes.* This relates to the previous point but also to the variables introduced by steric limitations of binding compared with the binding in free solution when the epitopes are fully 'exposed'. This is detailed in Chapter 7 (i.e. the difference between free and bound light chains).
3. *Fragments.* The most widely studied examples are those of free light chains. These problems are also considered in Chapter 7.
4. *Single (unique) versus multiple (broader population) epitopes.* This is quite analogous to the difference between monoclonal versus polyclonal antibodies. Probably the most carefully studied example for many years now has been quantitation of paraproteins in patients with myeloma. Traditionally, this has also been the most conspicuous example of the discrepancy problem. Given the various treatment options for patients with myeloma, it is of fundamental importance to be able to quantify the amounts of paraproteins, since this has a direct relationship with the number of malignant cells present. These cells are the targets of the therapies for myeloma.

There are various examples of these four points which will be considered in reverse order. After considering many different types of ostensibly 'quantitative' (e.g. nephelometric) assays, it eventually became clear that the most reliable method was electrophoresis, in which the peak height, or the area under the curve, of the myeloma spike was compared to the total area under the curve for the other proteins found on the same electrophoretic strip. The basic premises which underlies this assay, and therefore by implication a major source of the discrepancy problem, is the fundamental principle that like must be compared with like. The vast majority of people, who obviously do not have myeloma but do have a polyclonal distribution of their IgG population, can readily be compared with a standard source of human IgG, which in turn is polyclonal. In each myeloma patient there is, by definition, a unique sequence for the IgG molecule which is produced in excess

TABLE 6.6 Differences between polyclonal and monoclonal

Polyclonal	Monoclonal[a]
Broad	Narrow
Natural	Pathological

NB: monoclonal A from one person is very different from monoclonal B from another person
[a] Analogous to oligoclonal pattern

of any/all of the other members of the diffuse Gaussian distribution of polyclonal antibodies. To exacerbate the problem, not only does the individual clone differ from its polyclonal neighbours, but the monoclonal sequence in one patient with myeloma is very different from the myeloma clone in any other myeloma patient. This compounds the problem of measuring the amounts of myeloma protein from any given patient. One can therefore immediately see the analogous situation in patients with oligoclonal patterns, since the peaks of IgG are, in principle, very similar to a few monoclonal or myeloma patterns. Although the individual peaks or bands may not be as strikingly elevated as those found in myeloma, nevertheless the bands are ultimately derived from different clones in a patient with an inflammatory immune response. Clearly, these bands are (by definition/recognition) very different from the diffuse polyclonal background (see Table 6.6).

Concerning the study of free light chains (third point in the earlier list), people have been concerned for many years about the best choice of standards [1164]. The short answer, in the current context, is that the standards should be polyclonal rather than monoclonal because of all the above arguments. This question is considered in further detail in Chapter 7.

One more example of the discrepancy problem (second point above) relates to the demonstration of antigen-specific IgG. It is important to realise the steric availability versus the steric hindrance of the Ig molecule, which can be bound either in an optimal or 'exposed' manner with the Fc portion pointing up into the medium (see Figure 6.3A), or in a less exposed manner when the Fc portion is flattened against the nitrocellulose membrane as shown in Figure 6.3B. This can be practically illustrated in the figure prepared by Moyle et al. [781], in which the same volume of CSF is applied in both blots (see Figure 6.4). It is much more difficult to visualize the Fc in the isoelectric focusing of total IgG than it is in the antigen-specific immunoblot, where the Fc portion is more prominent (Figure 6.4A: Total IgG versus Figure 6.4B: herpes antigen-specific IgG).

The first point concerning molecular aggregates will be illustrated in the next Chapter 7.

To summarize the discrepancy problem: it is no surprise that in attempting to apply the same amounts of IgG from patients with oligoclonal patterns, rather

INTRODUCTION

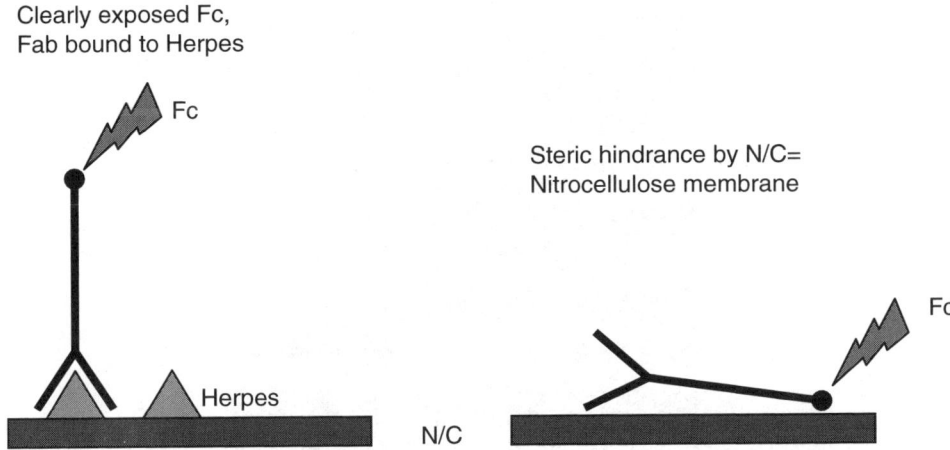

FIGURE 6.3 Cartoon showing how the Fc portion can be relatively hidden/exposed, depending on whether it is more in solution, versus tightly bound to the surface nitrocellulose (N/C).

haphazard results could be found, since individual oligoclonal bands do not, by definition, represent the same Gaussian distribution in molecular mixtures derived from the normal polyclonal IgG standard. This is compounded by the idiosyncratic banding patterns for individual patients.

Basically the problem is that we are not strictly comparing like with like. There can therefore be wide discrepancies among the various different quantitative methods, e.g. nephelometry, simple radial immunodiffusion and/or electro-immunodiffusion, latex beads or ELISA, etc. Results will also be erratic/idiosyncratic since they are due to the predominance of a unique sequence of IgG which is disproportionately elevated relative to the broad population mix. Figure 6.1 is a typical example of the discrepancy problem, where the first patient has a disproportionately low amount of IgG and the second has a disproportionately high amount, although ostensibly the same amount of IgG (50 ng) had been applied in each case.

To solve this problem, it therefore becomes important to have an assay with a very broad range which will be tolerant of all of these not uncommon (10–20 per cent), but nevertheless quite disparate, samples which do not fit into the normal distribution of IgG molecules. Reviewing the literature, the broadest range of encompassed IgG concentrations is probably found using horseradish peroxidase with the substrate ethylaminocarbazole. In our experience (as well as others), this spans a range of almost two orders of magnitude, from 20 to 1200 ng of total IgG. We therefore routinely apply 4 μl in parallel with a 1 in 400 dilution of serum using distilled water, to avoid the unnecessary addition of sodium chloride

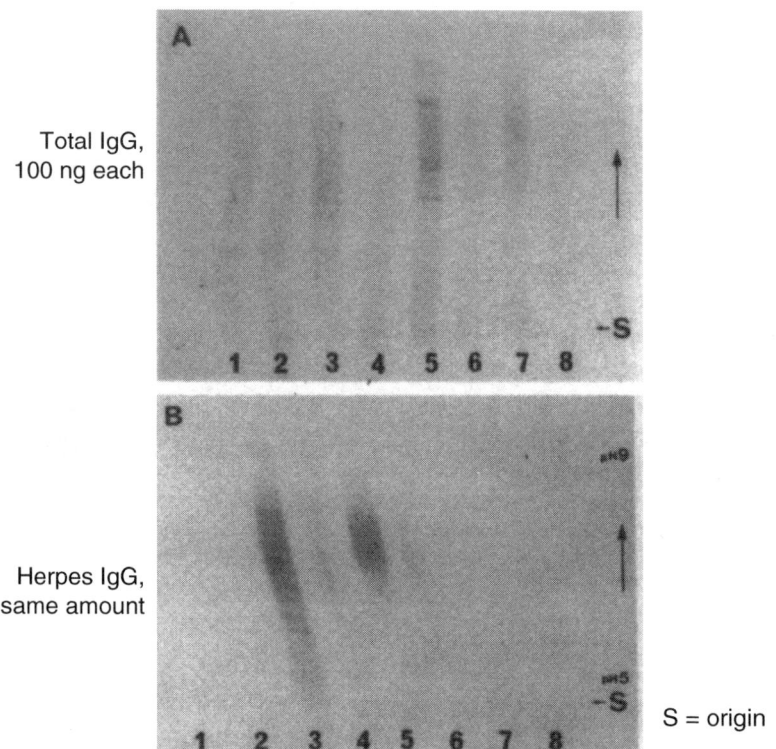

FIGURE 6.4 Herpes antibody (bottom) is more exposed by projecting into the solution, versus the total IgG (top) which is more tightly bound to the N/C membrane. Hence there is a greater amount of staining (bottom) in spite of the same amount of IgG and second antibody in both top and bottom. This is most striking in samples 3 and 5.

which would only increase the heat produced within the gel (because of the extra work required to shift these two counter-ions to their respective electrodes during focusing). In our laboratory, the incidence of repeats due to underload or overload (1 in 50) is clearly lower than that in studies using the silver method (1 or 2 out of 10). However, there is the additional bonus in the saving of time and money by not having to initially 'estimate' the amounts of IgG, yet without the satisfaction of finally avoiding the 20 per cent of problem samples. Certain other techniques have a much narrower range of tolerance, e.g. the 'identical' sandwich technique in which the antibody molecules are first trapped with CSF and serum in parallel, followed by the same antibody (although biotinylated) used as the second layer of the sandwich. Since both 'sides' of the sandwich are directed against the same epitope, this could explain the lower sensitivity of this assay. Sindic specific recommends 50 ng of IgG [1052]. The other practical difficulty avoided by immunofixation is a different kind of discrepancy: how to interpret and report any conflicting results between qualitative analysis and quantitative analysis. This may happen when the IgG index is elevated, but the oligoclonal pattern is negative. Given that the gold standard is isoelectric focusing, the final interpretation must weigh in favor of qualitative rather than quantitative analysis, since only the former can detect clonality. Fortunately the particular problem of false positives given by the IgG index is not a common occurrence. Other problems which have sometimes arisen, concern either the difficult issue of single bands or the other question of equivocal patterns, but these have usually been resolved through the use of kappa/lambda staining (Chapter 13, Figure 13.1A and B).

Conclusion

It is clear that the pros and cons of qualitative versus quantitative analysis have changed over the years, but more recent trends have continued to move more towards the qualitative side (in spite of all the caveats), due to its greater powers of discrimination which ultimately improve both sensitivity as well as specificity [1144].

PART 2

Normal CSF proteins, detailed discussion

CHAPTER

7

Methodologies and their limitations

Introduction
General methods
  Separation
  Visualization
  Quantification
Further examples of the discrepancy problem
Radiolabeled proteins

Introduction

There may be many pifalls in quantitative determinations without prior qualitative separation. Various pathologic conditions can lead to the problem of not being able to compare like with like. Two examples are: (1) the CSF form of a protein versus the serum form; and (2) the normal form versus a chemically altered form.

This chapter has three sections:

1. General comments on methods (for specific comments see the Appendix, Chapter 15).
2. The discrepancy between immunochemical and physical–chemical measurements.
3. The limited data on radiolabeled CSF proteins.

General methods

Ideally, CSF proteins should first be separated on the basis of their size and/or charge. They should then be visualized and finally quantified using their unique

TABLE 7.1 Three steps in the determination of protein levels

1. Separation
   Size: sodium dodecyl sulfate (SDS), Gradipore
   Charge: isoelectric focusing (IEF), Agarose
   Both: polyacrylamide gel electrophoresis (non-SDS)
2. Visualization (solid phase: nitrocellulose)
   Stains: Coomassie, naphthalene black, silver, nigrosine
   Peroxidase (or nitrocellulose discs: orthophenylene diamine)
   Autoradiographs
3. Estimation
   Scanning for quantification (see above 2) by densitometry
   Absorption: dye binding
   Fluorescence
   Chemiluminescence
   Counting: radiolabels
   Functional: C'3, factor D, ceruloplasmin (enzyme)

antigenic sites, by application of mono-specific antibody. Some of the individual options are given in Table 7.1.

Separation

The most discriminating method of separation is electrophoresis on polyacrylamide gels, where the sieving effect of acrylamide allows discrimination on the basis of molecular size in addition to the molecular charge (which is the major factor in other electrophoretic separations [200, 268, 1262]). The use of the denaturing detergent SDS will give the best estimate of the molecular weight. This requires the protein to be separated into its constituent polypeptide chains [874] for example, IgG is separated into heavy chains (55 kD) and light chains (22 kD). The net charge is covered by the sulfates of the detergent and hence one only resolves protein chain length or unfolded molecular size [919]. To obtain a better estimate of the molecular weight or molecular size under physiological conditions (without added detergent, therefore the protein is folded as 'normal'), it is more appropriate to use electrophoresis on Gradipore gels [272]. In these gels the degree of cross-linking is increased in a gradient fashion with consequent decreasing pore size, so proteins continue to migrate until they reach a pore size through which they cannot pass.

Separation on the basis of charge can be effected by isoelectric focusing [1260], but there are many inhomogeneities in the pH gradients produced by some ampholyte manufacturers and therefore artefactual bands (see also Chapter 15,

Figure 15.3) can readily be produced by this procedure [287]. Most nonsieving methods of electrophoresis, such as agarose, will separate proteins without artefacts, however, these techniques are not as discriminating as isoelectric focusing [548].

In isotachophoresis, 'spacer' buffers are used to make artificial 'steps' and thereby fractionate the gamma region [1179]. It is somewhat analogous to IEF (see Figure 15.3), but the individual investigator prepares the required 'gradient steps' rather than the ampholyte manufacturer producing them in bulk.

It is of some note that antigen–antibody complexes are not typically dissociated during IEF, unless they are subjected to a low pH (about 3.0) [697].

Separation by both size and charge can be performed using the now classical technique of two-dimensional gels [822]. This is typically done on denatured proteins using isoelectric focusing for the first dimension and SDS for the second. However, using 'one-dimensional' separation, namely polyacrylamide gel electrophoresis in the absence of SDS (and not Gradipore gels) allows the combined effect of size and charge to dictate the final resolution of the proteins. As a practical expedient, and especially in the setting of clinical service, this is by far the best method of separation.

Visualization

This could either be performed within the gel or by subsequent 'fixing' of the proteins to nitrocellulose and then performing solid phase immunodetection [1259]. Table 7.2 lists typical stains in the order of increasing sensitivity. The table also includes an ELISA method (HRP) as an index of the immunochemical sensitivity. Silver has been notoriously capricious, although later methods claim to be more reliable [388].

TABLE 7.2 Methods of detection in order of increasing sensitivity

Sensitivity of solid phase assays		Sensitivity of assays in solution	
Stains	Per band	Assays	Detection/dl
Naphthalene black	3.0 µg	Radial (Mancini)	1.0 mg
Coomassie	1.0 µg	Electro (Laurell)	0.5 mg
Nigrosine	10.0 ng	Nephelometry	0.1 mg
Silver	10.0 ng	Particle counting	0.1 µg
Enzymatic (HRP)	1.0 ng	Enzymatic (ELISA)	0.1 µg
Luminol	1.0 pg	Chemiluminometry	1.0 pg
Autorad	1.0 pg	Radioimmunoassay	1.0 pg

Quantification

The immunochemical sensitivities are also listed in Table 7.2.

Concerning Sindic's methods for free light chains, it should be noted that there are crucial differences in his method, namely he uses the same antibody twice, first as capture and second as detector after he has performed a biotinylation, whereas most people use two different antibodies for capture and detection.

Johnson [496] initially used Coomassie on concentrated CSF, but now most people use anti-IgG with no concentration. He used agarose gel electrophoresis whereas focusing gives a four-fold 'stretch' to the gamma region while actually causing sharper bands than one could prepare during electrophoresis. This is because diffusion occurs during the course of electrophoresis, as opposed to concertina-like 'sharpening up' of bands during focusing.

Chemiluminescence is perhaps the most sensitive of the assays for the antigen–antibody reaction, although it can sometimes be rather capricious (analogous to the silver stain). It also has a very steep slope, i.e. the rate at which the reaction takes place over the course of time, hence there is a very narrow window of opportunity to visualize this using photographic film. It is also rather difficult to make the assay quantitative, since once again relatively short differences in timing make dramatic differences in the numbers of silver grains which are precipitated by the photons which reach the photographic paper.

DELFIA (Dissociation-Enhanced Lanthanide Fluoro-Immuno-Assay) has the great attraction of having a very low background, although this may not be the most important issue if, in fact, one has a very weak signal. DELFIA is somewhat analogous to radioimmunoassay in that it depends upon the specific activity that one achieves with increasing numbers of labeled molecules per unit weight of indicator protein. But whatever this factor may be, the amount of light 'emitted' is only proportional to the strength of the 'excitation' light and the former will not change over the course of time. On the other hand, an enzymatic reaction will release an increasing amount of product over the course of time and this is analogous to using a radiolabeled ligand, which is left for long periods exposed to X-ray film in the same way that chemiluminescent substrates can be left for long periods exposed to photographic film. This enzyme-amplified fluorescent or chemifluorescent reaction can thus achieve, in principle, the same sensitivity as seen for chemiluminescence, but with the added attraction that the slope of the reaction over the course of time is less steep with chemifluorescence; therefore the reaction can be more quantitative since the timing is much less crucial than for the chemiluminescent reaction. The other technique for separating signal from noise in, for example, urine, is electrophoresis, in which the diffuse background noise is 'left behind', so the protein can migrate as a sharp band with a stronger signal.

Further examples of the discrepancy problem

If one uses an antibody to react with a heterogeneous antigen, the answers will vary with different heterogeneous subpopulations, namely intact versus fragments versus aggregates (see Table 7.3).

There are six types of discrepancy problems (see Table 7.4) that can be encountered using immunochemical assays to measure proteins, which have sustained physical–chemical alterations.

Using the analogy of the Mancini plate (see Figure 7.1), proteins may either be precipitated close to the origin (hence a falsely low value, compared to the 'standard' or unaltered protein), or far from the origin, i.e. well beyond the 'standard'

TABLE 7.3 Different sources which cause problems of discrepancy between immunochemical and physical–chemical estimates

1. Fragments
   Normal: 'enriched' by filtration on size basis: IgM monomer
   Normal: 'produced' by hydrolysis: tau protein
   Pathological: 'released': free light chains
2. Aggregates
   Normal: 'removed' by filtration on size basis
     IgM pentamer (see 1. Fragments for IgM monomer)
     Haptoglobin polymers
   Pathological: 'released': antigen–antibody complexes

TABLE 7.4 Discrepancy problems (numbered 1 to 6 for Figure 7.1). Due to altered exposure of homogeneous (Same) versus heterogeneous (Different) epitopes; the precipitates may be located either abnormally Close to or abnormally Far from the origin

Close (false low, true higher)
  Same   1. Cover homogeneous: IgA dimer
  Different   2. Uncover heterogeneous: free light chains

Far (false high, true lower)
  Same   3. Uncover homogeneous: IgM monomer
  Different   4. Minor sub-population: paraprotein (heterogenous)
  Different   5. Loss heterogeneous: tau (if not anti-NANA[a])
  Different   6. Cover heterogeneous (binding site):
                                        Hemoglobin antibody
                                        Antigen–antibody
                                        Alpha-2 macroglobulin enzyme?

[a] NANA = N-acetyl neuraminic acid (sialic acid)

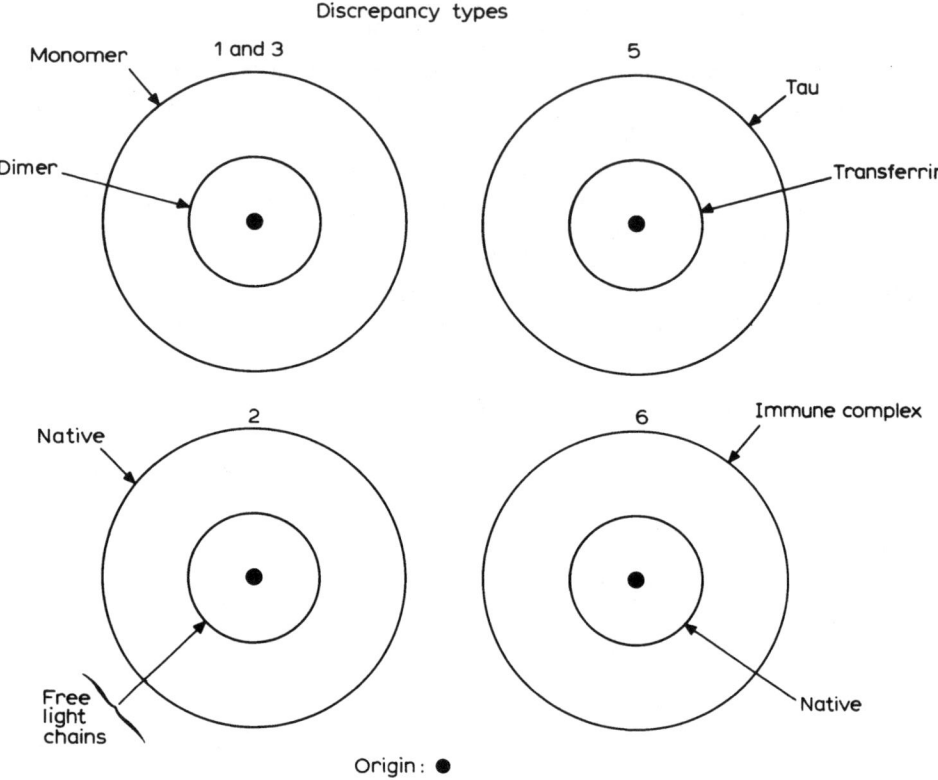

FIGURE 7.1 Different radial immunodiffusion of altered molecular forms, except type 4 (not shown, see Table 7.4 for details).

(hence a falsely high value). These are all variations on the theme of not comparing like with like. This problem was clearly elucidated by Laurell himself [623], who was responsible for one of the most widely used methods of immunochemical estimation. The same general problems apply to immunoturbidometric and kinetic immunonephelometric estimations (data not shown). These problems may not be as widely appreciated as they should be, because they occur less commonly in serum. However, in CSF, because of its physiological isolation (e.g. low total protein), these discrepancies easily impose significant limitations on the routine estimations of several CSF proteins. To give a specific example: a CSF sample from a patient with MS was estimated by radial immunodiffusion [700] to have an IgG to albumin ratio of 0.50 whereas by electrophoresis the value was 0.35. The result is still abnormal, but not as gross as one might think.

The earliest quantitative immunochemical investigations by Kabat [514] appear to show that in 4/16 patients with syphilis, there was more IgG than albumin. However Kabat's earlier studies had not demonstrated this when using electrophoresis [515]. Several other papers have reported IgG as comprising more than a third, or even half, the total CSF protein in MS [924, 1167]. We have not found this in our experience of over 25 000 samples of CSF electrophoresis with corresponding qualitative densitometric scans.

The oldest and best characterized problem is the immunochemical (falsely high) estimation of a paraprotein [786, 787]. Paraproteins can only be estimated correctly by electrophoresis followed by densitometry. The falsely high immunochemical results occur with paraproteins due to the fact that the 'standard' antigen mix (being polyclonal is directed against a large population of different epitopes), but it is then tested against a small subpopulation in which the 'test' or myeloma antigen predominates (thus only a small fraction of the population of precipitating antibodies can actually bind to the myeloma protein). To a lesser degree this would also apply to oligoclonal CSF bands, since they are analogous to, say, 3 to 5 monoclonal bands.

A number of antigens give different degrees of precipitation depending upon the relative exposure of their epitopes. In the case of oligomers and/or repeating subunits (domains) the epitopes can be either identical or different [1206].

In the case of IgA dimers [212], the results (per gram) show that the dimers are more easily precipitated than monomers and therefore dimers have shown a decreased radius (relative to monomer, see Figure 7.1, Type 1). The converse could equally be stated: the monomer is less easily precipitated than the dimer and therefore has an increased radius. If a standard curve is constructed using monomeric IgA and then 'tested' with dimeric IgA the results (falsely low) should be multiplied about $\times 2$. Conversely, if the 'standard' is dimeric IgA and the 'test' protein is monomeric, the results (falsely high) should be halved. This is analogous to serum standards (equivalent to more dimer), which are then tested with CSF (equivalent to more monomer), so that the CSF results should be halved. The same is also true of IgM, because the monomer diffuses further than the pentamer [785].

The IgA dimer diffuses less far per gram, presumably due to the production of hidden homogeneous antigens following auto-association of two monomers (see Figure 7.2). There are normally four potential binding sites (represented as circles in Figure 7.2) per single molecule of antigen (e.g. IgA). If two molecules of IgA are bound together as a dimer, then two of the eight epitopes may be buried or hidden and thus not able to bind to the antibodies during the diffusion procedure. Each dimer then combines with six molecules of antibody, which is equivalent to three molecules per bound monomer on the basis of weight. This is less than the four antibody molecules that can be bound by the free monomeric molecule.

A rather different situation applies in the case of heterogeneous epitopes (represented as circles, squares and triangles in Figure 7.3).

90　　　　　　　　　7. METHODOLOGIES AND THEIR LIMITATIONS

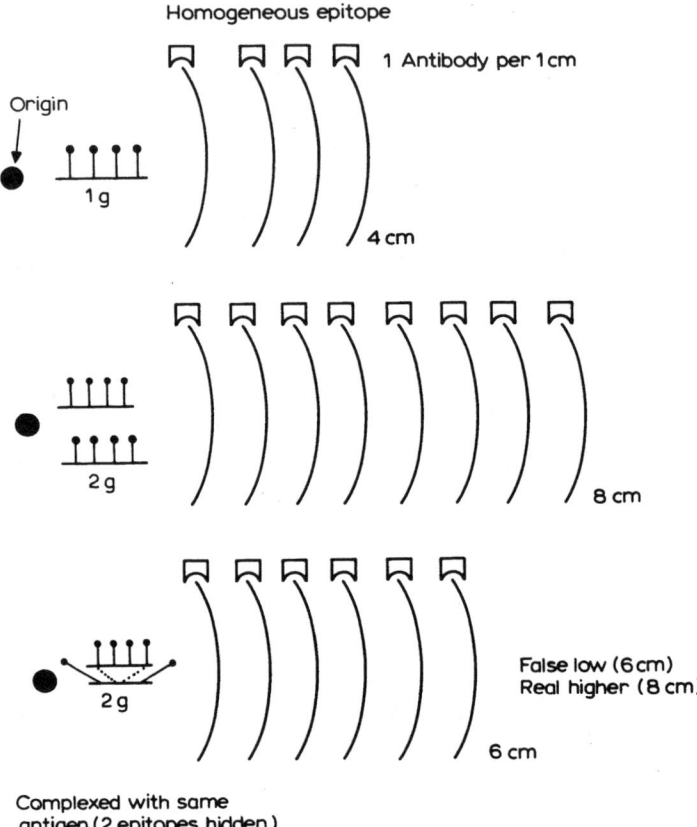

FIGURE 7.2 Precipitation during radial immunodiffusion of self associating units which are homogeneous.

Under normal circumstances (middle of Figure 7.3), only one type of epitope (indicated by the triangles) is buried or hidden due to the particular folding pattern of the protein in question (e.g. IgG).

Using circulating immune complexes as a second example, a separate subpopulation of epitopes may be buried or covered by the *in vivo* association with antibody (e.g. IgM antibody binds to the squares). When the complexed molecule diffuses from the origin (bottom of Figure 7.3), only one remaining class of epitopes (circles) would be available for binding of the anti-IgG antibodies in the immunodiffusion gel. However, the particular tertiary antigen (IgM, which is bound to the 'square' epitopes *in vivo*), unlike the case of IgA (above), is not a primary target for the antiserum that has been raised against the epitopes of the IgG molecule. Thus selective

FURTHER EXAMPLES OF THE DISCREPANCY PROBLEM

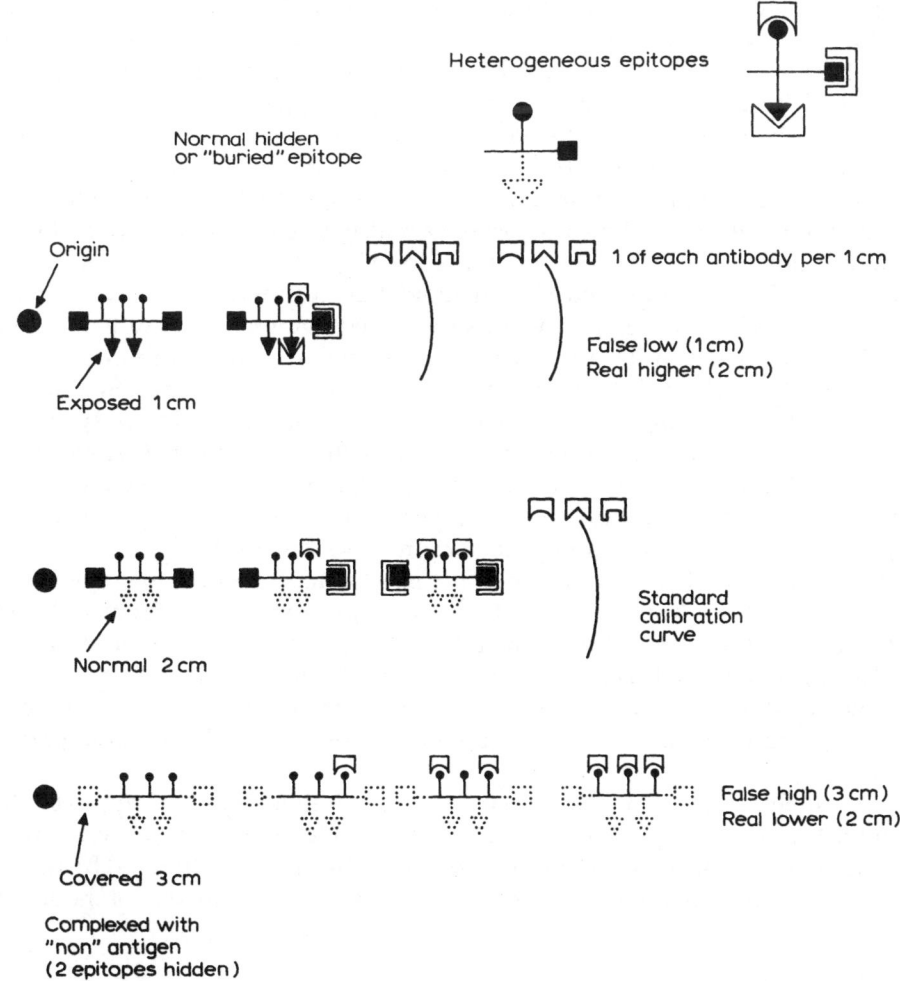

FIGURE 7.3 Precipitation during radial diffusion of antigen associated with a 'nonantigenic' protein, during diffusion from the origin (left to right) with covered or uncovered (exposed) antigenic sites. Three or more antibody molecules are required to bind to each molecule of antigen before a precipitate is formed.

removal or 'covering' of both subpopulations of antigenic sites ('normal' triangle epitopes and abnormal IgM complexed 'square' epitopes) on the IgG molecule means that the test substance (immune complex) will diffuse further into the gel.

Conversely, one can easily imagine a situation in which unfolded fragments of IgG (top of Figure 7.3) with newly exposed epitopes (the triangles) are now available to bind increased amounts of antibody per molecule of antigen (IgG fragment), and hence, will precipitate sooner or closer to the origin than in the normal case (where they are hidden from the 'anti-triangle' antibody subpopulation).

Although it has been known for many years that kappa and lambda light chains give double rings [76, 80, 82, 479], we may postulate that the inner ring represents the more unfolded, exposed, 'free' light chains, while the outer ring represents intact immunoglobulin or bound light chains (see Figure 7.1, Type 2). Unfolded or exposed fragments of myelin basic protein are also much more reactive and hence would precipitate closer to the origin (false low) [1284].

If haptoglobin is complexed with hemoglobin, it moves further and gives a falsely high level [622]. It might be thus presumed that oligomers of haptoglobin (association with self rather than with hemoglobin) would again mimic the situation with IgA where the monomeric species is enriched in the CSF and thus diffuses further requiring the CSF results to be halved. It may also be speculated that when alpha-2-macroglobulin is complexed with an enzyme it also gives rise to falsely high level [1008]. However, in some cases, the binding to a heterogeneous protein may induce a conformational change which results in the exposure of 'neo' antigens, i.e. precipitation nearer to the origin (false low) [623].

There is yet another possible reason for false high results, namely loss of sugar (NANA) from transferrin producing the tau protein (see Figure 7.1). Most studies of CSF transferrin have concluded that it is produced by 'local synthesis' (Table 7.5), although Felgenhauer noted briefly that CSF transferrin behaves anomalously due to the high level of percentage transfer and hence the possibility of local synthesis [290]. The falsely high amount of tau could explain in part the presumed 'local synthesis'.

Concerning the free light chains that appear as 'double rings' on kappa/lambda staining, the serum kappa chains typically circulate as monomers whereas the lambda chains are principally found as dimers. When these are filtered by the kidney (on the basis of their molecular size), there is a selective removal of the smaller

TABLE 7.5 Quantitative evidence for a transferring source within the CNS

$$\frac{\text{Transferrin quotient}}{\text{Albumin quotient}} = \text{Transferrin index ('local synthesis')}$$

Author (date)	Reference	Transferrin Index
Rosenthal and Soothill (1962)		1.50
Frick and Scheid-Seydel (1963)	[321]	1.50
Bock (1978)	[73]	1.46
Delmotte and Tourtellotte (1980)	[217]	1.40

kappa chains into the urine. The same size-effect may operate at the level of the various blood–CSF barriers.

Much careful work on affinity was performed by the triple Nobel laureate, Linus Pauling. The techniques were rather more long-winded in those days, in that they involved a dialysis membrane that contained a protein inside a sausage-shaped bag, and the ligand in question was allowed to diffuse throughout various stages of equilibration between the fluid inside and outside the bag. One of the earliest studies on antibodies was performed by Sips [205], and it soon became obvious that during the course of immunization the affinity for the antigen in question successively increased. This was termed the affinity maturation of the B-cell, and has usually been studied with IgG molecules due to their various intramolecular rearrangements. Contrast this with the IgM molecule, which tends to be relatively static in its affinities over the course of time. IgM has a much shorter life than perennial IgG, but the basic concept of affinity maturation reflects a kind of molecular Darwinism, in the sense of survival of the fittest, namely the higher affinity antibodies have a selective advantage and will be more likely to proliferate into fully fledged clones as mature plasma cells. This is in contrast to the initial, more primitive, so-called anamnestic response, in which the host presumably has innate memory before the initial exposure to the antigen in question, and therefore produces a polyclonal, non-specific response with low affinity. The hydrogen bonding, which is the molecular basis for the different degrees of affinity between antigen and antibody, ultimately reflects the number of shared protons (hydrogen atoms) linking the antibody and the antigen. This can, in effect, be titrated by using a competitor molecule which will also bind to protons, namely sodium thiocyanate (NaSCN), and thus the negatively charged SCN ion will act as a kind of chelator for the shared protons (i.e. positive charge) of the hydrogen bond (see Figure 7.4A and B). In Figure 7.4A we see two different epitopes, and in Figure 7.4B we can see how the 'correct' fit of antibody to its epitope will have a much higher number of hydrogen bonds than the 'incorrect' (low-affinity) binding to another epitope. Removing these protons (with successively higher concentrations of SCN) will release the antigen from its bond to the antibody. This differential effect of binding can be visualized quite easily by the titration of the different clones which bind to the antigen (see Figure 7.5 antibodies against albumin). This figure shows a decrease in polyclonal diffuse background as well as amongst the different clones, or bands. Some are released with a lower concentration of SCN, and therefore have lower affinity (Figure 7.5, top), others remain tightly bound at much higher concentrations of SCN and have a correspondingly higher affinity for the antigen (Figure 7.5, bottom). The original technique of equilibrium dialysis also involved using different concentrations of antigen–antibody and it is still possible to see the different slopes in titrations of standard curves produced from CSF and serum [206].

Another example of affinity studies which has been well documented is the binding of various drugs to albumin. One can distinguish, e.g. in the case of

7. METHODOLOGIES AND THEIR LIMITATIONS

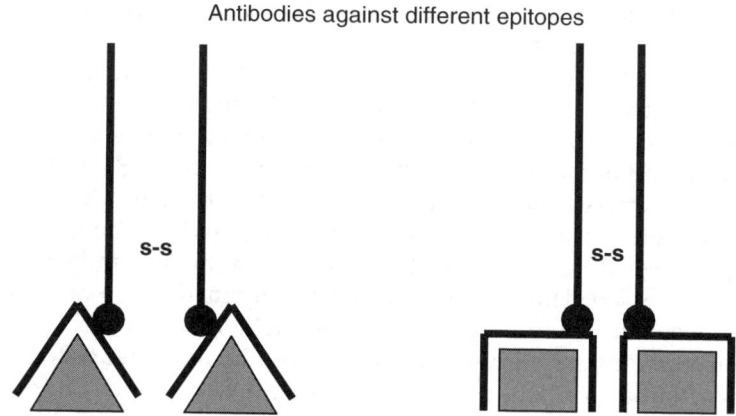

FIGURE 7.4A Cartoon of hydrogen bonds for 2 different shapes of epitope.

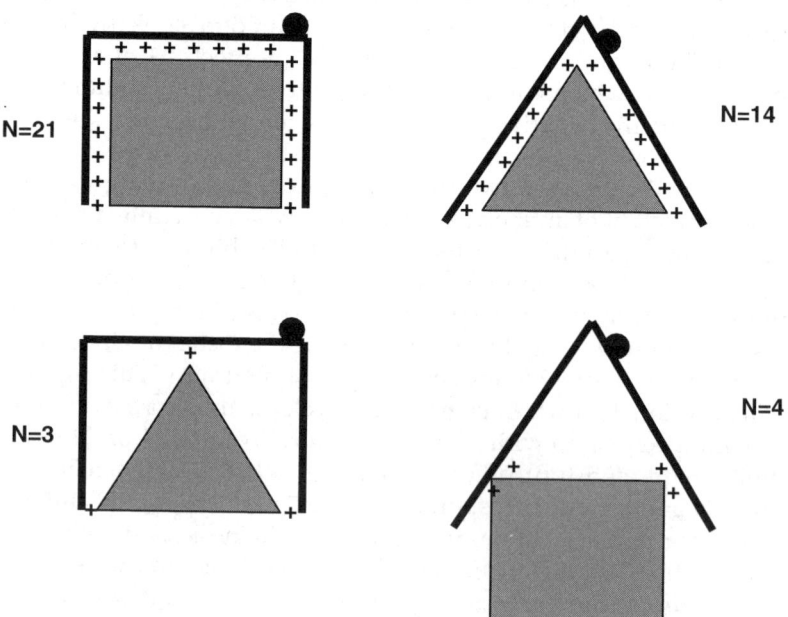

FIGURE 7.4B Cartoon to show increased number of hydrogen bonds with a better fit to either epitope.

FURTHER EXAMPLES OF THE DISCREPANCY PROBLEM

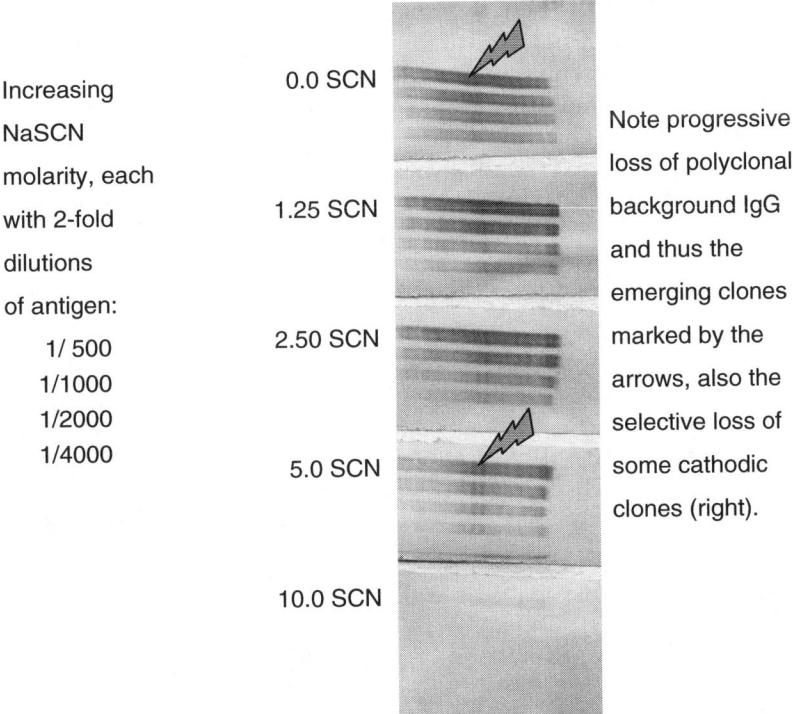

FIGURE 7.5 Increasing concentrations of SCN with the successive loss of the polyclonal background (arrow at top) versus the 'clear' bands which emerge (arrow at bottom) with the highest level of SCN.

anti-epileptic drugs, between the free and bound forms of the drug. In terms of delivering the drug to the appropriate brain receptor, there will be two-way competition between the relative affinity of the drug for albumin and its affinity for the receptor. There will also be local factors such as pH, which may modify the binding of drugs either to the receptor or to albumin. With regard to the binding of calcium to albumin, we can account for the free and bound levels of calcium, which are dependent not only on pH but also on the total concentration of albumin in the serum. Albumin has likewise been studied for two other important types of bonding, namely hydrophobic interactions (fats and/or bilirubin), and at the other extreme, electrostatic binding of drugs with a strong basic charge (e.g. pyridostigmine, which binds to the normally acidic albumin). For low molecular weight proteins/peptides, we know that 'small is beautiful' regarding ease of entry into the brain [279]. It is important to note that drugs which are fat soluble, bind easily to oligodendrocytes (myelin sheath) and will also pass readily through the endothelial cells of the blood–brain barrier (see Figure 7.6 for the oil/water partition coefficient for different drugs). Various drugs can also bind to albumin via hydrophobic, rather than electrostatic bonds.

Perhaps one of the most important practical consequences of a differential affinity is seen with the different types of penicillin that have been used to treat meningitis, encephalitis and/or cerebral abscess. This reminds us once again of this important element to the list of variables, namely the oil/water partition coefficient.

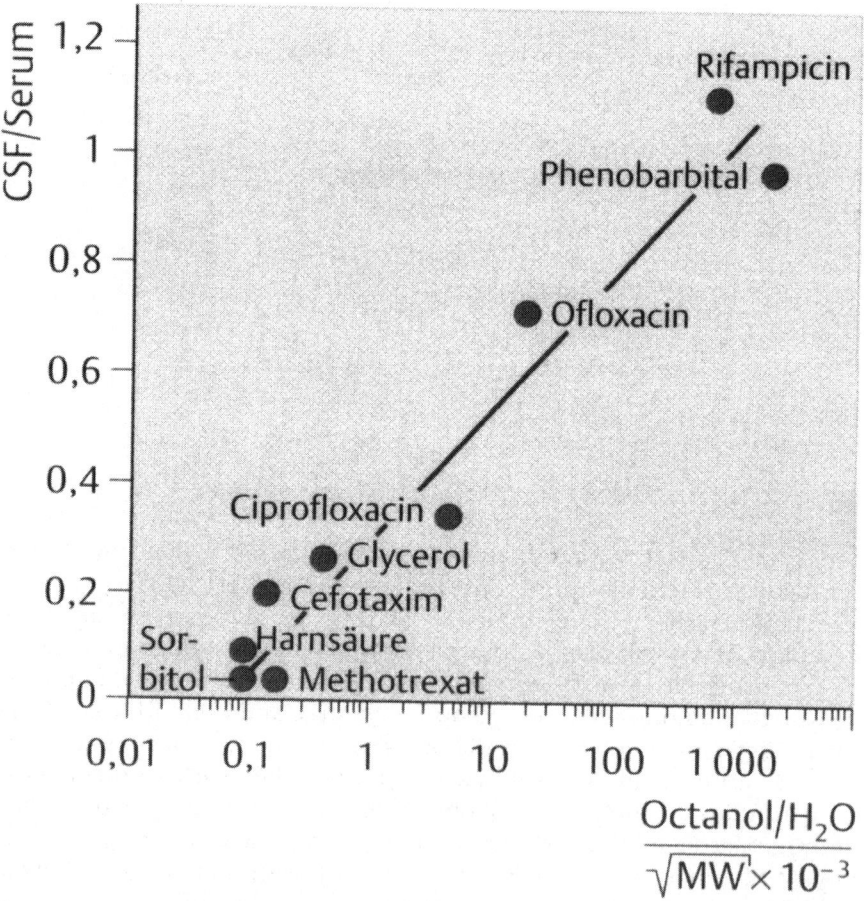

FIGURE 7.6 The correlation of greater ease of drug entry into the brain with increasing solubility in oil versus water (modified from several sources).

In practical terms, this means that the more fat soluble the drug is, the quicker it will pass across the cell membrane of the endothelial cell to enter the central nervous system.

It is worth reiterating that the classical experiments on the blood–brain barrier were performed with a dye which was strongly bound to serum albumin (Evans Blue), and hence the dye was relatively excluded from the brain as opposed to other organs. However, the normal concentration of albumin in CSF compared to blood is essentially a factor of 1 in 200. Therefore, there was not a complete absence of the dye from the brain, but rather 'by eye' there seemed to be no dye present, principally because of the two orders of magnitude difference in its concentration. It never ceases to amaze me that for many years now there has been a continuous trickle of otherwise highly knowledgeable colleagues who have asked me the simple question 'how do I get antibody into the brain?', to which I politely reply that the barrier is not absolute, but relative. The ratio in this case is 1 in 400 for IgG, as opposed to 1 in 200 for albumin, simply based on molecular size.

The importance of high affinity binding is apparent not only in the case of antigen–antibody, but also in the recycling of hemoglobin (due to the relatively short half-life of the red cell) and its binding to haptoglobin, which thus has the consequence of recapturing the iron atom. During evolution, the function of various hormones such as insulin as been improved, such that they only need to be present in a relatively low concentration, namely 0.1 nM/l. Thanks to its higher affinity, this is sufficient to achieve the desired effect.

In the laboratory, we have also taken advantage of the higher affinity binding between avidin–biotin, using it as a kind of molecular amplifier when we couple indicator enzymes, such as horseradish peroxidase. We can then visualize our marker antibodies either for specific bands on nitrocellulose membranes, or for specific parts of the brain under the microscope.

People sometimes speak dismissively about 'low levels of non-specific binding'. If one chooses an arbitrary example of, say, 5 per cent non-specific binding, it then follows that placing a twenty-fold higher concentration of such a binding agent, will produce almost complete (95 per cent) binding to the ligand in question. An initial report demonstrated that myelin basic protein would bind to IgG in the serum of patients with multiple sclerosis. It was therefore presumed that this molecule had some fundamental importance in the pathogenesis of the disease, not least because of many attractive aspects of the model disease, Experimental Allergic Encephalomyelitis (EAE). However, it was subsequently shown that the binding was not only low-affinity and non-specific but, what was much more striking was that it displayed a higher affinity for the common portion (Fc) of the molecule, rather than for the antigen-binding portion (Fab) [2, 820, 1056].

Another major concern with regard to low-affinity binding, which can be easily camouflaged by using higher concentrations, is in the use of antibodies which are ostensibly specific for 'free' light chains. Careful attention to the details of preparation of these reagents reveals the following sequence:

1. The animals are immunized with a Bence–Jones preparation, which is a genuine 'free' light chain, in that there are no bound heavy chains. Thus the animal produces antibodies against the normally 'hidden' epitopes of the light chains, as well as other antibodies against the normally 'exposed' epitopes. This has been considered previously in Chapter 6.
2. The serum is adsorbed several times with intact IgG or 'bound' light chains which display only the 'exposed' epitopes, and hence the presumption is that the antibodies that are not removed through this 'purification procedure' will represent the residual antibodies against the 'hidden' epitopes. The net result of all of these adsorptions is that, yet again, rather low-affinity antibodies are obtained, since the high-affinity antibodies were bound during the 'purification' procedure and are therefore unavailable. This will ultimately lead to multiple problems with cross-reactivities with other antigens.
3. The next important point relates to the three-dimensional configuration of the particular epitope. It is well known that when a protein is folded in its native state, the epitope that binds the antibody in question may not necessarily represent a simple linear sequence of five or six residues, but rather, contiguous turns in a long polypeptide chain, not unlike the two sides of a hairpin, which when straightened out would be a very long distance apart. Indeed, the original configuration of the epitope would be completely lost or at least split in half (see Figure 7.7).

Antibodies applied to a nitrocellulose membrane are (by definition) out of solution, and bind to nitrocellulose groups on the membrane via Van der Waals forces. We showed some years ago (as have others) that antibodies against 'free' light chains will then also bind to 'bound' or the intact light chains of a 'complete' IgG molecule [1260].

The other example of affinity, which unfortunately was not directly measured but only indirectly inferred, was from a study reported by Bell *et al.* [53]. We were fortunate to receive CSF and serum samples which had been kept frozen at the original Herpes Reference Centre in Manchester, UK by Professor Maurice Longston. These patients had not been treated with aciclovir since, at the time, this drug did not hold the pre-eminent position it holds today. Although the number of patients was relatively small, nevertheless a significant proportion had the more primitive polyclonal response, and thus they were more likely to die. Conversely, those who had at least mounted an early monoclonal response to the herpes antigen, survived. Therefore, considering the technique of Eastern blotting against Herpes simplex virus, this again demonstrated that the fittest (literally) survived because they had, for their molecular basis, the clonal selection of antibodies directed against the target, and these stood out well above the diffuse polyclonal background.

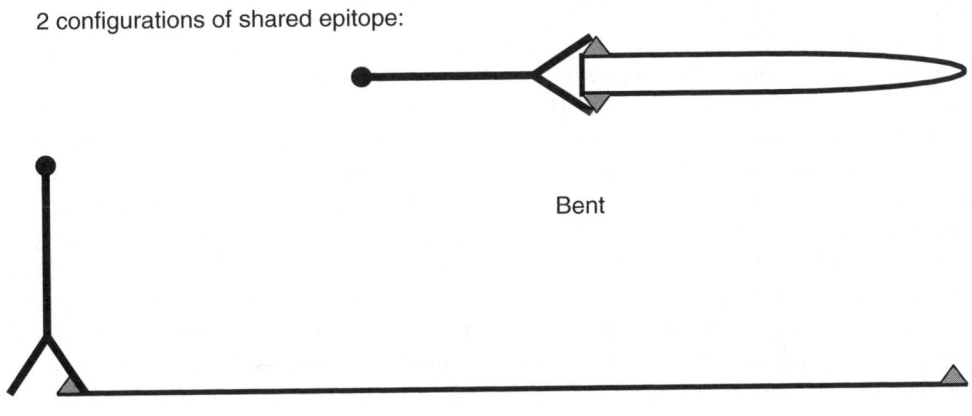

FIGURE 7.7 Cartoon to show two different configurations as an antibody can 'span' two adjacent loops of the polypeptide chain, but then the full binding is lost with the change in the 3D structure of the protein.

One final question about affinity: what is the best standard calibrant for measuring the free light chains? This is analogous to the classical argument between the monoclonal and the polyclonal antibody. This is not just a question of the standard, which can be simply monoclonal versus polyclonal, but the original antigen used for raising of the monoclonal antibodies in question, which will reflect a single idiotype distinct from the thousands of light chains that have quite different sequences. It will therefore bind with lower affinities to light chains whose sequence differ from the original immunizing epitope [796].

Radiolabeled proteins

Injecting radiolabeled substrates into patients does not simply solve the problem of local synthesis, indeed it may create its own problems, e.g. if non-equilibrium conditions exist and/or if the labeled substrate does not behave exactly like the unlabeled molecule. Of course, the discrepancy problem still applies to any 'cold' assays, i.e. as used in the calculations to produce the specific activity data. Specific activity (SA) is given by the radioactive counts per minute (cpm) divided by the amount of protein (mg). Thus the SA = cpm/mg protein. Although there are relatively few studies of radiolabeled CSF proteins in humans, they have attracted a wide audience because of ethical difficulties with some aspects of such studies from current perspective. Much of the earlier work was written in German by

Frick and Scheid-Seydel [318–321] and taken together represents the most extensive study of radiolabeled proteins in the CSF. One of the annoying problems with this particular work is the relatively large number of mistakes in simple arithmetic, which is enough to put off all but the most dedicated academicians. The net effort is worthwhile, however, since such a large number of patients with different disorders was investigated. To aid others with the mathematical understanding of the results, Figure 7.8 details the simple algebraic manipulations performed on the data. In Frick's experiments, 130 per cent of radioactive counts orginated from serum transferred into CSF (hence they multiplied the radioactive count by 0.78 to correct for this (see Table 7.7), but in Tourtellotte's case more than 90 per cent of the counts were for albumin crossing from serum into CSF, thus they needed no correction factor. The two formulae are therefore essentially identical. It could be argued that the fixed 'C' value (Figure 7.9) represents a systematic error, e.g. aggregates. The lower half of the same figure gives a representation of what could be expected if clear *de novo* synthesis occurred during the course of the experiments. In that particular case, one would have to exclude a shorter half-life for IgG in the CSF, but the available figures to date show parallel slopes for decay curves in CSF and serum from individual patients [1180]. The formula for determination of the proportion of locally synthesized IgG (i.e. the fraction of the total IgG which is not derived from the serum) is essentially the same method as that used by Felgenhauer [275, 290] and Reiber [902] (see Figure 7.10).

It is necessary to ensure that the system is at equilibrium, otherwise a rather transient phenomenon is being studied. Table 7.6 shows that the time to reach equilibrium is of the order of four days for albumin and six for IgG. Although the early workers made this point quite clear, several of their lumbar punctures were performed at a time when equilibrium had not yet been reached. Because of the dynamics of the CSF flow, equilibrium will be reached first in the ventricles and then in the cistern and finally in the lumbar region [1110].

The relative specific activity for albumin and IgG (Table 7.7) appears to be slightly higher in CSF than in serum. However, the lower specific activity of transferrin was used as an argument for its 'local synthesis'. The immunochemical determination of transferrin (including beta-2 transferrin) has its own special problems of discrepancy as noted above.

The rates of transfer from serum into the CSF are given in Table 7.8 and show quite a wide disparity for the values for IgG transfer (hundred-fold difference between the highest and lowest values). If the typical rate of CSF formation is 18 ml/h, then given a normal concentration of 2 mg/dl, 0.36 mg per 18 ml could be expected, so that the correct result is of the order 0.3 mg/h. This would also be consistent with the value of 0.9 mg/h for albumin. In order for the median IgG value to be correct, there would have to be a selective IgG 'sink' that would destroy 90 per cent of what is transferred. This seems rather unlikely since the half-lives of both albumin and IgG are of the order of two weeks, which is also quite comparable

1. Frick's calculations for radiolabelled IgG:

Divide

$$\frac{SER}{CSF} = RSA \quad \text{i.e. Relative Specific Activity}$$

For electrophoresis

$$\frac{\text{Gamma area}}{\text{Total area}} = \%\,G \quad \text{i.e. percentage Gamma}$$

then use a constant (e.g. 0.78, or 78 per cent)

$$\frac{\%\,G}{RSA} \times 0.78 = C \quad \text{i.e. CSF Gamma}$$

then

$$100 - \frac{C}{\%\,G} \times 100 = \text{local synthesis}$$

However % G will factor out, viz:

$$100 - \left(\frac{\frac{\%\,G}{RSA} \times 0.78}{\%\,G} \right) 100$$

$$100 - 78 \frac{\%\,G}{RSA} \times \frac{1}{\%\,G}$$

$$100 - \frac{78}{RSA}$$

$$RSA = \frac{SA\ SER}{SA\ CSF}$$

$$RSA = \frac{\frac{cpm\ SER}{mg\ SER\ G}}{\frac{cpm\ CSF}{mg\ CSF\ G}}$$

$$RSA = \left(\frac{cpm\ SER}{cpm\ CSF} \times \frac{mg\ CSF\ G}{mg\ SER\ G} \right)$$

$$\frac{1}{RSA} = \left(\frac{cpm\ CSF}{cpm\ SER} \times \frac{mg\ SER\ G}{mg\ CSF\ G} \right)$$

$$100 - 78 \times \left(\frac{cpm\ CSF}{cpm\ SER} \times \frac{mg\ SER\ G}{mg\ CSF\ G} \right)$$

$$1 - 0.78 \times \left(\frac{cpm\ CSF}{cpm\ SER} \times \frac{mg\ SER\ G}{mg\ CSF\ G} \right) = \text{local synthesis}$$

FIGURE 7.8A Frick's calculations, which are continued on the next page.

7. METHODOLOGIES AND THEIR LIMITATIONS

2. Tourtellotte's calculations for radiolabelled IgG:

$$1 - \frac{B}{A} = C = \text{local synthesis}$$

$$1 - \left(\frac{\frac{\text{cpm CSF}}{\text{mg CSF G}}}{\frac{\text{cpm SER}}{\text{mg SER G}}} \right) \frac{\text{mg CSF G}}{100 \text{ ml}}$$

$$\frac{\text{mg CSF G}}{100 \text{ ml}} - \left(\frac{\text{cpm CSF}}{\text{mg CSF G}} \times \frac{\text{mg SER G}}{\text{cpm SER}} \right) \times \frac{\text{mg CSF G}}{100 \text{ ml}}$$

$$\frac{\text{mg CSF G}}{100 \text{ ml}} - \left(\frac{\text{cpm CSF}}{\text{cpm SER}} \times \frac{\text{mg SER G}}{100 \text{ ml}} \right)$$

$$1 - \left(\frac{\text{cpm CSF}}{\text{cpm SER}} \times \frac{\text{mg SER G}}{\text{mg CSF G}} \right) = \text{local synthesis}$$

FIGURE 7.8B Specific activity is given for serum and CSF by the above calculations. Frick's calculations are essentially the same as Tourtellotte's (but Frick had a constant of 0.78).

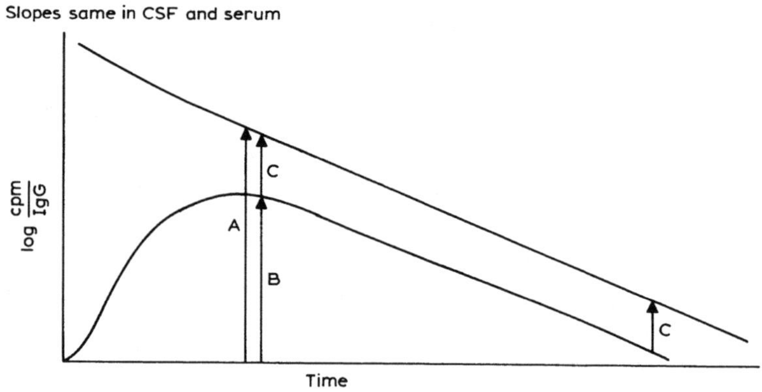

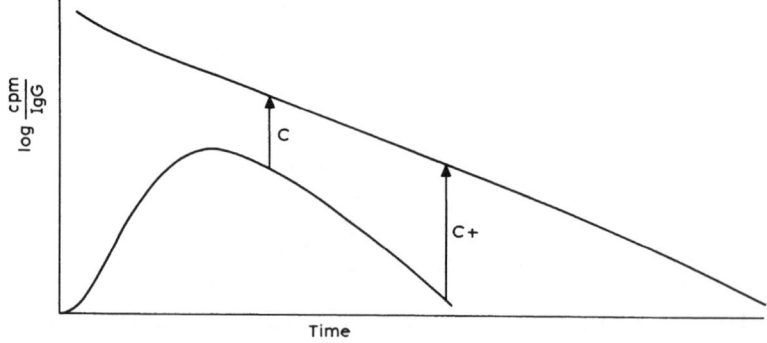

FIGURE 7.9 Decay curves of static (top) and falling (bottom) slopes.

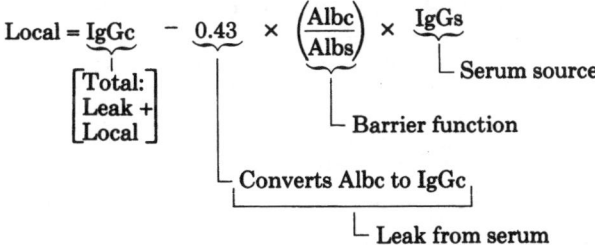

Local = Total − Leak

FIGURE 7.10 Formula to correct for the individual serum level (Serum source) and CSF total protein level (Barrier function).

TABLE 7.6 Time to reach equilibrium in CSF (serum injection)

Author (date)	Reference		Time taken
Albumin			
Frick and Scheid-Seydel (1958a)	[318]		4 days
Fieschi and Agnoli (1964)	[294]		< 6 days (except 1/5 patients)
Tourtellotte et al. (1980)	[1180]		2 days
		Alb: median	4 days
IgG			
Tourtellotte et al. (1980)	[1180]		2 days
Frick and Scheid-Seydel (1958b)	[319]		7 days (lumbar)
Frick and Scheid-Seydel (1958b)	[319]		3 days (cisternal)
Cutler et al. (1967)	[187]		2 days (ventricular)
Cutler et al. (1967)	[187]		6 days (lumbar)
		IgG: median	6 days (lumbar)

TABLE 7.7 Relative specific activity in CSF/serum

Author (date)	Reference	Specific activity
Albumin		
Frick and Scheid-Seydel (1958a)	[318]	1.2
Fieschi and Agnoli (1964)	[294]	1.1
Tourtellotte et al. (1980)	[1180]	0.9
IgG		
Frick and Scheid-Seydel (1958b)	[319]	1.3
Cutler et al. (1967)	[187]	1.1
Transferrin		
Frick and Scheid-Seydel (1963)	[321]	0.85
Frick ('beta') (1960)	[320]	0.78

TABLE 7.8 Rates of transfer from serum into CSF

Author (date)	Reference	Rate of transfer	
Albumin			
Fieschi and Agnoli (1964)	[294]	0.90 mg/h	
IgG			
Frick and Scheid-Seydel (1958b)	[319]	2.00 mg/h	
Lippincott et al. (1965)	[657]	0.04 mg/h	
Cutler et al. (1967)	[187]	3.60 mg/h	
Cutler et al. (1967)	[187]	0.30 (mg/h, but only 200 ml/day of CSF actually formed	
		IgG: median	2.0 mg/h
		'Calculated'	0.3 mg/h

TABLE 7.9 Half-life in the CSF and serum for albumin and IgG

Half-life in CSF				
Author (date)	Reference		Half-life	
Albumin				
Frick and Scheid-Seydel (1958a)	[318]		14 days	
Fieschi and Agnoli (1964)	[294]		12 days	
Tourtellotte et al. (1980)	[1180]		10 days	
		Alb: median	12 days	
IgG				
Frick and Scheid-Seydel (1958b)	[319]		16 days	
Cutler et al. (1967)	[187]		11 days	
Tourtellotte et al. (1980)	[1180]		14 days	
CSF injection				
Lippincott et al. (1965)	[657]	IgG: 'fast'	0.5 day	
Lippincott et al. (1965)	[657]	IgG: 'slow'	6 days	
		IgG: median	14 days	
Half-life in serum (same as CSF except)				
Serum injection				
Lippincott et al. (1965)	[657]		19 days	
CSF injection				
Lippincott et al. (1965)	[657]		11 days	

to their half-life in the plasma (Table 7.9). The only atypical results are again those of Lippincott et al. [657], in which a comparison of his Tables 1 and 2 reveals that those cases with injection into CSF (versus serum) had quite a significant increase of total protein value following the CSF injections. Thus, there must have been a significant breakdown of blood–CSF barriers.

CHAPTER

8

Fitting the data to curves

Introduction
Mathematics of the multiple sources of CSF proteins
Conclusion

Introduction

Each CSF protein should be considered in the first instance to have a primary serum source and to have a concentration gradient which drives it across various barriers, whose degree of filtration or effectiveness can readily change. However, proteins can be synthesized and/or modified within the confines of the theca. These three factors must always be included in any equation: serum level, function of barrier(s), and local synthesis (modification). For those who estimate local IgG synthesis, I have derived a formula (and simple plot) to allow for the change in percentage transfer with damaged barriers. Note that in simple, untransformed linear plots the slope allegedly changes from 0.7 to 1.0 as total protein values increase.

Fitting of data to a curve is the mathematic counterpart of the concept of multiple blood–CSF barriers (refer to Chapter 5).

Mathematics of the multiple sources of CSF proteins

To evaluate the levels in CSF of any given protein properly one needs to know the serum level (and its variability) the degree of filtration (percentage transfer) and whether the percentage transfer changes (and by how much) when CSF total protein levels are elevated (change in selectivity and/or different loci as 'new' sources of CSF proteins become available under various pathological conditions).

The data can be reported using a formula or an index which can also be depicted graphically. We must first clarify the fundamental concepts of how to express results, before going on to discuss the specific details of normal values (Chapter 9).

When one obtains numerical values for amounts of IgG, the first question to be addressed is: how does one decide on an interpretation of these values in the light of a specific diagnosis (e.g. MS). In general terms, the best measure of any test is its efficiency, or its ability to yield the correct answer in the face of possible false negatives or, much more worrying, false positives. Clinicians will not be put off making a diagnosis if they know that only 3–5 per cent of patients will have a false negative result (indeed with a relapsing and remitting disease such as MS they would probably choose to repeat the test at a later date when it may well have become positive).

We have seen (Chapter 3) that it is better to interpret the individual quotients for each patient's Albc/Albs levels as a final average, rather than first averaging all the Albc levels, then averaging all the Albs levels and finally averaging the quotient of these two values. Once again, for the most reliable conclusions, the individual index for each patient (IgGc/IgGs)/(Albc/Albs) should be summed and then averaged rather than dividing a single average value of all IgGc/IgGs quotients by the single average value for all Albc/Albs quotients.

The additional advantage, however, of plotting individual IgGc/IgGs against Albc/Albs, is that any changes in slope with increasing levels of Albc/Albs (i.e. barrier breakdown) will be seen at once, whereas by calculating the average index, this information would not be apparent. One other alternative (not usually performed) would be to plot individual index values against individual Albc/Albs levels. Yet again, the intuitive value of comparing IgG with albumin is much to be preferred. The fact that changing slope is best accommodated by log–log transformation will be demonstrated. It is clear from inspection of any of the plots, that the value of any quotient (or ratio) does not allow distinction between a higher denominator and a lower numerator (or some combination thereof). Likewise any index value will, by definition, not distinguish which of the four parameters (and/or combinations) has been altered. To give a simple example: if a patient had a marked drop in his serum IgG level (e.g. immunosuppressive drug therapy) this could give a falsely high value for the IgG index.

Because the normal range of CSF total protein is so wide (20 to 55 mg/dl) [1170] and since it varies with age [1163] (and perhaps to a lesser degree with the amount of fluid withdrawn [921]) it is difficult to give an absolute value for the upper limit of normal for IgG. The very first quantitative study of CSF IgG suggested that the IgG should be divided by the value of the total CSF protein [514]. Although 'age-correction' factors have been introduced for very young children, a similar factor has not yet been developed for the much more common (and increasing) older age population [338]. The 'rule of thumb' for CSF total protein was: the patient's age is approximately the level in mg/dl.

CSF IgG is primarily derived from the serum and therefore a larger amount of serum IgG being transferred to the CSF might be misinterpreted as being local synthesis. To correct for this relatively higher proportion of serum IgG/total serum protein compared with the CSF IgG/total CSF protein, and thereby to compensate for the serum IgG, the CSF IgG was divided by the serum IgG. In addition, to correct for any barrier changes, the CSF albumin was also divided by the serum albumin (as it is well known that the latter can also decrease in various disease states) see Figure 8.1.

Basically, the idea of correcting for differing serum levels of IgG and albumin (as well as the degree of permeability of the barrier by using the albumin ratio, since this protein cannot be synthesized within the CNS) was first used by Tourtellotte in 1970 [1167], and later modified [1168]. The formula is felt by some to be rather more complicated than simple ratios. However, as we shall see, this is not strictly the case.

The most straightforward of these ratios was originally proposed by Delpech and Lichtblau in 1972 [219], and has now come to be called the IgG Index. Unfortunately, the same problem plagues both the index and the formula (or any of their multiple variations) namely the assumption that the degree of protein transfer (the 'slope' of Felgenhauer [290]) does not change over the range of increasing barrier damage. In fact, careful examination of the data of Ganrot-Norlin [336] shows that as the barrier function declines (i.e. as the total protein increases), there is proportionally more transfer of IgG and less selective filtration. To reiterate: this is a normal function with increasing age [338].

The formulation of Delpech and Lichtblau [219] was designed primarily to correct for changes in serum levels and not specifically for changes in barrier function. In fact any cases of abnormal barrier function were excluded, using agar electrophoresis to define any samples which showed a 'transudative' pattern. Each axis depicts the relative amounts of two different proteins i.e. IgG/albumin, thus being analogous to IgG as a percentage of total protein. The ratio of IgG/albumin is plotted for each of the two fluids (CSF and serum) on the two axes (see Figure 8.2, top). As one proceeds further right along the X axis there is either elevated serum IgG or depressed serum albumin or a mixture of both. These are typically mirrored in the changes along the Y axis. The degree of barrier damage can not be deduced from analysis of the graphic points. This mode of expression can be termed 'proportional composition' (IgG/albumin) and can be contrasted with 'percentage transfer' (i.e. IgGc/IgGs). In this latter formulation, (see [335]) analysis of the graph shows the degree of barrier changes rather than serum levels and thus presents the converse of the Delpech and Lichtblau formulation. In the case of the Ganrot and Laurell formulation the two axes depict quotients of the same protein (Albc/Albs or IgGc/IgGs) on either side of the blood–CSF barriers. This results in a dimensionless number or a fractional transfer (percentage of the particular serum protein which is transferred or filtered across the different barriers into the CSF, see Figure 8.2, bottom).

108   8. FITTING THE DATA TO CURVES

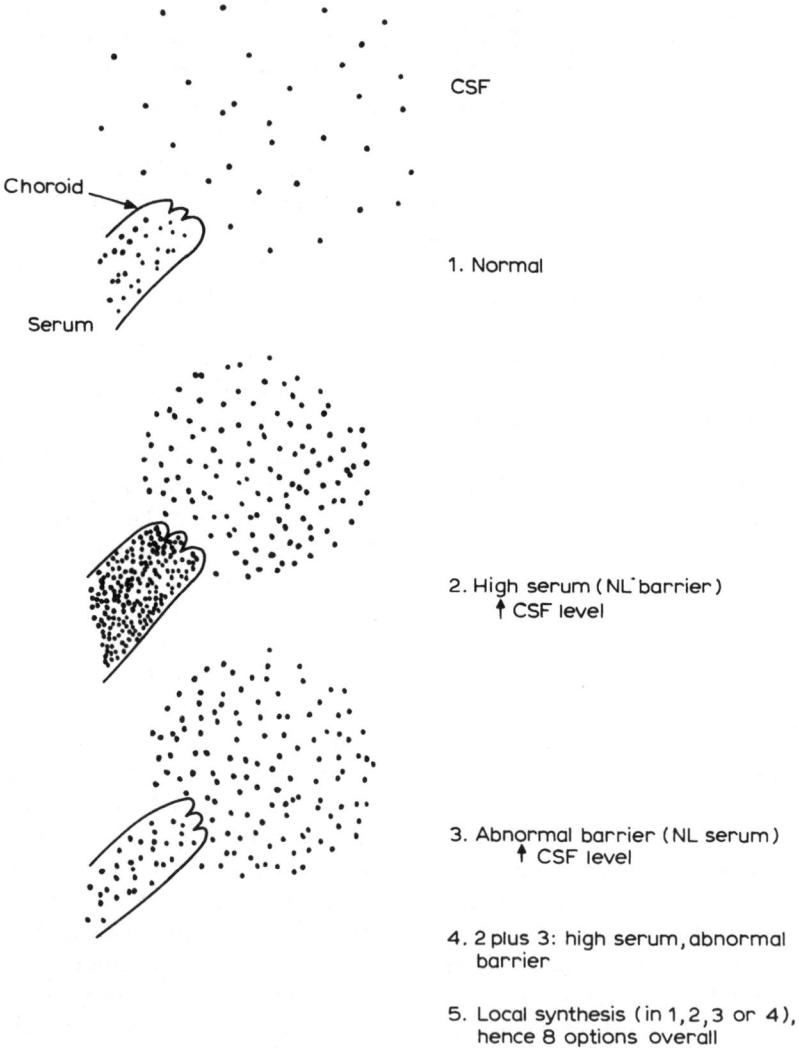

FIGURE 8.1 The separate and combined influence of serum levels and/or barrier function, which both hamper estimation of the third source: local synthesis.

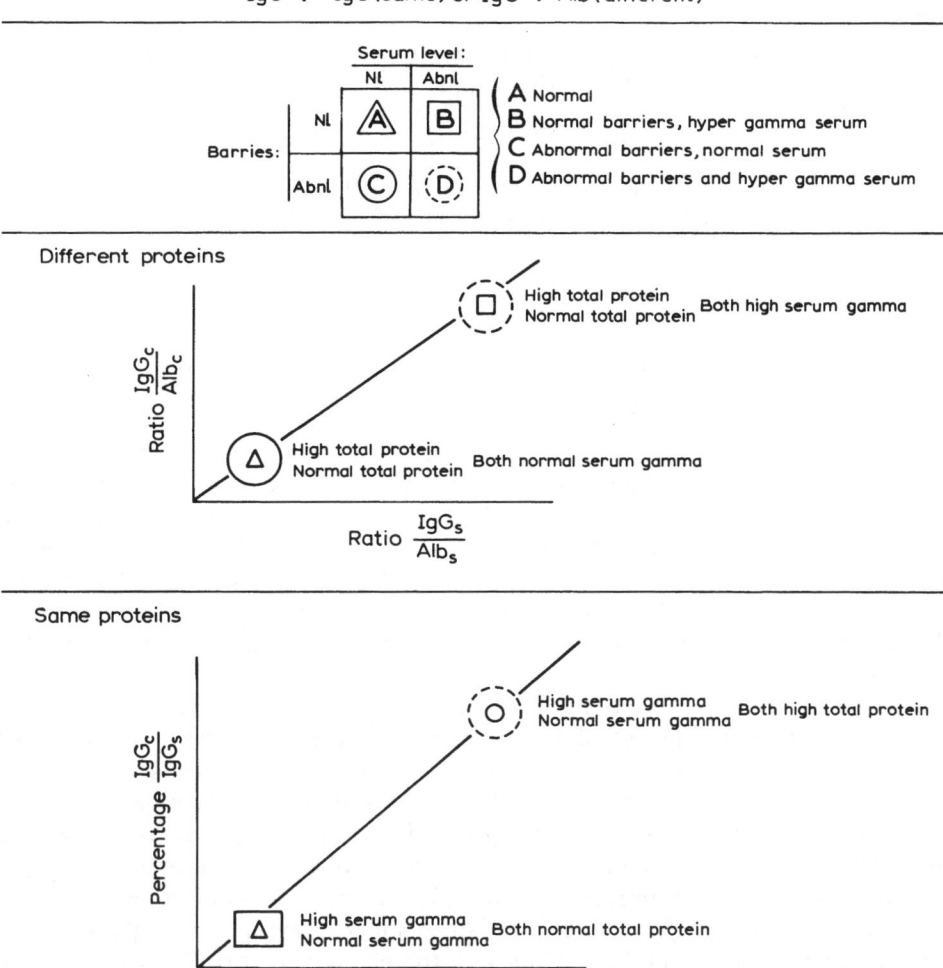

FIGURE 8.2 Two methods of plotting with the primary emphasis on either serum (top) or barrier function (bottom).

As one proceeds to the right along the X axis there is successively more impairment of barrier function i.e. elevation of the total CSF protein level. When data is presented in this way the relative proportion of serum IgG to serum albumin (or corresponding proportion of CSF) cannot be deduced from analysis of the graphic points.

Comparison of the two modes of graphic presentation reveals an obvious algebraic similarity and an identical index value.

Delpech and Lichtblau [219]

$$\frac{(IgGg)/(Albc)}{(IgGs)/(Albs)} = \frac{IgGc}{Albc} \times \frac{Albs}{IgGs} = \frac{IgGc}{IgGs} \times \frac{Albs}{Albc}$$

Ganrot and Laurell [335]

$$\frac{(IgGc)/(IgGs)}{(Albc)/(Albs)} = \frac{IgGc}{IgGs} \times \frac{Albs}{Albc}$$

Conversely, there is an obvious difference not just in the axes but also in the slope. This is readily seen in the data of Eickhoff [236, 240] where the Delpech and Lichtblau slope is 0.53 and the Ganrot and Laurell slope is 0.32. The average index values of the two plots are the same, namely 0.47 (recalculated from the plotted data points of the two different formulations). In their original publication the data of Delpech and Lichtblau [219] yielded a slope for their ratio plot of 0.63 but the index was 0.52.

The Ganrot and Laurell plot [335] has a certain intuitive appeal, since the X axis essentially mirrors the level CSF total protein, and the value of 1.0 would reflect complete equilibration of CSF with serum. A slope of, say, 0.5 reflects the larger molecular size of IgG versus albumin. In simple terms, for every molecule of IgG which finds its way across the various blood–CSF barriers, two corresponding molecules of albumin (smaller protein size) also cross. The relative amounts of serum IgG to serum albumin seem to have less intuitive appeal to the practising neurologist and although a value of 1.0 (or greater) is possible (i.e. IgG is greater than 50 per cent of the total protein level), the fundamental differences between the two proteins (e.g. size, charge, etc.) seem to limit the appeal of the Delpech and Lichtblau plot.

In conclusion, there are three separate approaches to the same end (i.e. to correct for differences in serum levels and differences in barrier efficiency):

1. Slope of proportional transfer (ratio IgG/Alb for CSF versus serum).
2. Slope of percentage transfer (quotient IgG versus quotient albumin).
3. Index.

The Tourtellotte [1168] formula may not appeal, since it offers numeric results in milligrams of IgG synthesized per day. However, there was no graphic plot of this formulation and the basic data used to generate the Ganrot and Laurell plot [335] and the Delpech and Lichtblau plot [219] can be expressed as mg/day as well as by the index. This point was used by Reiber to calculate amounts of locally synthesized IgG [902]. The perceived complexity of the Tourtellotte formula possibly hindered its ready comprehension and therefore acceptance, but in principle it is actually no different from either of the two graphic presentations of the index. The presumptive difference lies in the identical assumption of 'equal molar transfer' which leads to further mathematical manipulation (as will be detailed later). Indeed, this identical assumption is quite implicit in the index calculations. However, the net result can be seen to be essentially the same as either of the two index calculations. Thus the concept of 'equal molar transfer' presumes that for each molecule of albumin moving across the barriers an equivalent molecule of IgG travels with it. In any case, the constant so derived (0.43) represents the quotient of the molecular weight of albumin divided by the molecular weight of IgG. This correction factor is derived from Tourtellotte's presumed molecular weights of:

$$\frac{60\,000}{150\,000} \frac{\text{Alb}}{\text{IgG}} = 0.43$$

There are two slight difficulties with this assumption:

1. The serum molar concentration ratio is in fact 10 to 1 (Alb:IgG) and molar equivalence of transfer would thus rely on a kind of Maxwell's demon to exert a positive selection for IgG ten times more frequently than albumin from the serum pool.
2. To convert albumin into IgG-equivalence, one should in theory divide by the molar constant (0.43) rather than multiply, i.e. the constant 0.43 has the 'units' albumin/IgG (0.43 albumin/IgG). When one algebraically multiplies an albumin value by the constant, the result would be albumin squared divided by IgG (albumin × albumin/IgG). However, if one were to divide the albumin value by the constant, the net result would be to convert albumin into IgG equivalence (albumin divided by albumin/IgG = IgG).

Happily, neither of these issues causes any problems in practice, since the constant 0.43 actually turns out to be the slope of the equation described in Figure 8.3A.

The Schuller formula is quite similar to the Tourtellotte formula [1168], except that the Albs value is a constant, rather than a variable. However, this assumption of invariant serum albumin is rather difficult to justify (see Figure 8.3B).

8. FITTING THE DATA TO CURVES

The Tourtellotte formula was given in units of mg/100 ml. One could thus multiply by 5 to give mg/day (i.e. $5 \times 100 = 500$ ml CSF formed per day). Putting the factor of five to one side, we have the original formula:

$$IgGc - \frac{IgGs}{369} - \left(\frac{Albc}{1} - \frac{Albs}{230}\right)\left(\frac{IgGs}{Albs}\right)0.43$$

This can be changed by simple algebra

$$IgGc - 0.43\left(\frac{IgGs}{Albs}\right)\left(\frac{Albc}{1} - \frac{Albs}{230}\right) - \frac{IgGs}{369}$$

$$IgGc - 0.43\left(\frac{IgGs}{Albs}\right)Albc + 0.43\left(\frac{IgGs}{Albs}\right)\frac{Albs}{230} - \frac{IgGs}{369}$$

$$IgGc - 0.43\left(\frac{Albc}{Albs}\right)IgGs + 0.43\left(\frac{IgGs}{230}\right) - \frac{IgGs}{369}$$

$$IgGc - 0.43\left(\frac{Albc}{Albs}\right)IgGs + 0.002\left(\frac{IgGs}{1}\right) - 0.003\frac{IgGs}{1}$$

$$IgGc - 0.43\left(\frac{Albc}{Albs}\right)IgGs - 0.001\,IgGs$$

The Reiber formula [697] is given in its original form as:

$$IgGc - \left[0.43\left(\frac{Albc}{Albs}\right) + 0.001\right]IgGs$$

This can be changed by simple algebra

$$IgGc - \left[0.43\left(\frac{Albc}{Albs}\right)IgGs + 0.001\,IgGs\right]$$

$$IgGc - 0.43\left(\frac{Albc}{Albs}\right)IgGs - 0.001\,IgGs$$

This is thus essentially identical to the Tourtellotte Formula.

FIGURE 8.3A The Reiber formula is essentially identical to the Tourtellotte formula.

The IgG index does not specify a particular cut-off value *per se*, but if one can determine the appropriate slope, any IgG present above the cut-off value is presumed to be due to local synthesis. This is the approach used by Reiber [902] (see Chapter 7) in which his formula effectively approximates to the proportion of serum leakage of IgG, and thus one derives the difference (from total IgG) which represents locally synthesized IgG. Felgenhauer uses a method which is also essentially equivalent in determining local synthesis [290, 989]. The method includes the additional determination of alpha-2-macroglobulin to allow for any changes

The Schuller formula is given as:

$$\text{IgGc} - \left[\frac{(\text{Albc} - 240)\,\text{IgGs}}{60\,000} + 30\right]$$

This can be changed algebraically to

$$\text{IgGc} - \left[\frac{(\text{Albc} - 240)\,\text{IgGs}}{6 \times 10^4}\right] - 30$$

$$\text{IgGc} - \left[\left(\frac{\text{Albc}}{245} - \frac{240}{245}\right)\frac{\text{IgGs}}{\sqrt{6} \times 10^2}\right] - 30$$

$$\text{IgGc} - \left[\left(\frac{\text{Albc}}{245} - 0.98\right)\frac{\text{IgGs}}{245}\right] - 30$$

$$\text{IgGc} - \left[0.41 \times 10^{-2}\left(\frac{\text{Albc}}{245} - 0.98\right)\text{IgGs}\right] - 30$$

$$\text{IgGc} - \left[0.41\left(10^{-2}\frac{\text{Albc}}{245} - 0.98 \times 10^{-2}\right)\text{IgGs}\right] - 30$$

$$\text{IgGc} - \left[0.41\left(\left(\frac{\text{Albc}}{2.45 \times 10^4}\right) - 0.010\right)\text{IgGs}\right] - 30$$

$$\text{IgGc} - 0.41\left(\frac{\text{Albc}}{2.45 \times 10^4}\right)\text{IgGs} + 0.004\,\text{IgGs} - 30$$

FIGURE 8.3B The Schuller formula is very similar to the Tourtellotte formula, but there is no ability to vary the serum albumin level, as it is assumed to be a constant.

in barrier selectivity (hence equivalent to change of slope in percentage transfer of IgG over that of albumin). This would seem to be the preferred method in principle, however in practice, probably due to rather large scatter of alpha-2-macroglobulin values, Felgenhauer stopped using this method. It also implied that there is no need to use Felgenhauer's 'correction factors' for the change in slope with elevated levels of percentage albumin transfer [275], which were so clearly defined in his earlier work [290, 989].

There has been some disagreement about the particular value of the slope constant. Reiber [902] simply asserted that it should become 1.0 (the value for serum) at levels above 0.7 per cent albumin transfer. Ewan and Lachman [261] used the value 0.43 but questioned whether it should not be 1.0 (like serum) or 0.66 (representing the normal ratio of gram amounts on either side the barrier). As we shall see in the ensuing discussion, the more fundamental question is not related to the problem of choosing a particular value for the constant but the much more important problem of whether indeed it can be construed as a constant at all, perhaps being better as a variable. In any case, the notion of 'equal molar transfer' is not supported by the following facts:

1. At normal CSF protein levels, the ratio of IgG to albumin is much lower in CSF (e.g. IgG = 10 per cent of total protein or IgG/albumin = 0.17) whereas in the corresponding serum the ratio is higher (e.g. IgG = 20 per cent of total protein or IgG/albumin = 0.34) [1167].
2. At elevated CSF proteins levels, the ratio of IgG/albumin (although higher than in the normal CSF range) is not as high as in serum [336].
3. Finally, most important is in fact a change of slope between the lowest and highest CSF protein values (see examples here).

In summary, therefore, there are two quite fundamental difficulties with the assumption of 'equal molar transfer':

1. The constant (0.43) is either too high or too low (Table 8.1); and
2. The constant is not in fact a constant but a variable (Table 8.2).

There can be different values for the 'constant', depending on differing degrees of barrier damage. This particular difficulty is not to be construed as relating primarily or exclusively to the Tourtellotte formula [1180], as it clearly applies equally to any index. In fact the later publication of the extended data of Ganrot-Norlin [336] showed quite clearly in her Figure 2 that the slope changed with elevated CSF total protein: 0.45 in the normal range, 0.75 at 4 per cent albumin transfer and 0.75 at 8 per cent albumin transfer. Although using slightly different axes, the original data of Felgenhauer [290] also showed (1976) that the slope of IgG relative to albumin changed from 0.35 to 0.70 with increasing levels of CSF total protein, but never actually reached 1.0, i.e. complete equilibration with the serum levels.

TABLE 8.1 Different slopes for IgG versus albumin

1. Slopes: assumed				
Molecular weight	0.43	(67 kD/155 kD)	Tourtellotte	[1180]
Molecular size	0.68	(Stokes radius)	Felgenhauer	[271]
Serum ratio	1.0	(no barrier)	Reiber	[902]
CSF ratio	0.62	(normal barrier 230/369)	Ewan and Lachman	[261]
Urine ratio	< 0.40[a]	(greater sieving)		
2. Slopes: used				
Tourtellotte	0.43			
Low index	0.55			
Mean index	0.69			
High index	0.88			

[a] Ewan speculated > 1.00 but see text re: problem of any given constant 'correction factor'

TABLE 8.2 Change in slope (% increase) at various levels of albumin transfer

Protein	Author (date)	Reference
1. IgG		
50 per cent: 0.40 becomes 0.63 (at 2 per cent transfer)	Felgenhauer (1976)	[290]
50 per cent: 0.50 becomes 0.68 (at 2 per cent transfer)	Felgenhauer (1983)	[275]
50 per cent: 0.45 becomes 0.75 (at 8 per cent transfer)	Ganrot-Norlin (1978)	[336]
50 per cent: 0.31 becomes 0.45 (at 1 per cent transfer)	Eickhoff (1977)	[240]
2. IgA		
100 per cent: 0.37 becomes 0.76 (at 10 per cent transfer)	Felgenhauer (1983)	[275]
350 per cent: see his graph	Felgenhauer (1976)	[290]
3. IgM		
400 per cent: 0.13 becomes 0.67 (at 10 per cent transfer)	Felgenhauer (1983)	[275]
4. A2M		
50 per cent: 0.23 becomes 0.33 (at 2 per cent transfer)	Reiber (1980)	[902]

Felgenhauer went so far as to publish a 'correction factor' in the form of a curve, to correct for the discrepancy [290]. Although the later Felgenhauer data showed a slightly smaller change in slope (0.5 to 0.67) with the higher level of CSF total protein, only three patients had albumin percentage transfer levels of greater than 3 per cent (or six times the normal level), thus approximating to a total protein level of 300 mg/dl (see Figure 1 in [275]). It is still clear from this figure that the lower values for CSF total protein are found more to the right (i.e. a lower slope of 0.5). It is also apparent from his 1976 article [290] that the main change in slope was in the first few doubling elevations of the total protein. Since the data is presented on an inverse and log scale, it is sometimes difficult to see differences, especially with elevated levels of percentage transfer. IgA has changes in slope which are even more striking than those of IgG, and again Felgenhauer included a special graph to correct the IgA values obtained [290]. In spite of these changes in the IgG and IgA slope, Felgenhauer claimed that there were no changes in the slope of alpha-2-macroglobulin [290]. This may be because the degree of 'scatter' or biological variation is larger for alpha-2-macroglobulin than for IgG (2.7 versus 2.1, see Chapter 9). Inspection of the 'normal' slope for alpha-2-macroglobulin reveals a rather large variation (see his Figure 3 [291]). Our own data also reveals a wider variability in alpha-2-macroglobulin slope, as does the data of Livrea [665]. Nevertheless, Reiber [902] found a statistically significant difference in the slope of alpha-2-macroglobulin when comparing the slope in the normal range (0.19) and at elevated levels of total CSF protein (0.33) using 15 patients in each group. Unfortunately he assumed (without any data or statistics) that the slope

TABLE 8.3 Successive changes in slope with increasing barrier damage

Barrier damage	Range of percentage transfer (Alb)	Slope	Cases
Normal	0–0.7	0.49	214
Moderate	2–5	0.72	50
Severe	5–60	0.85	17
Entire range	0–100	0.96[a]	284

[a] Extrapolation shows negative intercept, see text

of IgG would suddenly become 1.0 (from 0.4) at a level of 0.7 per cent albumin transfer.

When our data is plotted according to percentage transfer of IgG versus percentage transfer of albumin, it is clear by statistical analysis (see Table 8.3) that there is successive increase in slope as the barriers are progressively broken down.

The negative intercept for the entire range of data implies a curvilinear function. When the same normal data ($n = 214$) was plotted according to ratio (IgG/Alb) for CSF versus serum, the slope was 0.35 (versus 0.49 as shown earlier). However, when the index was calculated for each patient i.e. (IgGc/IgGs)/(Albc/Albs) and the mean taken, it was found to be 0.55. General reasons for these differences are as discussed earlier. In this particular case, while the values for the two slopes are quite different, the value for the index (0.55) is clearly closer to the slope of the percentage transfer (0.49) than to the slope of the ratio plot (0.35). In any case, the mathematical formulation in Figure 8.3 shows that the 'constant' (e.g. 0.43 for Tourtellotte [1180] or Reiber [902]) which is to be multiplied by Albc/Albs, is the straightforward equivalent of the slope of the percentage transfer plot. Since we suspected that there could be more than one barrier which would be breached in the many different CSF samples, we decided to divide them into broad diagnostic categories. We then found (Table 8.4) that there were two pathological groups which could be clearly distinguished by statistical methods as being different from normal i.e. having mild barrier damage (slope = 0.56) and these are termed: capillary transudate as in Guillain–Barré syndrome (GBS) or tumors, versus arterio-venous exudate (slope = 0.70), namely more severe damage as in meningitis and Froin's syndrome.

We also looked at the slope of alpha-2-macroglobulin, a much larger molecule, against albumin and found yet another group (Table 8.5) which was distinguishable on the basis of statistical analysis. The use of alpha-2-macroglobulin also obviates any possible difficulties due to local synthesis of IgG in the various diseases.

Thus, GBS and tumor were indistinguishable from each other, the Froin's syndrome was clearly different, as it was different from meningitides, the 'capillary' group and from the normals.

TABLE 8.4 Different slopes of IgG versus albumin using percentage transfer data in different diseases

Presumed source	Slope	Cases
Normal filtrate	0.49	214
Capillary transudate		
Guillain–Barré syndrome	0.55*	13
Tumor	0.58	14
Arteriovenous exudates		
Meningitis	0.68*	17
Froin's	0.71	6

* Different ($p < 0.01$) from other groups (and from normal) but not different within group

TABLE 8.5 Different slopes of alpha-2-macroglobulin versus albumin using percentage transfer data in different diseases

Presumed source	Slope	Cases
Normal filtrate	0.19	214
Capillary transudate		
Guillain–Barré syndrome	0.30*	10
Tumor	0.37	10
Arterial (direct) exudate		
Meningitis	0.44*	11
Venous (indirect) exudate		
Froin's	0.62*	6

* Different ($p < 0.01$) from other groups (and from normal) but not different within group

In summary, there are four sources of CSF protein: one normal and three pathological, each with a possible anatomical basis. These may represent normal filtration and the following three types of leak: (1) capillary; (2) arterial (direct); and (3) venous (indirect).

There may be some grounds for agreement with this anatomical/pathological hypothesis when one scrutinizes the data of Livrea *et al.* [665]. They state in the text that the pathological types are 'significantly different from each other but they did not [differ] from controls'. Nevertheless, their data clearly shows no overlap of 95 per cent confidence limits between the controls and GBS or between controls and meningitis. Their different slopes may also represent differences in case selection from our own. Their variations in individual diseases (as well as in our own data) are not due to differences in the level of Albc/Albs, since this clearly increases in parallel with increases in slope. In any case, it would seem that our general conclusions are in agreement: different slopes for different diseases with presumably different sources of CSF protein. One practical limitation follows: if different diseases have their different slopes, how can one judge a given patient's result as greater than a particular cut-off value (slope) unless one knows the diagnosis (i.e. which slope to use as pathological control)? This is clearly a tautological situation.

We therefore decided to pursue the matter by using a log–log transformation of the data since it seemed 'by eye' to approximate to such a function. The same mathematical function seems appropriate with the exception of Felgenhauer's 'correction factors' which have been clearly plotted for the changes in IgG slope albeit principally for the normal range [290, 989]. To our pleasant surprise, the relation clearly became linear when this approach was adopted (Figure 8.4). Although no

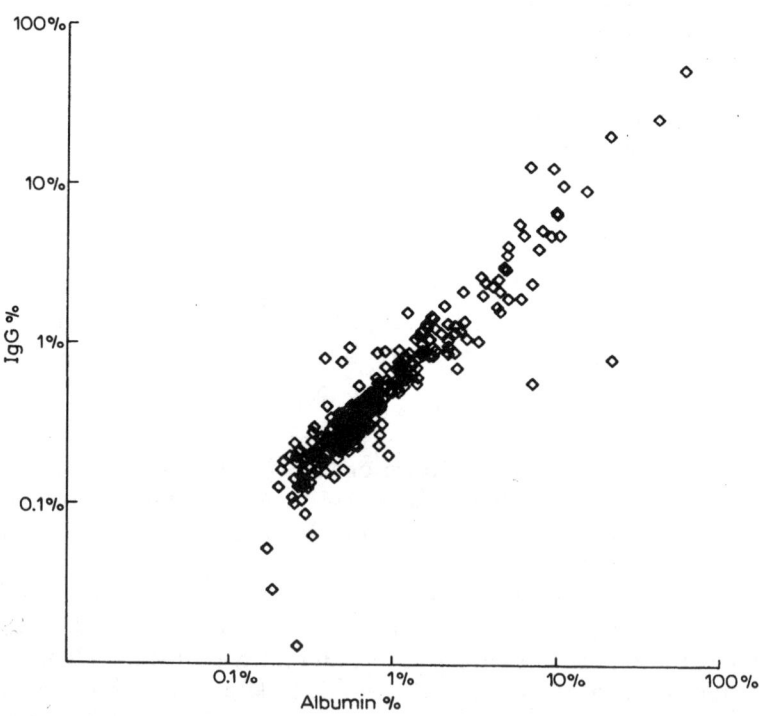

FIGURE 8.4   The logarithmic method of plotting IgG barrier function.

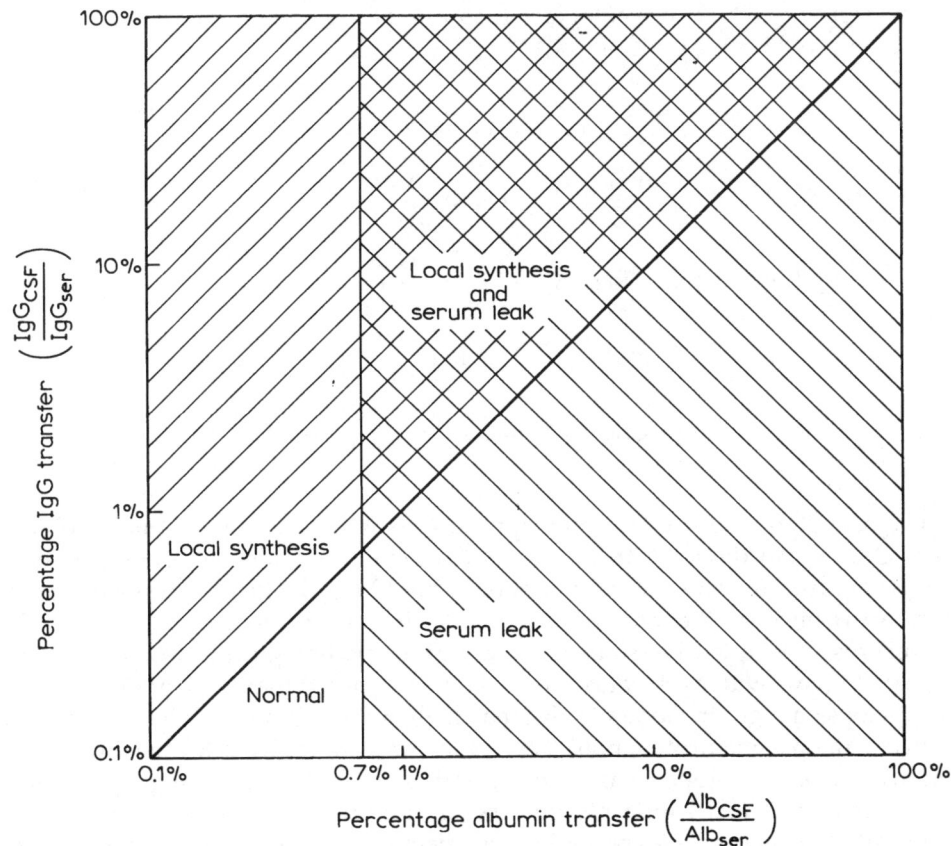

FIGURE 8.5 Use of the logarithmic plot to distinguish IgG barrier abnormalities (serum leak) from local synthesis.

amount of mathematical manipulation will change the overlap of normal samples with those which were found to be oligoclonal positive (see Figure 10.1), a much more mathematically suitable as well as clinically practical method is given in Figure 8.5.

When statistical testing was applied to the data (Kolmogorov–Smirnov, goodness of fit test), it was found that the best fit for the data was given by the double log transformation. Not only was the fit better than the original untransformed data, but it was also better than the exponential transformation of the same data [683].

The formula which is analogous to those given earlier, and which allows calculations of locally synthesized (LS) IgG is as follows:

$$\log y = m \log x + b$$

$$\log \frac{\text{IgGc}}{\text{IgGs}} + \text{LS} = m \log \frac{\text{Albc}}{\text{Albs}} + b$$

$$\log (\text{IgGc} + \text{LS}) - \log \text{IgGs} = m \log \frac{\text{Albc}}{\text{Albs}} + b$$

$$\log (\text{IgGc} + \text{LS}) = m \log \frac{\text{Albc}}{\text{Albs}} + b + \log \text{IgGs}$$

$$\log (\text{IgGc} + \text{LS}) = 0.997 \log \frac{\text{Albc}}{\text{Albs}} - 0.68 + \log \text{IgGs}$$

The next step is to use the same approach as with the alpha-2-macroglobulin data. Once again the logarithmic model fits the data better than does the linear model (Figure 8.6). This suggests that one should try the log–log transformation in comparison with the simple linear model for any protein which may be filtered across multiple diverse barriers in various pathological states. This model appears to have wider relevance, since the linear model makes no allowances for the obvious changes in slope which occur with any protein which has its primary source in the plasma protein pool. The precise anatomical basis for the differential selectivity changes (slopes) has yet to be clarified in various neurological diseases. However, that there are multiple barriers (or multiple sources of proteins) for discrete pathological processes seems beyond dispute, i.e. changes in various diseases which result in elevation of total protein levels are not simply due to a monolithic process, such as filtration on the basis of an unchanging 'mole for mole' transfer process.

There seems to be a general consensus for the idea of a change in slope for most proteins, as the blood–CSF barriers are successively broken down. In any case there is an abundance of data to show that the slope clearly changes for IgG, hence any fixed index or formula will be seriously compromised by the false assumption of a constant slope when it is obviously a curvilinear function (Figure 8.7).

A sweeping theory has been proposed by Reiber, namely that a hyperbolic formula can explain both normal and pathological results, and thus it has some kind of fundamental biological relevance [904]. The simplistic notion that flow rate dictates the level of total protein, and particularly that blocked flow gives rise to a high total protein, is manifestly untrue in the case of hydrocephalus. Indeed, in benign intracranial hypertension (BIH) there are strikingly low levels of CSF albumin. The article on BIH by Inshasi [473] presented data from a retrospective study

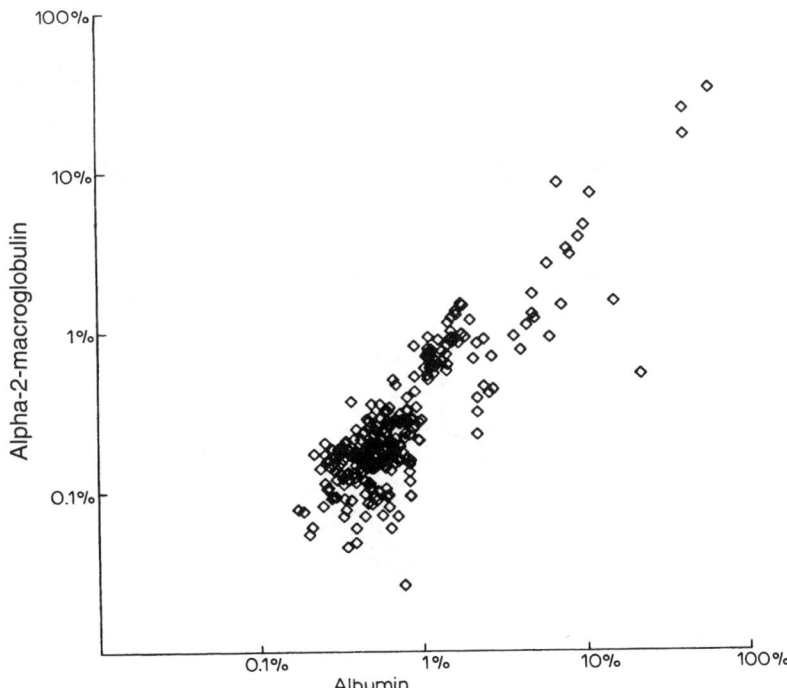

FIGURE 8.6 The logarithmic method of plotting alpha-2-macroglobulin barrier function.

on 10 patients, published previously [733], and an additional 11 new patients studied prospectively by Inshasi, giving a total of 21 patients. The main finding was that the patients had very low CSF total protein, specifically, low CSF albumin. The highest single value was 225 mg/l, followed by one at 185 mg/l and all the remaining 19 patients had < 166 mg/l. The mean of all the values of the 21 patients was 111 mg/l. In the first series, 3/4 patients who had an elevated IgG index also had a decreased CSF albumin (< 91 mg/l). There was a fourth patient who did not have an elevated IgG index who also had a low CSF albumin. Thus, 4/10 of these patients had low CSF albumin. In the second series, 1/2 patients who had an elevated IgG index also had a decreased CSF albumin (< 81 mg/l), and two more patients who did not have an elevated IgG index also had a low CSF albumin, thus 3/11 patients had low CSF albumin. Overall, therefore, 7/21 patients had a CSF albumin < 91 mg/l.

The other major problem with simplistic explanation by just different flow rates is the high level of CSF total protein in Guillain–Barré syndrome (GBS), where there

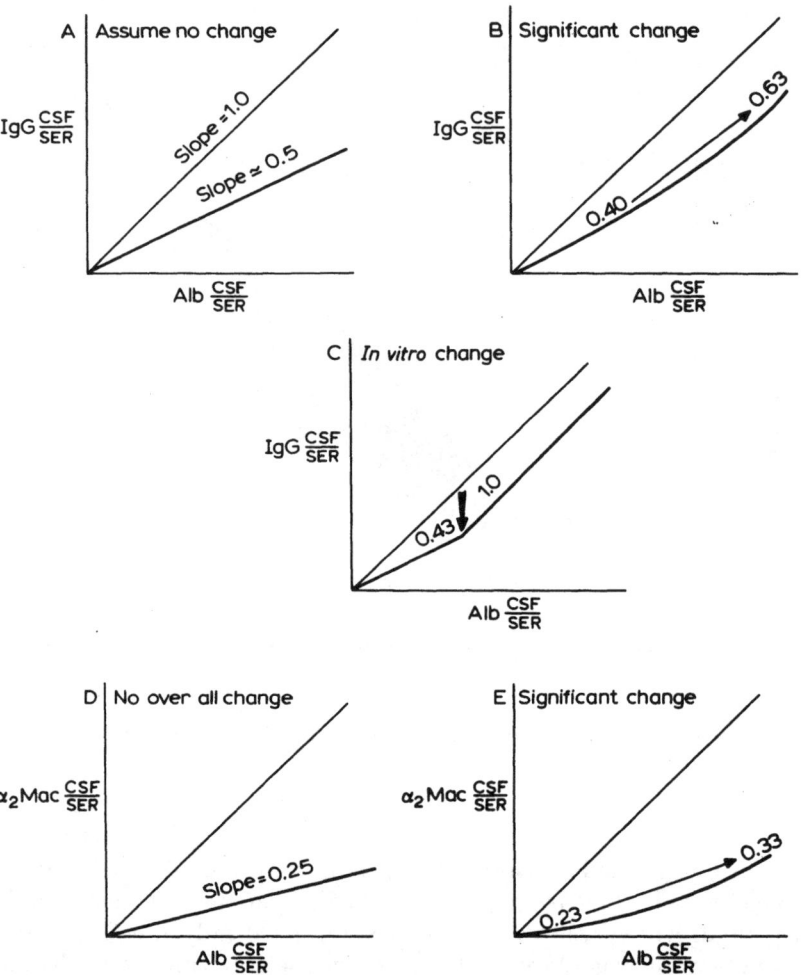

FIGURE 8.7 Options for slope changes with different barrier functions: (A) no change of IgG; (B) gradual change of IgG; (C) assumed sudden change of IgG (arrow) at 0.7 per cent albumin transfer; (D) no change of alpha-2-macroglobulin; (E) gradual change of alpha-2-macroglobulin.

are dramatic elevations in the spinal cord region but perfectly normal levels in the ventricles. The Queckensted maneuver does not show any evidence of blocked flow. Conversely, inflammation at the level of the spinal roots may cause a further increase in permeability of the dorsal root ganglia, which would also produce a dramatic increase in spinal cord CSF total protein levels. The idea that there are major sites of resorption at the level of the roots, which selectively become blocked in GBS, i.e. profuse, local arachnoid villi, is rather fanciful.

The various attempts to define a mathematical index reveal a number of caveats. Results can be thought of as either intuitive or counter-intuitive. Initial calculations of Delpech and Lichtblau proposed a ratio of IgG to albumin in CSF on one axis, with the ratio of IgG to albumin in serum on the other axis. They were concerned with the possibility of hypergammaglobulinemia and hence chose to 'normalize' this high IgG factor by dividing IgG in serum by albumin in serum. Laurell's greatest contribution was the strong intuitive approach of dividing IgG in CSF by IgG in serum on the Y axis versus albumin in CSF divided by albumin in serum on the X axis, hence as one progressed from left to right along the X axis the increase in total protein levels in CSF was obvious.

The next step was the use of the curved line as opposed to the straight line as a further attempt to normalize the multiple parameters. Felgenhauer first plotted an empirical curved line between what he called 'observed' versus 'expected' IgG in serum divided by IgG in CSF. Rather like Delpech and Lichtblau, this perhaps clouded the true meaning whereas the converse ratio of IgG in CSF divided by IgG in serum is much more attractive intuitively, since again it is analogous to an increase in CSF total protein. Reiber initially proposed a linear formulation with a dramatic 'kink' to change to the slope of 45° at an arbitrary point on the level of increasing total protein at a value of 'x' mg (7×10^{-3}) quotient of Albc/Albs along with the arbitrary choice of the slope being 1 (or a 45° angle). Fortunately the collaboration between Felgenhauer and Reiber gave rise to the hyperbolic formulation which is now widely used [906] and although it is somewhat invidious and I am perhaps biased, it is interesting that they plot a log–log relationship, which is what we originally described before 1988. One must also comment that Reiber's data is not Gaussian in distribution [904], especially for IgM. Also, Ohman's data shows that IgM 'cross-over' or contamination from a traumatic tap can have a profound influence on CSF levels. It is arbitrary, or even 'circular', to remove those points 'above' 3SD of the line when compared with IEF as the gold standard [904]. It is also interesting that the 'hyperbolic' plot (which again is practically 'log–log') does not show the data points changing their slope, so as to 'curve' more toward the Y axis at low values of X, but the data points actually head 'straight' for a negative Y intercept [1267].

As mentioned before (Figure 8.2), in an historical perspective, the formulation of the data has been approached in one of two ways: (a) percentage transfer (IgG/IgG); or (b) proportional composition (IgG/albumin).

Percentage transfer is a dimensionless ratio whereas proportional composition relies on two different proteins. The final common calculation (to produce the index, see Figure 8.2) will bring the two alternative methods together to produce the final dimensionless index, but it is instructive to consider them separately. If the data is presented graphically, then the axis labels will be different in the two cases, namely in case (a) percentage transfer, it will be a dimensionless number, and for case (b) proportional composition, it will typically be grams of IgG per gram of albumin (the two proteins still have their individual and quite different molecular weights, net charge and Stokes radius). Further examination of case (a) reveals that while the axes are presented as percentage transfer (IgG/IgG versus albumin/albumin), since the barrier is normalized for each of the two proteins on an individual protein basis (IgG or albumin on either axis), the slope of the line will reflect any differences in relative transfer of the two proteins e.g. more IgG transferred per gram of albumin transferred, at different points along the X axis i.e. with increased total protein. Conversely in case (b) where the axis is presented in proportional composition (IgG/albumin) there is no correction for the barrier status *per se* (on either axis). There is a further difficulty in the presentation of ratios of IgG/albumin by this method, namely that the elevated ratio can be due either to an elevated level of IgG (serum hypergammaglobulinemia as originally proposed) or to a depressed level of albumin (e.g. the many causes of impaired liver function etc.) or to a mixture of both processes. It is difficult to estimate the time lag between, say, an acute systemic insult followed by lowered serum albumin and the lowering of CSF total protein (since there is less albumin to be filtered across the blood–CSF barriers). In both cases (a and b) of percentage transfer and proportional composition, the final quotient (or IgG index) will normalize both barrier function and the serum levels. Working backwards from the final values, if one knows only the index value (without the primary data upon which it is based) it is not possible to make any deductions about either the state of the barriers or the serum IgG (or albumin) levels. As noted earlier, knowledge of the IgG/albumin ratio (again without access to the primary data) allows no deduction as to whether the IgG is high, the albumin low or if there is a mixture of the two abnormalities. The percentage transfer has strong intuitive attraction, but given the problems of kinetics of blood–CSF filtration, e.g. a rather debilitated patient with low serum albumin, how quickly does the CSF total protein level drop? Likewise, if the patient is immunosuppressed, does a drug (e.g. ACTH) act on the systemic lymphoid tissue before the CNS?

Conclusion

The presentation of a single, simple fact of IgG index says nothing directly about barrier function and/or serum levels (by definition). The ratios of IgG/albumin do not distinguish between the two individual proteins. For the best presentation of the data therefore, all primary values (IgG and albumin in CSF and serum) should be given. If a graph is to be shown, the percentage transfer method is preferred to proportional composition since the result of Albc/Albs is directly related to

CSF total protein. However, the clearest method of graphical presentation for the clinical laboratory is logarithmic namely, log–log (Figure 8.5). On balance, it is still desirable to do qualitative (versus quantitative) analysis, but best to do both qualitative as well as quantitative analysis, and to express the results in a fashion which allows for the pathological fact of change in slope (change in percentage transfer) under diverse pathological conditions.

CHAPTER

9

Normal blood–CSF barrier values

Introduction
Expressions of values for normal data
Degrees of biologic variability
Critique of CSF immunoglobulin levels

Introduction

A shortened list of normal values is given in the Appendix (Chapter 15). There is still considerable variability, if not disagreement, among some investigators as to what constitutes normal values.

The discussions of normal values have been left until this stage since they still pose several difficulties, namely different patient populations, age and different commercial sources of antiserum. Qualitative examinations of proteins have been quantified and these values are also listed. This chapter has three sections:

1. Expressing the data (mainly IgG) in relation to other proteins (e.g. albumin or the total protein level);
2. The comparatively wide range of normal variability for CSF protein levels (plus the increase with older age) and
3. Points of possible discrepancy in dealing with normal versus abnormal values for IgG.

Expressions of values for normal data

Barriers can be assessed in absolute or relative terms. Since IgG can be synthesized within the CNS (unlike albumin), barrier corrections for IgG usually involve albumin as the denominator (see Table 9.1).

TABLE 9.1 Measures of barrier function and local IgG synthesis

Barriers	
1. Total protein	(Absolute)
2. Albc/Albs	(Quotient × 100 = percentage transfer)
3. IgGc/IgGs	(Quotient × 100 = percentage transfer)
4. IgAc/IgAs	(Quotient × 100 = percentage transfer)
5. A2Mc/A2Ms	(Quotient × 100 = percentage transfer)
Intra-CNS synthesis of IgG	
1. IgG	(Absolute)
2. IgGc/total protein	(× 100 per cent of total)
3. IgGc/Albc	(Ratio)
4. IgGc/Albc/IgGs/Albs	(Index)
5. Tourtellotte = Reiber	(mg/day)
6. Schuller	(no serum albumin)

Since the bulk of the CSF proteins are derived from serum, it is relevant to begin by discussing their normal ranges in serum. Although individual laboratories must always establish their own absolute values, it is of interest to note some of the normal biological variations seen in a large series (22 000 samples) drawn from the normal screening applied by Ritchie's laboratory [930].

Degrees of biologic variability

Transferrin and albumin have the most stable values and haptoglobin, IgA and IgM, the most variable. IgG is at the lowest of the intermediate range of variability and orosomucoid is at the highest.

The data can be analyzed by taking the value for the mean plus two standard deviations and dividing it by the value of the mean minus two standard deviations (Table 9.2).

This bears some relation to the coefficient of variation (usually used to decide how much variation a particular analytical test carries when the same sample is re-tested many times) and such a quotient can be considered as follows: if a test has a coefficient of variation (standard deviation divided by the mean) of less than 10 per cent it probably has a methodologic variability less than the biologic variability of the population to be studied. It can be seen that albumin or transferrin are the most stable proteins and hence are good 'denominators' (reference proteins or internal controls) against which one can see changes in other proteins such as IgG. As a denominator, IgG has less variability than IgA or IgM. Alpha-2-macroglobulin is rather more variable than albumin or transferrin and others have

TABLE 9.2 Biological scatter of serum proteins given as High/Low (+2SD/−2SD) and CV (SD/Mean)

	High/Low	CV %
Transferrin	1.70	13
Albumin	1.72	14
IgG	2.1	18
Alpha-1-antitrypsin	2.4	21
C'3	2.6	22
Alpha-2-macroglobulin	2.7	22
Apo-B-lipoprotein	3.8	28
C'4	4.0	32
Orosomucoid	4.7	35
IgM	7.6	38
IgA	8.7	41
Haptoglobin	10.0	45

also found this to be the case [665], since they have found a much wider scatter when alpha-2-macroglobulin is used as the denominator or reference protein.

Historically, the most extensively used denominator for data on barrier functions has been the level of CSF total protein. It is widely known that this increases with old age [338, 1163]. The range in a larger series of strict normals (medical students) who were of relatively young age (22 to 26 years) was based upon 149 subjects [1170] and gave values of 36.4 mg/dl + 9.2 or 18 to 55 mg/dl resulting in a population variance (PV different from CV) of 25 per cent. Although the PV is not strictly comparable, it is used because of its intuitive relevance. Although Link [1163] had a smaller number of patients (said to be 'not abnormal clinically') in this age range ($n = 21$), he found a PV of 17 per cent which increased progressively with age to 25 per cent. If we wish to compare any other parameter with total protein, we must first look at its biologic or population variance (PV). Several such examples are given in Table 9.3.

The next most commonly used indicator of barrier function is the level of CSF albumin. This has a PV of about 27 per cent which rises with age to 32 per cent. If one attempts to improve the situation by dividing the CSF albumin by the serum albumin (to correct for any variation in liver function), the situation does not change i.e. the PV is again about 27 per cent rising to 31 per cent with age. Either of the CSF albumin values (absolute or percentage transfer) shows a lower correlation with age than does the total protein: $r = 0.54$ for the total protein versus $r = 0.42$ for CSF albumin and essentially the same value $r = 0.46$ for percentage transfer of albumin [1163]. A similar value, $r = 0.41$, for percentage transfer of albumin correlated with age was found by Laurell [335]. The distribution of IgG values

TABLE 9.3 Rank order of population variance (PV = SD/Mean) for IgG index versus other blood–CSF barrier parameters

Serum	PV %
Na	2
Ca	5
K	9
Albs	14
IgGs	18

CSF and/or serum	PV %
Ganrot [335] N = 54	
IgGc	68
IgGc/Albc	23
Index	12
Albc	41
Reiber [902] N = 334	
IgGc	34
IgGc/IgGs	38
IgGc/Albc	28
Albc	28
Index	25
Albc	20
Albc/Albs	30
Link [1163] N = 93	
IgGc	33
IgGc/IgGs	30
IgG/total protein	25
IgGc/Albc	19
Index	13
Albc	30
Albc/Albs	30
Total protein	20

is no tighter, as one might expect from the Ritchie data for serum levels [930]. It would therefore seem preferable to reduce any age dependence to nil if possible, in order to normalize all values for each patient. This happens with the IgG index i.e. (IgGc/IgGs) divided by (Albc/Albs) in that the PV dropped to 13 per cent as well as remaining unchanged between the ages of 17 and 77 [1163].

Turning from the question of barrier function to that of IgG levels in the CSF, it was noted earlier that the absolute level following normalization was interpreted more accurately when divided by the total protein [514]. However, the same general questions about population variance should again be examined: the PV of absolute CSF IgG levels is 24 per cent rising to 35 per cent with increasing age [1163]. Division by serum IgG levels makes no appreciable change as the PV again rises from 26 per cent to 35 per cent. There are two possible options regarding the CSF IgG level: either divide by total protein as Kabat suggested [514] with the PV rising from 17 per cent to 25 per cent ,or divide by CSF albumin where the PV is also rather variable (from 24 per cent to 16 per cent). One is hard pressed to decide which to adopt but, as noted before, the better solution is probably the IgG index with the smaller PV of 13 per cent. The consensus of most authors' data (full details given in Table 9.4), is that given the wide variance of IgG it is better (i.e. there is less variation) if one divides CSF albumin level than by total protein, and best to use the index.

Having compared the relative degree of scatter for the different ratios, quotients and indices, it is worth noting that the patient populations will always differ and therefore some results will have a wider scatter (e.g. Ganrot and Laurell's index had a variance of 12 per cent PV versus Reiber [902] index with 25 per cent PV) but within the rank order for each laboratory, the index will typically have the tightest PV.

Turning next to the particular normal values for the three laboratories listed in Table 9.3, with additional values collected from other laboratories, we see that the result for normal index can vary from 0.38 to 0.58 with an approximate average of 0.48, while the cut-off level for an abnormal index (mean + 2 standard deviations) may also vary from 0.55 to 0.88, again with an approximate mean of 0.70 (see details in Table 9.5).

The so-called range of variability (RV) works out at 10 per cent which is a relatively low value for scatter. The next important question about the slope of the IgG/albumin line is: does it change (a) in the normal range and (b) in the abnormal range i.e. with damaged barriers.

Looking first at the normal range, data from Felgenhauer [290] shows that it changes by a factor of two overall, with an increase in slope associated with increasing barrier permeability (less selectivity) and most of this change takes place in the normal range. Although most other authors do not show any change within the normal range, inverse and logarithmic plotting of the data by Felgenhauer [290] effectively expands the normal range relatively more than the abnormal range.

The only other suggestion of possible change in the normal range is the data of Ganrot-Norlin [336], which gave a negative Y intercept, suggesting that if she had examined samples from patients with lower protein values, the slope would have been less than 0.45.

In the abnormal range, again Felgenhauer's data [290] shows the continuing change noted earlier for the normal range, but with an overall 50 per cent increase in slope. The data of Ganrot-Norlin [336] also clearly shows a 50 per cent increase

TABLE 9.4 Normal values from individual papers on blood–CSF barrier parameters. Listed are mean, standard deviation (SD), upper limit of normal (M + 2SD), coefficient of variation (CV = SD/Mean), number of cases (N), method of analysis, and source of antibody (Aby)

Authors (date)	Reference	Mean	SD	M + 2SD	CV %	N	Meth	Aby
1. Delpech and Lichtblau (1972)	[219]							
Index IgG		0.85	± 0.16	0.85	31	43	RID	Pasteur
2. Ganrot and Laurell (1974)	[335]							
Index IgG		0.45	± 0.05	0.55	12	54	EID	'own'
3. Olsson and Pettersson (1976)	[836]							
Index IgG		0.46	± 0.10	0.66	22	44	EID	'own'
4. Felgenhauer et al. (1976)	[291]							
Quotient Alb		0.0038	± 0.0011	0.0060	30	20	EID	Behring
Quotient Cp		0.0032	± 0.0009	0.0050	29	20	EID	Behring
Quotient A2M		0.0011	± 0.0004	0.0019	37	20	EID	Behring
5. Tibbling et al. (1977)	[1163]							
Quotient Alb		0.0046	± 0.0014	0.0074	30	93	RID	Behring
Quotient IgG		0.0022	± 0.0007	0.0036	30	93	RID	Behring
Index IgG		0.46	± 0.06	0.58	13	93	RID	Behring
6. Bock (1978)	[73]							
Quotient Alb		0.0048	± 0.00042	0.0056	9	22	RID	Dako
Quotient IgG		0.0026	± 0.00020	0.0030	8	22	RID	Dako
Quotient A2M		0.0002	± 0.00009	0.0004	45	22	RID	Dako
Quotient IgA		0.0009	± 0.0001	0.0011	13	22	RID	Dako
Quotient Trf		0.0066	± 0.0005	0.0076	8	22	RID	Dako
Quotient Pre		0.0720	± 0.0046	0.0812	6	22	RID	Dako
Index IgG		0.58	± 0.04	0.66	7	22	RID	Dako
Index A2M		0.12	± 0.02	0.16	17	22	RID	Dako
Index IgA		0.19	± 0.03	0.25	33	22	RID	Dako
Index Trf		1.46	± 0.10	1.66	7	22	RID	Dako
7. Ganrot-Norlin (1978)	[336]							
Index IgG		0.45	± 0.05	0.55	12	90	EID	'own'
8. Ahonen et al. (1978)	[5]							
Quotient Alb		0.0050	± 0.0015	0.0080	35	30	RID	Orion
Quotient IgG		0.0017	± 0.0006	0.0029	30	30	RID	Orion
9. Al-Kassab et al. (1979)	[8]							
Quotient Alb		0.0047	± 0.0010	0.0067	21	36	Neph	'own'
Quotient IgG		0.56	± 0.10	0.76	18	36	Neph	

(continued)

TABLE 9.4 (Continued)

Authors (date)	Reference	Mean	SD	M+2SD	CV %	N	Meth	Aby
10. Poloni et al. (1979)	[876]							
Index IgG		0.45	±0.13	0.70	29	?	RID	Behring
11. Reiber (1979)	[901]							
Index IgG		0.41	±0.10	0.62	25	334	RID	Behring
12. Reiber (1980)	[902]							
Quotient Alb		0.0046	±0.0014	0.0074	30	334	RID	Behring
Quotient IgG		0.0020	±0.00075	0.0035	38	334		
Index A2M		0.23	±0.05	0.33	23	15		
13. Delmotte and Tourtellotte (1980)	[217]							
Quotient Alb		0.0034	±0.0014	0.0062	40	46	Neph	Behring
Quotient IgG		0.0016	±0.0008	0.0032	47	46	Neph	Behring
Quotient A2M		0.0007	±0.0003	0.0013	40	46	Neph	Behring
Quotient IgA		0.0019	±0.0009	0.0038	50	46	Neph	Behring
Quotient Oro		0.0030	±0.0017	0.0064	39	46	Neph	Behring
Quotient Trf		0.0047	±0.0018	0.0083	39	46	Neph	Behring
Quotient Pre		0.0455	±0.0100	0.0655	22	46	Neph	Behring
Index IgG		0.51	±0.12	0.75	24	46	Neph	Behring
14. Livrea et al. (1981)	[664]							
Index IgG		0.38	±0.15	0.68	39	24	RID	Behring
15. Kamp et al. (1981)	[528]							
Quotient Alb		0.0049	±0.0016	0.0081	34	40	RID	Behring
Index IgG		0.51	±0.09	0.69	18	40	RID	Behring
16. Sun et al. (1981)	[1106]							
Index IgG		0.58	±0.15	0.88	26	30	RID	Behring
17. Killingworth (1982)	[563]							
Index IgG		0.53	±0.14	0.81	26			
18. Felgenhauer (1982)	[275]							
Index IgG		0.69[a]	} @ Albumin quotient 0.005					
Index IgA		0.37						
Index IgM		0.13						
Index IgG		0.67	} @ Albumin quotient 0.100					
Index IgA		0.76						
Index IgM		0.67						

[a] Graph shows 0.50 (thus IgG slope changes)

TABLE 9.5 IgG Index values (authors taken from Table 9.4)

Authors	Mean	Mean + 2SD
1.	0.52	0.85
2.	0.45	0.55
3.	0.46	0.66
5.	0.46	0.58
6.	0.58	0.66
7.	0.45	0.55
9.	0.56	0.76
10.	0.45	0.70
11.	0.41	0.62
13.	0.51	0.75
14.	0.38	0.68
15.	0.51	0.69
16.	0.58	0.88
17.	0.53	0.81
Range	0.38–0.58	0.55–0.88
Median	0.48	0.70

TABLE 9.6 Normal values for the formula of Tourtellotte (authors taken from Table 9.4)

Authors	Mean	Mean + 2SD
6.	0.12	0.16
12.	0.23	0.33
Behring antibody	0.23	0.33

in slope. Finally, although with smaller numbers and over a narrower range of barrier damage, Eickhoff's data also shows a 50 per cent increase in slope [240]. A further discussion of similar changes in slope (increase) with other proteins has been presented in Chapter 8. In spite of this caveat, values are presented for the normal range of alpha-2-macroglobulin in Table 9.6.

Thus far we have seen that the most biologically stable value for the normal barrier function is the ratio of ratios for IgG to albumin in CSF versus serum, the so-called IgG index which is independent of changes with age. What about the other extreme of age: the developing child? It is known that a baby can have a very high level of CSF total protein and still be considered normal [1277]. It appears that the expected value for total protein in a young adult is reached at about eight months, but the IgG and alpha-2-macroglobulin levels appeared to become stabilized as albumin mainly peaks at two months [1083].

The normal values for CSF IgG expressed by the Tourtellotte formula are given in Table 9.7.

Although this has been shown to be essentially equivalent to the IgG index (see Chapter 8), the values are still used in many laboratories. One can easily derive the main values for barrier function i.e. percentage transfer of albumin and percentage transfer of IgG from the Tourtellotte formulation and these are shown in Table 9.8.

Although we have noted earlier that the percentage transfer of albumin has more biological variability than the IgG index, nevertheless the normal range can vary from 0.34 per cent to 0.5 per cent with an average of about 0.4 per cent, and abnormal values (namely the mean + 2 standard deviations) can vary from 0.56 per cent to 0.8 per cent, again with an average of about 0.7 per cent (Table 9.9).

The analogous values for percentage transfer of IgG are as follows: normal range of 0.16 per cent to 0.26 per cent with an average of 0.2 per cent and the mean + 2 standard deviations of 0.29 per cent to 0.36 per cent with an average of 0.3 per cent. Again, relative to albumin, the average quotients (IgG divided by albumin) also yield a value of 0.5 which is quite close to that which was calculated

TABLE 9.7 Normal values for the formula of Tourtellotte

	Mean	Mean + 2SD	Cases
Tourtellotte (1970) [1167]		0.00	67
Tourtellotte (1975) [1166]	−3.3	3.3	70
Ewan and Lachman (1979) [261]	−7.7	15.3	39 (Neurol)
	−6.0	11.0	8 (Non-neurol)
Mandler et al. (1979) [701]	−0.5	5.5	27
Valenzuela et al. (1982) [1205]		3.5 (added in their proof)	
Delmotte and Tourtellotte (1980) [217]		6.6	46
	Median	6.0 mg/day	

TABLE 9.8 Values for percentage transfer from Tourtellotte formula

	Mean for Albumin %	Mean for IgG %
Tourtellotte (1975) [1166]	0.43	0.27
Mandler et al. (1979) [701]	0.59	0.29
Delmotte and Tourtellotte (1980) [217]	0.34	0.16

TABLE 9.9 Values for percentage transfer from serum into CSF (authors from Table 9.4)

Authors	Mean %	Mean + 2SD %
Percentage transfer of albumin		
4.	0.38	0.60
5.	0.46	0.74
6.	0.48	0.56
8.	0.50	0.80
9.	0.47	0.67
12.	0.46	0.74
13.	0.34	0.62
15.	0.49	0.81
Range	0.34–0.50	0.56–0.81
Median	0.42	0.68
Percentage transfer of IgG		
5.	0.22	0.36
6.	0.26	0.30
8.	0.17	0.29
12.	0.20	0.35
13.	0.16	0.32
Range	0.16–0.26	0.29–0.36
Median	0.21	0.33
Percentage transfer of IgA		
6.	0.09	0.11
13.	0.19	0.38
Behring antibody	0.19	0.38
Percentage transfer of A2M		
4.	0.11	0.19
6.	0.02	0.04
13.	0.07	0.13
Median	0.07	0.13

for the IgG index on the basis of individual patients. This is also analogous to the slope of 0.5 between the percentage transfer ratios of albumin and IgG.

Critique of CSF immunoglobulin levels

A number of specific articles deserve mention with regard to certain critical points to be highlighted:

1. Bouloukos *et al.* (1980) claimed to find abnormalities in patients with stroke and other diseases not typically thought of as being associated with an immune response [88]. The values for normal patients included IgG as a percentage of total protein up to a level of 21 per cent and normal IgA values were claimed to be up to 25 per cent of total protein. Clearly the total protein values must have been too low (method not given) and thus the falsely low denominator would give rise to falsely high values throughout.
2. Grubb (1974) showed oligoclonal bands in CSF with immunofixation [403]. This demonstration, 30 years ago, has not been given sufficient credit.
3. Riddoch and Thompson (1970) found 2/205 patients whose IgG comprised 50 per cent of the total protein [924]. This seems excessive and is discussed under the discrepancy problem (Chapter 7). Schuller and Tompe (1973) showed good correlation between electrophoretic densitometry and Laurell rockets for IgG [1004]. This also deserves wider recognition.
4. Sindic *et al.* (1984) showed that IgE was most elevated in TBM (using a cut-off value of 0.6 for IgE index). This may aid differential diagnosis [1063].
5. Prasad (1983) claimed elevated levels of IgD in myasthenia gravis CSF [884]. This may be relevant to pathogenesis.
6. Sindic *et al.* (1984) showed that dimeric IgA (index 0.12) was more typical of local synthesis than was monomeric IgA (index 0.44) [1061].
7. Measurement of IgM levels is highly variable and a list of some of the main contributions is given in Table 9.10.
8. Measurement of unbound or 'free' immunoglobulin light chains:

It is difficult to summarize the results for free light chains (versus bound or intact immunoglobulins) not least because the methodologies are so diverse as well as authors freely admitting various inconstant findings (see cross index of references). However, the main problem occurs because the majority of the assays do not distinguish between free and bound light chains by using physical–chemical techniques. The most discriminating technique for free light chain determination involves two sequential steps, namely: (1) electrophoretic separation of bound (intact) IgG or IgA from free light chains; and (2) specific immunofixation of kappa and lambda as well as gamma and alpha i.e. heavy chains. After examining several hundred CSF samples in this way, we found an association between elevated numbers of free light chains and recent exacerbations in MS [1203].

TABLE 9.10 Normal levels of IgM

First author (Date)	Reference	Total protein %	'Upper' limit M+2SD %	Index	'Upper' limit M+2SD %	CSF/SER %	'Upper' limit M+2SD %
Mingioli (1978)	[766]	0.01	0.08				
Nerenberg (1975)	[801]	0.03	0.046				
Williams (1978)	[1295]	0.02	0.10				
Kobatake (1980)	[587]	0.06	0.11				
Brouwer (1983)	[103]			0.04	0.08		
Forsberg (1984)	[307]				0.061		0.03
Sindic (1982)	[1054]			0.02	0.079	0.0009	0.03

The Bence-Jones proteins are by-products of myeloma, hence all the caveats which were discussed previously apply equally to these free light chains. This contrasts with polyclonal IgG (as in chronic TB), where there is excessive production of free light chains, but these are certainly not monoclonal.

The important point about standards, specifically in the choice of free light chains, reflects back to the problem of only selecting a single Bence-Jones protein (as opposed to a wider population). Commercial standards are sometimes prepared from a single patient since they may have received a large volume of urine from just one case. The best standard would not be just a mixture of different patients with myeloma, but of different patients with chronic TB, where there would in effect be polyclonal free light chains [1116].

PART 3

Individual diseases, and the future

CHAPTER 10

CNS immunoglobulin synthesis

Introduction
Efficient diagnosis
Differential diagnosis
Specific antigens
Technology
   General methods for protein separation/visualization
   Separation by isoelectric focusing
   Antigen-specific immunoblotting
   Dot-blots
   Relative specific antibody on Eastern blots
   CSF versus MRI
Serum source of CSF antibodies
   Focusing of serum and parallel CSF
   Differential diagnosis of serum bands
   Serum oligoclonal bands in MS
   Paraneoplastic serum and CSF antibodies
Conclusion

Introduction

The presence of oligoclonal bands in CSF is not pathognomonic for MS. However, if it can exclude other antigens with the appropriate ancillary tests, the diagnosis of MS is very strongly indicated, since about 98 per cent of clinically definite patients do show this abnormality. Quantitative measures such as the IgG index are less reliable since they can yield both false negative and false positive results.

It is clear that antibodies of various classes can be produced by lymphocytes resident within the thecal spaces (local synthesis) and five questions should be answered:

1. What is the most efficient method for the lab to diagnose this pathological entity?
2. What is the 'laboratory's' differential for the causes of local synthesis?
3. What major antigens can be bound to these antibodies?
4. What is the best technology?
5. What is the role of serum?

Efficient diagnosis

As noted previously there are two general methods for making the diagnosis of local immunoglobulin synthesis (i.e. within the CNS):

1. Quantitative analysis, e.g. of IgG and albumin.
2. Qualitative analysis, e.g. polyacrylamide gel electrophoresis (PAGE) and isoelectric focusing followed by immunofixation (for oligoclones).

The best discrimination is obtained by quantifying the qualitative separation and thereby having the best of both methodologies.

Examination of Table 10.1 illustrates the efficiency in terms of improved results in studies in which both qualitative and quantitative analysis have been performed on the same patient population (see Chapter 7 for details on how to compare tests).

Our own data shows the discrepancy between qualitative and quantitative levels in Figure 10.1. Both false positive and false negative results can easily be discerned regardless of the value of the chosen cut-off slope. The same problem was also documented previously using the technique of kappa and lambda immunoblotting [1151].

The laboratory diagnosis of local synthesis should prompt the clinician to exclude other known antigens before finally settling for the clinical diagnosis of MS. Table 10.2 shows a list of known antigens as well as other laboratory tests for individual diseases and some idea of the percentage of positive findings in each disease (since clinically definite MS is more than 98 per cent positive).

In the interpretation of known antigenic responses (e.g. measles for SSPE) within the CNS, the important caveat is always, does the antibody derive primarily from the systemic circulation and only secondarily enters the CSF? This leads to three other specific questions:

1. What is the serum level (titer) of antibody?
2. What is its clonal disposition (polyclonal versus oligoclonal)?
3. What is the state of the blood–CSF barrier (abnormal leakage)?

TABLE 10.1 Qualitative versus quantitative tests ranked in order of efficiency (higher in MS than OND) using sensitivity values

Author (date)	Reference	Test	Cut-off	Abnl	#
Olsson & Pettersson (1976)	[836]	Agar	(Napht)	85%	33
		Index	(> 0.66)	88%	(18% OND)
Link and Tibbling (1977)	[654]	Agar	(Napht)	88%	59
		Index	(> 0.58)	86%	
Christensen et al. (1978)	[142]	Agar	(Napht)	83%	27
		Index	(> 0.85)	77%	
Thompson et al. (1979)	[1150]	PAGE	(Coom)	94%	32
		IgG	(> 12%)	75%	
Poloni et al. (1979)	[876]	IEF	(Coom)	90%	30
		Index	(> 0.70)	60%	
Johnson et al. (1980)	[498]	Agarose	(Coom)	81%	78
		IgG	(> 12%)	61%	
Hershey and Trotter (1980)	[445]	Index	(> 0.66)	91%	46
		Agarose	(Coom)	90%	41
		IEF	(Coom)	83%	41
		Formula	(> 4.5)	78%	46
		IgG/Alb	(> 0.27)	61%	46
Gerson et al. (1981)	[346]	Agarose	(Napht)	78%	54
		IgG	(> 18%)	35%	
Bloomer and Bray (1981)	[69]	Agarose	(Napht)	94%	118
		Formula	(> 6)	75%	
		IgG/Alb	(> 0.25)	67%	
Sun et al. (1981)	[1106]	Agarose	(Coom)	75%	16
		Index	(> 0.88)	60%	
		'M-C'	(> 0.001)	80%	(25% OND)
Livrea et al. (1981)	[664]	IEF (fixation)		99%	64
		Formula	(1.0)	70%	
		Index	(> 0.68)	63%	

Abbreviations: OND = Other neurological disease; Napht = Naphthalene black stain; Coom = Coomassie stain; Formula = Tourtellotte's; IEF = Isoelectric focusing

We have discussed the first and last questions previously (see Chapter 8) i.e. the so-called 'index' or (better) the relative specific antibody (i.e. the percentage of total IgG which was specific for the antigen in question). The more important question, i.e. the qualitative nature of the IgG should again be answered in both qualitative ('by eye') and quantitative fashion (by scanning densitometry) [845, 1156].

In SSPE the clonal patterns are different in the serum and CSF, yet more than 90 per cent of the CSF antibody is measles-specific [1228]. In herpes encephalitis

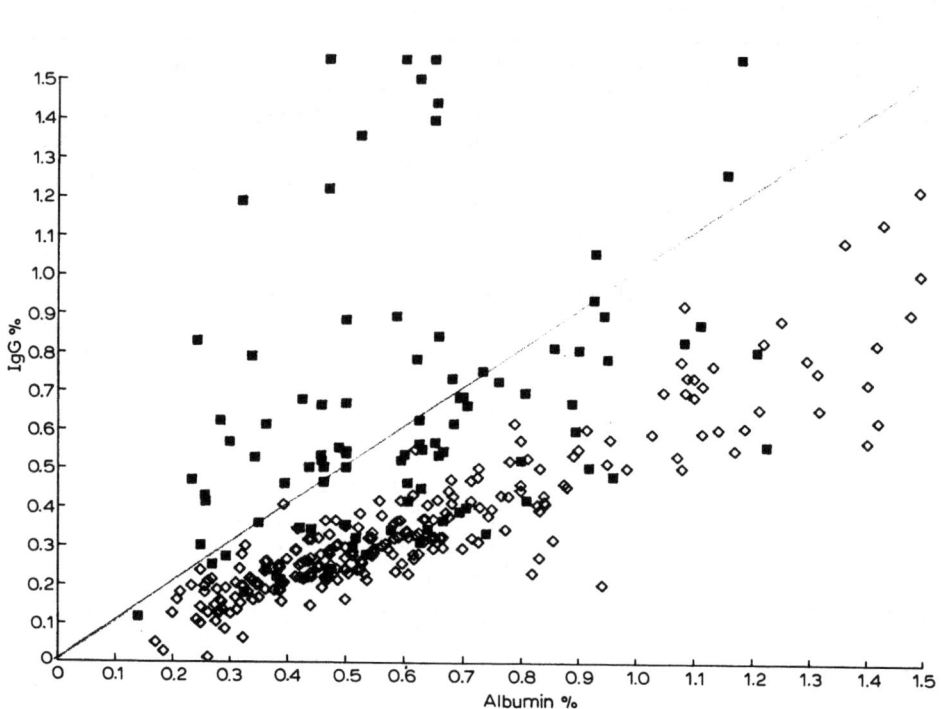

FIGURE 10.1 Qualitative versus quantitative analysis in the detection of local synthesis. Filled squares were oligoclonal positive. Open squares were oligoclonal negative by isoelectric focusing.

the serum is polyclonal and the CSF is oligoclonal. The relative specific antibody is greater than 1.0 [286].

Differential diagnosis

Given an oligoclonal pattern, what other antigens should be excluded before being confident of a diagnosis of MS (see Table 10.2). It should be noted that infections do not typically show oligoclonal patterns in the first 7 to 10 days. Also, in our experience, strokes are not associated with oligoclonal bands, unless there is secondary infection or arteritis. The actual number of cellular clones which appear to respond in a 'non-specific' manner are typically < 1 per cent or even < 0.01 per cent of the total lymph node drainage cells [880].

TABLE 10.2 Differential diagnosis for oligoclonal CSF

Sensitivity %	Diagnosis	Test
98	MS	Visual evoked potentials
100	SSPE	Measles antibody titer
100	Neurosyphilis	Treponemal antibodies
80	AIDS	Anti-HIV antibody
80	Lyme disease	Anit-Borrelia antibody
50	Lupus	C'3 + anti nuclear Factor
20	Behcets	C'3 + CSF polymorphs
60	Ataxia telangectasia	Low serum IgA
100	Adrenoleukodystrophy	X-linked, long chain fats
60	Harada's disease	Meningitis-uveitis
< 5 if < 1 wk	Encephalitis	Viruses: herpes, mumps
< 5 if < 1 wk	Meningitis	Bacteria: high lactate, CRP
< 5	Tumor	CT scan
< 5	Sarcoid	Kveim

Although some have argued that the occurrence of oligoclonal bands bears no relation to the pathogenesis of MS [718], [852] nor perhaps to the disease itself [661], this is not our experience. We find that the amounts of the major oligoclonal bands on PAGE (gamma 5 peak height relative to the internal standard of transferrin) show a definite correlation with the severity of the disease (Kurtzke score) [1262].

Recent work on CSF antibodies has shown that IgG is not due to random nonsense production, but they derived from what would be expected from a normal maturation in response to a typical antigenic stimulus [893].

A correlation was seen between the level of the IgG index and the number of bands in the data derived from Specialty Labs (Guy Grimsley, personal communication, 2002) and this has also been shown by Livrea [663].

We also found a significant correlation between elevated numbers of free light chains and recent exacerbation [1203], which was a retrospective study of the patients' notes examined in a 'blind' fashion, before any attempt at correlations between the two assessors of the clinical/immunochemical data.

In addition we have found an association between the transudation of haptoglobins with the progressive form of MS [1262]. These findings are all summarized in Table 10.3.

Felgenhauer compared the relative amounts of IgG, IgA and IgM in various neurological diseases and claimed that this had predictive value in differential diagnosis [275]. There were four broad categories:

1. Those with an increase mainly in IgG without concomitant increase in IgA or IgM (i.e. MS or encephalitis).

TABLE 10.3 Clinical–pathological correlations of MS CSF

Clinical	Immunochemical	p value
Disability (Kurtzke)	Gamma 5 (area ratio)	< 0.001
Recent exacerbation	High number free light chains bands	< 0.004
Long duration	Low number free light chains bands	< 0.010
Progressive form	Excess haptoglobin 2-1	< 0.001

2. Those whose greatest increase was in IgA but who also had an increase in IgG and IgM (i.e. meningitis of bacterial or tuberculous origin).
3. Those in whom the greatest increase was in IgM but also had raised IgG and IgA (i.e. syphilis or Lyme disease).
4. Those whose greatest increase was in IgG but who also had raised IgA and IgM (i.e. herpes encephalitis).

In a similar differential diagnostic approach, Sindic [1053] did not find these broad categories; but he used fixed levels of cut-off for abnormal index values while Felgenhauer [275] used higher cut-off levels for higher degrees of barrier damage (see Chapter 9).

Specific antigens

The detection of antigen-specific IgG has several advantages over simply detecting total IgG as in MS, but it also imposes further requirements on the assay (see Table 10.4).

To correct for any changes in either (a) serum levels or (b) barrier function, it is preferable not to express the results as an index with albumin but to compare like

TABLE 10.4 Rank order for increasing diagnostic specificity

1. IgG as percentage of total protein
2. IgG Index
3. Antigen Index
4. Antigen Relative Specific Antibody (RSA) ELISA
5. IgG immunoblot
6. Antigen immunoblot
7. Antigen immunoblot (RSA)
8. Antigen IgM and IgA (IgG above)
9. Affinity titration

with like, i.e. antigen-specific IgG as a percentage of total IgG on both sides of the barrier:

$$\frac{\text{Antigen IgG CSF}}{\text{Total IgG CSF}} \bigg/ \frac{\text{Antigen IgG SER}}{\text{Total IgG SER}} = \text{Relative Specific Antibody}$$

This can be termed the relative specific antibody (RSA) i.e. the specific antibody is the amount of antigen-specific IgG expressed as a percentage of the total IgG for each compartment. The 'relative' implies CSF divided by serum and thus relative specific antibody is derived. The basic concept for pathophysiology is that the specific antibody is divided (e.g. anti-herpes) by the total IgG, thus, giving a value for specific antibody or percentage of total IgG on each side of the blood–CSF barriers. By comparing like with like (specific versus total IgG) the result is not influenced by either the barrier state or the level of serum IgG. Additionally there is no problem with any changes of slope or 'degree of selectivity' of the barriers. Analogous to percentage transfer (for barrier function) we have percentage of total IgG (for immune function). It must also be noted again that there is no need for a determination of albumin in either CSF or serum to correct for any barrier alterations.

This has the appeal that, if the percentage of antigen-specific IgG is higher in CSF than in serum, local synthesis is strongly implied. If the percentage is the same in CSF and serum, then passive forces have produced equilibrium in the two compartments. If the percentage is lower in CSF, then either there has been selective removal, e.g. 'precipitating out' and/or local synthesis of other antigen specific IgG, i.e. not derived from the serum pool. The same assay should therefore be employed (or as close as possible conditions and/or amounts) for detecting the antigen-specific IgG as for detecting the total IgG.

Although a value of greater than 1.0 for relative specific antibody should mean local synthesis [286] as with any test, one should determine the mean plus and minus two standard deviations. Muller [790] determined the range (0.5–2.0) for 24 normals using syphilis-specific IgG, therefore greater than 2.0 was judged abnormal. Ukkonen *et al.* [1201] used mumps-specific IgG and simply stated that greater than 2.0 was abnormal. Since Johnson [507] found a case with a normal value of 1.2, the value of 1.0 is probably not a reliable cut-off value and perhaps the value 2.0 should be more widely tested to ascertain whether or not the normal scatter extends beyond 2.0 [771]. Looking further ahead, it would be even more informative to distinguish between non-specific (polyclonal) responses and antigen-specific (oligoclonal) responses.

Early work insisted on the empirical 'four-fold' difference in the titers of specific versus 'non-specific' antibody titers. There is however, no clear evidence that these assumptions have a statistically verified basis, particularly since there can be an anamnestic response.

Although the early data of Vandvik [1228] showed that there was more measles antibody in CSF than in the corresponding serum of SSPE patients, the relative specific antibody was 5.0 and MS patients had a relative specific antibody of 2.0. Thus, the net difference between two diseases, expressed as a quotient, was 2.5. We have found that the relative specific antibody for SSPE (using scans of nitrocellulose blots) is greater than 90 per cent (see Figure 10.2) and the corresponding value for

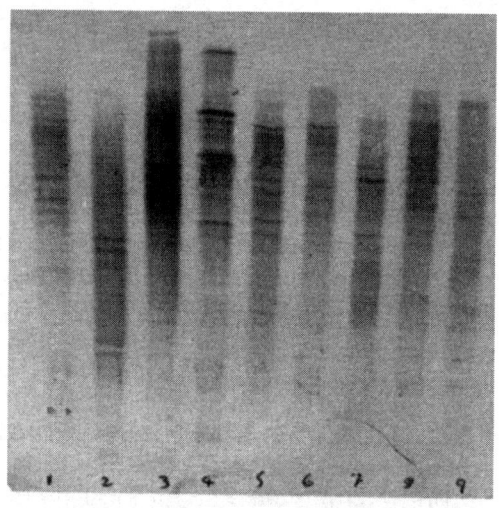

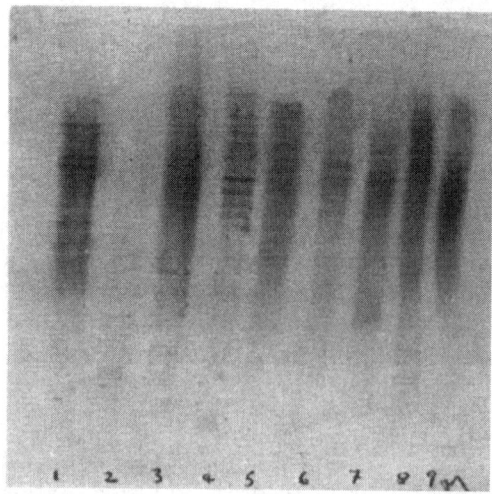

FIGURE 10.2 Antigen specific immunoblotting with measles virus (bottom) compared with total IgG (top). Number 2 is paraprotein, others are subacute sclerosing pan encephalitis (SSPE).

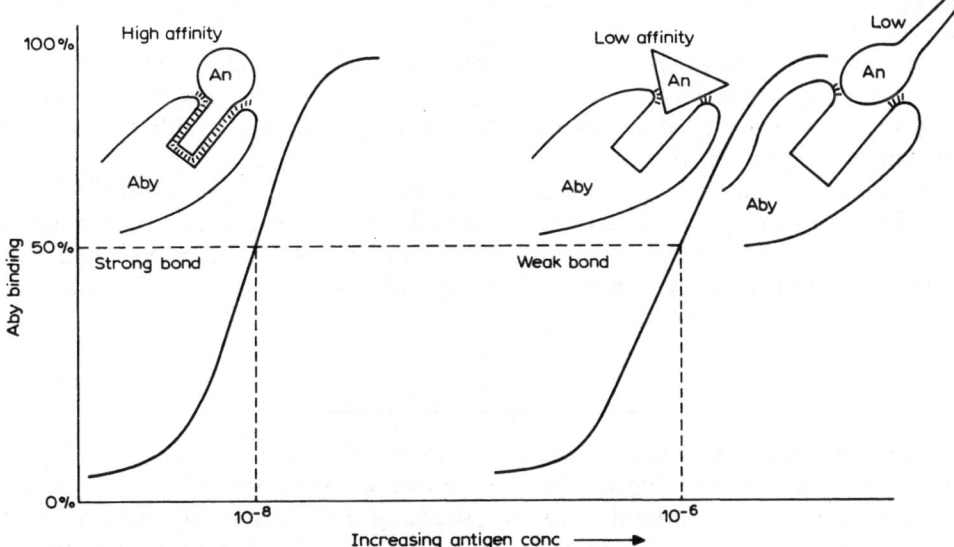

FIGURE 10.3 High affinity (specific) versus low affinity (non-specific) binding of antigen(s) to antibody.

MS is less than 1 per cent [782]. The net quotient for the two diseases therefore approaches a ratio of 100 to 1.

One can titrate both the amounts of CSF IgG and antigen. This has been termed the dual-dilution phenomenon [206]. Analogous to the Michaelis–Menton equations for half-saturation of substrate to enzyme binding site, the Sips plot [205] can show that non-specific, i.e. low-affinity antibody binds at a hundred-fold higher concentration than the high affinity, specific binding antibody. This is seen with myelin basic protein where specific binding occurs at 10^{-8} molar levels while non-specific binding occurs at 10^{-6} molar levels [205] (see Figure 10.3). Using the same amounts of CSF IgG from an MS patient as from an SSPE patient, it is possible to use differing amounts of measles antigen as a form of titration curve. We have found that the amount of measles antigen which is sufficient to bind more than 90 per cent of the IgG from patient with SSPE will bind less than 1 per cent of IgG from MS patients. Therefore, although there are small amounts of IgG which will bind to several viruses [1237] (or even myelin basic protein) [846, 1056] the major proportion of the CSF IgG from patients with MS does not bind to any known antigen. These arguments are based on 'anamnestic' responses which imply that essentially all the CSF IgG can be accounted for, but unfortunately, are not borne out by the quantitative data. Indeed, a careful analysis of the anamnestic data shows that the proportion of such non-specific response is also of the order of only a few per cent or less [880].

As was mentioned earlier, very large amounts of antigen (e.g. measles virus) can show binding to specific clones (almost a paraprotein-like pattern) but due to the large excess of antigen this probably reflects non-specific or low-affinity antibody binding (see Figure 10.3).

In similar circumstances, it has also been observed that up to one-third of the CSF IgG from MS patients can bind to measles. Again the amounts of antigen are so high that it is probably non-specific binding [1174]. In neurosyphilis about 60 per cent of the CSF IgG is bound by antigen [1236].

Technology

The method for detection of antigen binding must be: (1) sensitive (1 ng) to autoradiography levels; and (2) rapid, so it can be performed within one working day, if, e.g. it has to be diagnosed (assuming the polymerase chain reaction (PCR) result was negative) and early treatment has to be started for e.g. herpes encephalitis.

The PCR can amplify a single strand of DNA from a virus such as herpes simplex, and can be used to make a rapid diagnosis, although the neurologists will usually start treatment with aciclovir as soon as they form a clinical impression that herpes simplex infection is a reasonable part of the differential diagnosis. Nucleic acids from a number of different viruses and other organisms have been studied and general reviews are given [208, 1162].

General methods for protein separation/visualization

There are a number of general methods which are still very useful before focusing on specific proteins, including viruses and other pathogens. For many years we had screened CSF with a general stain, Coomassie Blue, which not only shows all proteins but also has a proclivity for basic proteins such as immunoglobulins. In neurological hospitals it is particularly important to screen for paraproteins, especially if there is any question of neuropathy [516]. When there are limiting volumes of CSF it is important to be able to use a very sensitive gold stain following the blotting of the agarose gel, subsequent to electrophoresis (silver stain and nitrocellulose are not a happy mixture). This technique requires only 4 µl of CSF [553]. The gold staining of CSF protein is shown in Figure 10.4. However, we have now shifted to staining all serum proteins with Coomassie, using 2 µl of undiluted serum. The Coomassie staining of serum proteins is shown in Figure 10.5 with 3 paraproteins (2 IgG on either side of an IgA).

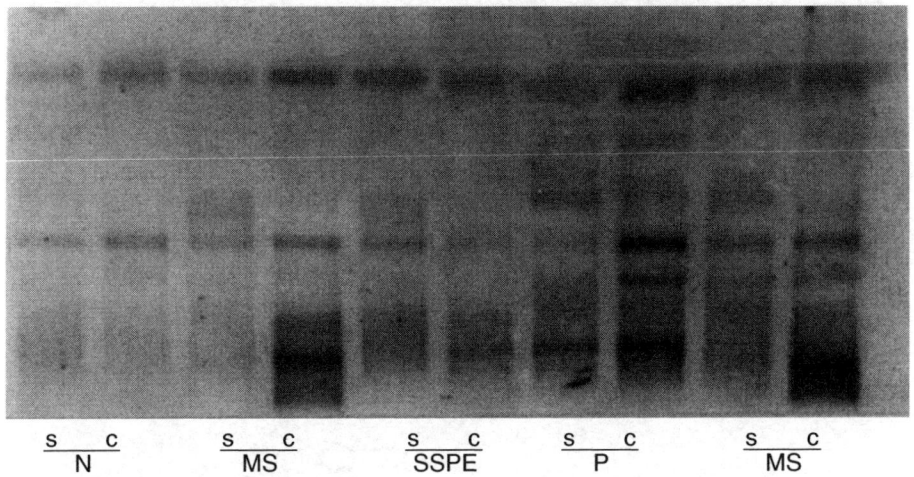

FIGURE 10.4 Gold stain on nitrocellulose using 4 μl of CSF as a general screen.

Separation by isoelectric focusing

Although we have found some differences in the quantitative ratios of kappa/lambda light chains, once again it is the qualitative analysis which is much more discriminating, as with the determination of IgA [629, 1028] and IgM. We use kappa/lambda staining as a routine 'tie-breaker' to discriminate between patterns which are difficult to interpret by IgG immunostaining. An example is shown later in Figures 13.1A and 13.1B (Chapter 13). Figure 13.1A shows a single band reacting to both kappa and lambda (which may represent an immune complex) as opposed to the unique lambda band seen in Figure 13.1B. This is lacking on the kappa stain and hence, is a clear example of intrathecal synthesis, although the whole question of prognosis for single bands is a separate study [198], see details in Chapter 13. Nevertheless some of the more traditional examples of oligoclonal bands are shown here.

Differences in commercial ampholyte mixtures are derived in no small part from the fundamental differences in the chemical reactions by which they are produced [549]. Basically there remains only one reliable source. These differences are reiterated here in Figures 10.6 and 10.7. Figure 10.6 shows a commercial kit from Perkin Elmer based upon polyacrylamide with in-gel fixation followed by silver staining of all proteins (as well as, once again, the discrepancy between the 3rd versus the penultimate track, and they are all supposed to have the

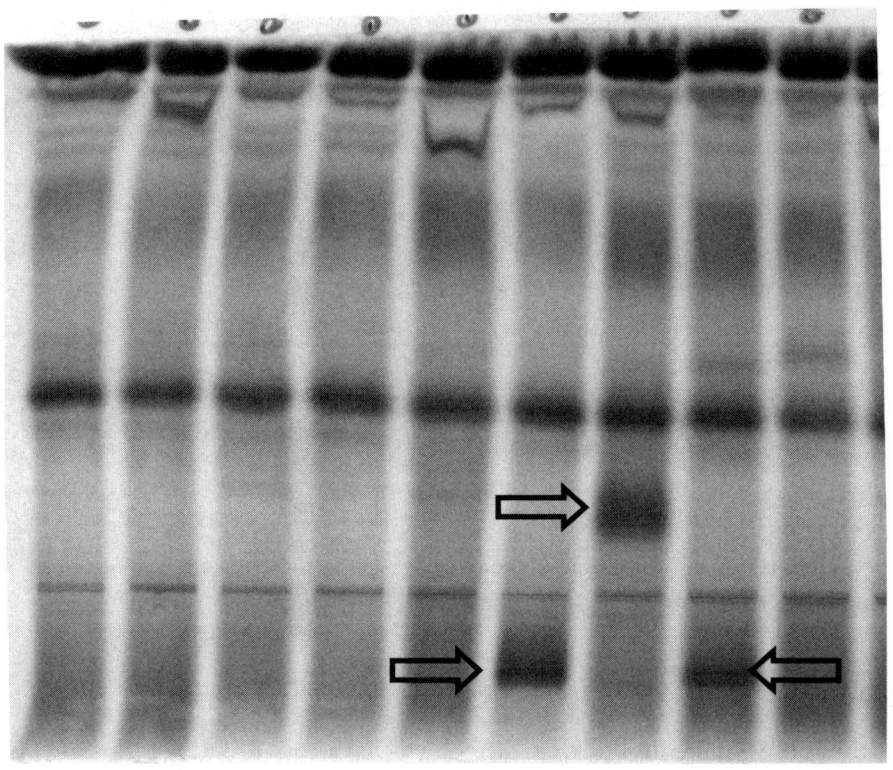

Serum paraproteins: Two IgG (bottom) and one IgA

FIGURE 10.5 Coomassie stain of serum as a paraprotein screen.

same 50 ng of IgG). Figure 10.7 shows a different commercial kit from Helena, in which the ampholytes are added to the agarose gel solution before electrophoresis and after separation the proteins are blotted onto nitrocellulose. Specific immunostaining with antibody against the heavy chain (Fc) portions of the IgG molecules precedes the incubation of the coupled antibody attached to the enzyme horseradish peroxidase (HRP) which amplifies the underlying IgG over a wide range of 20–1200 ng of the patient's IgG in parallel CSF and serum. The difference between the Figures 10.6 and 10.7 is the much more prominent artefactual banding pattern in the former, which produces the so-called 'common' bands which are exactly the same in all patients and have no physiological or biological relevance. They only serve to make the overall interpretation much more difficult.

TECHNOLOGY   153

Inhomogeneity of ampholytes → artefactual 'common' bands

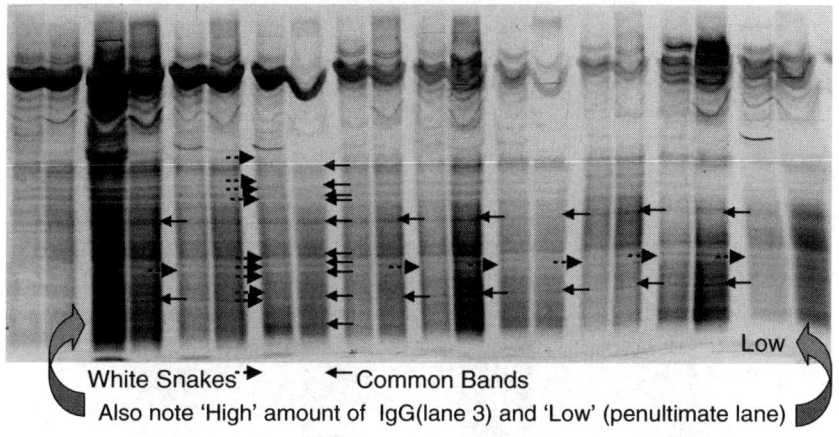

FIGURE 10.6   Artefactual or 'common' bands on silver stain.

Common (artefactual) bands and 'white snake'
shown by the double headed arrows

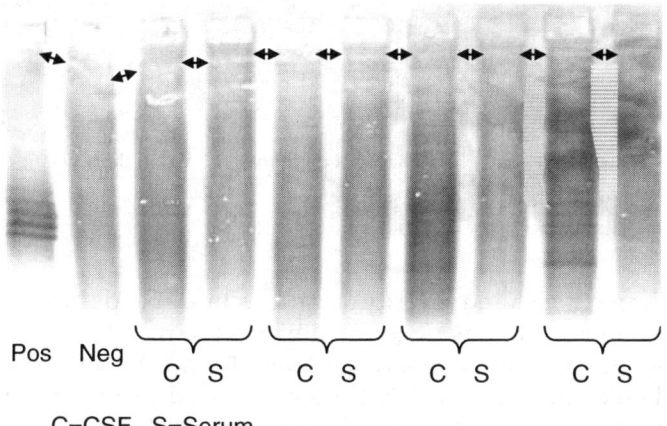

C=CSF   S=Serum

FIGURE 10.7   HRP immunostain ('one' artefactual band).

In practical terms, using the silver stain produces a very high level of 'noise' which unfortunately obscures any of the pathological (oligoclonal) bands with the large number of 'pseudo-bands' which simply reflect the serious inhomogeneities in the underlying ampholytes (see Figure 10.6). This is also seen in the staining of the ampholytes themselves in the absence of any protein (see Chapter 15, Figure 15.3).

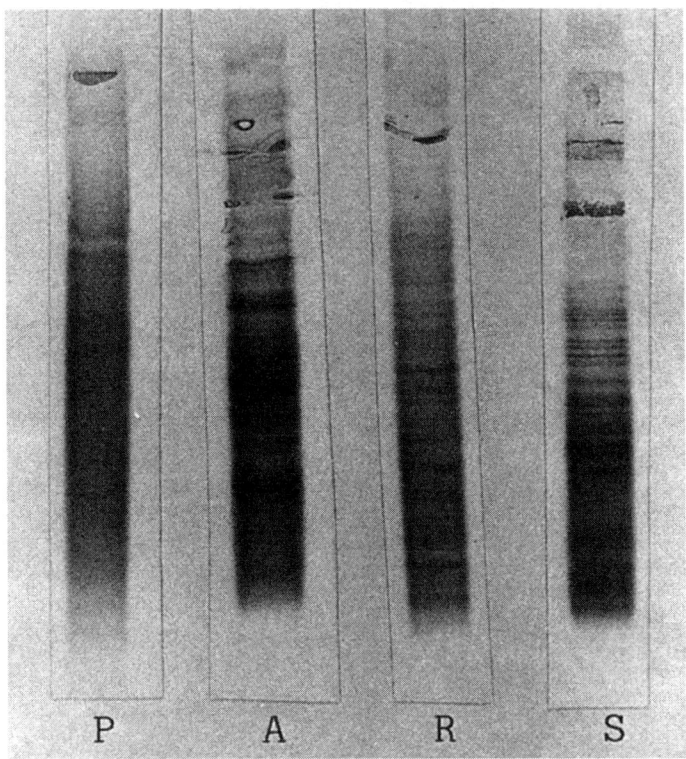

FIGURE 10.8 Smooth distribution of Pharmalytes in the IgG region (left) versus the other 3 rough patterns.

Conversely, the incidence and severity of the 'common' bands 'noise' in Figure 10.7 is clearly much lower and any oligoclonal bands can readily be seen. To put this in more physiological terms, the normal pattern for IgG is diffuse and polyclonal, which is easily seen in Figure 10.7, but this Gaussian distribution is unfortunately grossly distorted in Figure 10.6, thereby making the final diagnosis of the presence of oligoclonal bands in the latter much more difficult to ascertain. This major difference in the types of ampholytes is also shown in Figure 10.8, where the much more evenly distributed pattern is shown on the left (Pharmalytes) whereas the next three (from other commercial sources) are all much less satisfactory, due to the coarse irregularities which are present [549]. The basic chemistry of the synthesis is quite different, namely, the Pharmalytes are prepared from different substituted amino acids, whereas the others are all prepared from substituted acrylamide (e.g. LKB). Silver stain will of course react with Pharmalytes, but not the others, hence many kits have to include these very 'rough' patterns of the IgG region (e.g. Perkin Elmer). The most recent International Consensus requires immunofixation of IgG, and thus the general silver stain is no longer recommended [316].

Antigen-specific immunoblotting

In the diagnosis of viral or other brain infections, the two methods which are relevant to the study of proteins are Western and Eastern blots (as opposed to the study of nucleic acids, which involve Southern and Northern blots).

One of the best examples of Western blots was the study from Felgenhauer's laboratory [803] (see Figure 10.9). Here the proteins from different viruses such as varicella zoster or herpes simplex were fractionated according to their molecular weights and then two parallel tracks of CSF and serum are exposed to approximately equal amounts of IgG (100 µl of CSF). It can be seen very clearly that proteins with differing molecular weights are differentially stained by the IgG from the two compartments of CSF and serum. Note also the 'loss' of higher molecular

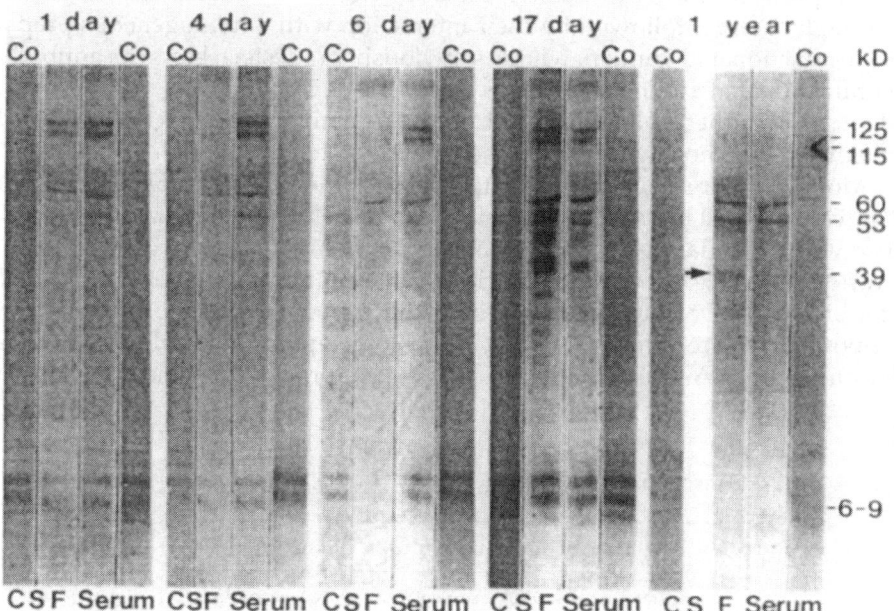

FIGURE 10.9 Western blots for different viral proteins (SDS) for CSF versus serum (same amount of IgG).

weight bands on day 4 and 6, with emergence of lower molecular weight bands on day 17 and at 1 year. This is unequivocal evidence for local synthesis of these specific antibodies in the central nervous system. Concerning Eastern blots, our techniques have shown separation of antibody in CSF and parallel serum according to their net molecular charge, via isoelectric focusing followed by binding of approximately the same amounts of IgG from the two compartments to the unfractionated viral preparations from different viruses such as measles [782] and herpes simplex [781]. At about the same time as we published our method [780], our fundamental techniques were also being modified along the same lines by investigators in the laboratory of Ter Meulen [228], but this article clearly acknowledged the use of our earlier basic methodology [1259], since our new method had then been reported at a number of International meetings [1129] before we had submitted the technique to a refereed journal [780].

What is the meaning of the historical origin of the term, 'Eastern' blots? In Eastern blots, separation by isoelectric focusing is based on the antibody (molecular charge). This is then probed with unseparated antigen. 'Western' blots are quite different. Along another axis (North/South versus East/West) the latter represent the binding of total (unfractionated) serum or CSF antibodies following the separation of antigens on the basis of their molecular weights in SDS gel electrophoresis. Thus we propose the term 'Eastern blots' in which antibodies are fractionated according to charge, followed by their interaction with a homogeneous preparation of unfractionated antigen, whereas obviously in Western blots the antibody is also unfractionated and homogeneous (see Table 10.5).

The term 'Southern blots' is based upon the name of the original author, Ed Southern, [1080] who bound complementary DNA to DNA which is analogous to what Mother Nature does during mitosis. In Southern blots, separation by electrophoresis is based on the DNA primary sequence (molecular size). This is then probed with secondary or 'complementary' DNA.

'Northern' blots were used to describe a different binding not of DNA to DNA, but the binding of DNA to mRNA, which is the first step towards protein synthesis (as opposed to cell replication). In Northern blots, separation by electrophoresis is based on the mRNA (molecular size). This is then probed with the DNA primary sequence.

TABLE 10.5 North/South/East/Western Blots

	Separation	Probe
Southern	DNA primary	DNA secondary
Northern	mRNA	DNA primary
Western	Antigen	Antibody
Eastern	Antibody	Antigen

In Western blots, separation by electrophoresis is based on the antigen (molecular size). This is then probed with unseparated antibody.

Dot-blots

Dot-blots are prepared as single strips to conserve precious antibody and commercially produced antigen. Dots are produced by the intersection after a 90° rotation between antigen and antibody. Specifically, a plastic mold with eight long troughs is put on top of a piece of nitrocellulose to form a water-tight seal so that solutions of different antigens can be applied to each of the troughs. This is then allowed to incubate overnight, during which time the antigens attach by noncovalent bonding to the nitrocellulose paper. The mold is then rotated through 90° and eight different samples of CSF are applied in the troughs such that each sample crosses over the eight different antigens. Thus one can see whether or not a single antigen will bind the CSF antibody. Multiple antigens may bind, as in the case of multiple sclerosis. A single antigen will typically give the diagnosis of infection with the antigen in question.

In the dot-blot, CSF alone is studied as a screen, since the final diagnosis will be made when a single antigen is positive. At that later stage the diagnosis will involve isoelectric focusing of parallel CSF and serum, which is obviously much more discriminating than the (unfractionated) 'dot' of CSF which was applied during the screening test. Once again this reinforces the idea that the binding of antibody to any of the particular antigens is dictated primarily by the affinity of the antibody for the specific antigen. This could also be detected in principle by performing parallel titration of the antibody in CSF versus serum to see if the two lines were indeed parallel (although they may be different in concentration), but given the case of a difference in slope, the affinities would be different between CSF and serum. Therefore, the only way to distinguish between a large amount of low-affinity antibody and a small amount of high-affinity antibody is to use the independent technique of thiocyanate titration [686].

Antigens used as pathological controls include the most common neurotropic viruses and other agents, namely, measles, herpes, rubella, mumps, toxoplasma, syphilis, tuberculosis and CMV (see additional antigens in Table 10.6). In cases of a specific viral infection, e.g. measles in the case of SSPE, only one of the antigens (anti-measles) will be positive and all the others will be negative. Here we see a case (see Figure 10.10) with only CSF antibodies against varicella (arrows), and then on the bottom, the focusing pattern of the same parallel CSF/serum from this case shows bands specific for varicella in the CSF. However, in cases of multiple sclerosis, there will typically be several different antigens which will be positive. This hypothesis was originally proposed by Felgenhauer as a possible diagnostic test for multiple sclerosis, and the title of his article ended in a query when first

TABLE 10.6 Antigen-specific IgG in CSF

Type	Organism
Bacteria	S. pneumoniae
	H. influenzae
	N. meningitidis
	M. tuberculosis
	T. pallidum
	B. burgdorferi
Viruses	Measles
	Herpes simplex type I, II
	Varicella zoster
	Rubella
	Mumps
	Cytomegalovirus
	JC papova virus
	ECHO virus
	EB virus
Protozoans	Toxoplasma
	Malaria
Other	Kveim
	Aspergillus

submitted to the referees 'Cerebrospinal fluid virus antibodies: A diagnostic indicator for multiple sclerosis?' [289]. This subsequently became modified into the so-called 'MRZ' test, namely Measles, Rubella, herpes Zoster (or varicella). This is still used, mainly in Germany, following on from the hyperbolic formula which was introduced by Reiber in an attempt to explain the co-variance of the albumin quotient versus the IgG quotient. Due to a number of factors, probably associated with the discrepancy problem (referred to previously in Chapter 6), Reiber then interpolated back to the idealized form of the hyperbola, since otherwise in some patients there could be 'too much' IgG. This was based on the fact that the intersection point of QIgG/QAlb could be quite elevated from the line of the hyperbola. Felgenhauer also noted this originally, although he expressed it as the difference between the 'observed' and the 'expected' IgG levels [291]. In any case, QLim represents mathematical manipulation to try to 'make sense' of the fundamental difficulties posed by this discrepancy problem whereby the oligoclonal IgG does not behave in the same fashion as the polyclonal standard.

Relative specific antibody on Eastern blots

Since the MRZ test does not strictly compare like with like (IgG/Alb) and also due to the discrepancy problem (hence the need for QLim), the better approach

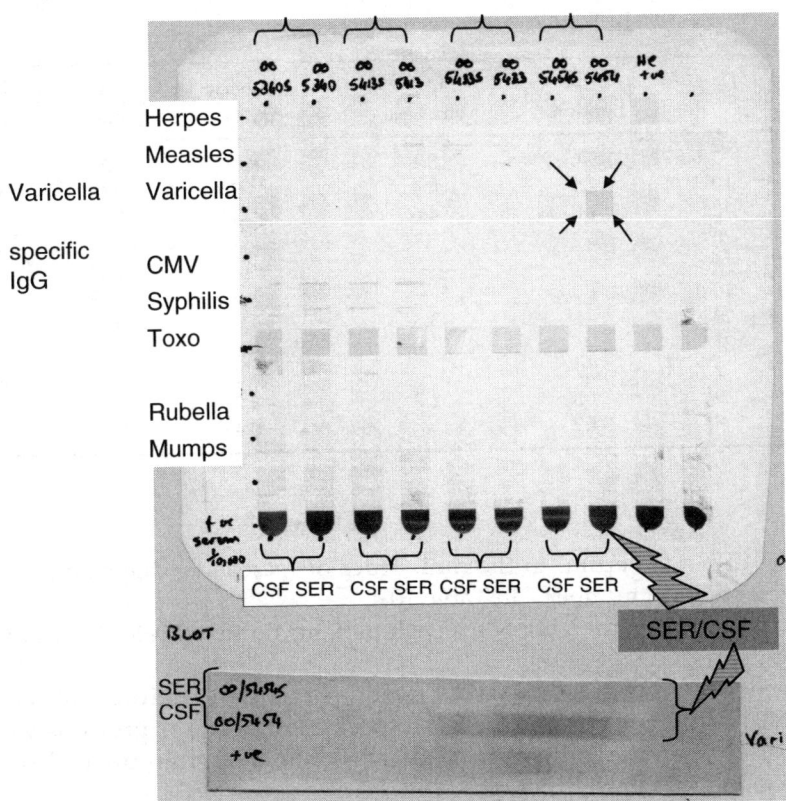

FIGURE 10.10 Dot-blot screen, followed by Eastern blot (IEF) for CSF versus serum against varicella (same amount of IgG).

is the use of dot-blots (as a screen) followed by Eastern blots (relative specific antibody). It should be stressed that we use the dot-blot test primarily to look for brain infection as the MRZ test is less sensitive and specific than IEF in the diagnosis of MS. In any case, the MRZ test is said only to be positive in a proportion of those who have oligoclonal bands in their CSF, and thus it does not find additional MS patients who might just be negative for local synthesis of IgG.

Several patients with Herpes encephalitis were studied by Bell et al. [53] and the relative specific antibody (RSA) is given in Table 10.7 which shows the effect of integrating the area under the curve for the total amounts of IgG in CSF and parallel serum to give one quotient. A second quotient is calculated by the division of the area under the curve for the antigen-specific (in this case herpes simplex) antibody

TABLE 10.7 Measures of herpes-specific and total IgG

Patient	Day	RSA London	RSA Manchester	Antibody index	IgG index	Outcome (sequelae)
JS	12	9.6	8.1	8.8	1.11	Minor
SH1	10	2.1	6.4	4.5	0.69	None
SH2	18	1.8	4.6	10.3	2.20	
WC	20	36.3	5.7	12.4	2.19	Minor
PK1	4	–	–	–	0.42	Died
PK2	10	1.0	–	7.8	1.69	
TM1	3	–	3.8	1.0	0.30	Severe
TM2	21	–	2.0	4.4	0.36	
SB1	2	0.7	–	3.4	–	Severe
SB2	19	1.3	1.7	4.7	2.70	
EB	8	1.9	–	0.4	0.91	Died

in parallel CSF and serum. In the rank order of increasing diagnostic specificity this represents level number 7 in Table 10.4.

First column gives the initials for each patient; numbers refer to either the 1st or 2nd puncture.

The second column gives the day of puncture relative to the onset of symptoms.

The third column is the RSA (relative specific antibody) as previously defined. This is based upon the four scans of total IgG in CSF/serum divided by herpes simplex-bound IgG in CSF/serum.

The fourth column shows a similar calculation based upon the ELISA results performed in Manchester using a calculation analogous to that for the RSA in the previous column.

The fifth column is the antibody index based upon the ELISA from Manchester of herpes antibody in CSF/serum divided by the Albumin quotient in CSF/serum.

The sixth column is the traditional IgG index.

The seventh column is the clinical outcome. The first three patients, all of whom had an early serum monoclonal response, had a good outcome [53]. The final four patients had a primitive polyclonal response in serum, and a poor outcome. The qualitative clonality is thus the only parameter which can correctly distinguish the two groups. This was statistically significant, and further details are given in Chapter 12.

The next step (level 8 in Table 10.4) is shown by immunostaining for IgM which is specific for the antigen in question. In this case we have an example of tuberculous meningitis which was successfully treated. The antigen blot showed that in parallel with the decrease in the amounts of bound IgM antibody there was a continuing increase in the levels of IgG (see later Chapter 12, Figures 12.1 and 12.2).

TABLE 10.8 Sensitivity and specificity of IEF for MS

Author	Reference	Total number of patients	MS #	%
A. Sensitivity				
Kostulas et al.	[597]	1114	58	100
McLean et al.	[733]	1007	82	95
Ohman et al.	[829]	558	112	96
B. Specificity				
Beer et al.	[50]	189	98	87
Paolino et al.	[847]	44	26	86

CSF versus MRI

The MRI technology can be contrasted with CSF in the diagnosis of multiple sclerosis (see Table 10.8).

Although there have been many retrospective studies comparing these two modalities in the diagnosis of multiple sclerosis, there are very few that are not only prospective but which also calculate what difference CSF analysis would make to the final diagnosis. There are four main points to make:

1. Sensitivity was slightly less, 97 per cent versus 99 per cent, for IEF versus MRI.
2. The specificity was much greater, namely 98 per cent versus 87 per cent in favor of IEF.
3. Molecular mechanism – focusing deals with IgG, which is the molecular basis for inflammation/infection, as opposed to water which is bound to varying degrees in the MRI sequences for a broad spectrum of neurological diseases.
4. MRI scans change with normal aging ('a man is as old as his arteries') whereas with IgG all that increases with age (typically over the age of 65) is the benign paraproteins which have a characteristic non-oligoclonal pattern.

Serum source of CSF antibodies

Another fundamental perspective is that of looking at the main source of CSF antibodies, i.e. the serum proteins which are transferred passively (according to molecular size) into the CSF. These proteins can have rather important effects within the brain, relate to their function (normal or pathological) or to their role in diagnosis of brain diseases.

Focusing of serum and parallel CSF

It could be argued that the overwhelming majority of spinal fluid samples are collected to determine whether or not there is intrathecal IgG synthesis.

Naturally, one must always determine the level of total protein, however one could (provocatively) make the point that the determination of albumin in CSF is rather a distraction from the main event, namely IgG. For many years now there have been complicated formulae which relate the amounts of albumin to the amounts of IgG, but it is now almost universally accepted that the gold standard is the parallel isoelectric focusing of CSF and serum immunostained for IgG. This bears no formal reference to the determination of albumin. Any changes in barrier function (which of course are reflected by the levels of albumin in CSF and serum albumin) are in effect corrected for by the appropriate dilutions of the parallel CSF and serum samples, such that visual inspection of the two patterns shows comparable amounts of IgG 'by eye'. As has been argued previously, the determination of 'abnormal' (oligoclonal) IgG is fraught with many difficulties, myeloma being the classic example. One could even go further and argue that all the added expenditure in the determination of serum and CSF albumin, including all the extra manipulations and attendant labor costs, will only serve to introduce yet more unnecessary variables, which only produce more noise within the system. A study of the history of the diagnosis of brain infections neatly illustrates this point. Specifically, the original determinations were derived from serum assays, e.g. complement fixation, hemagglutination etc, in which optical densities and/or serial dilutions were used to determine the final titer of antibodies. A practical dogma evolved in which a four-fold increase (i.e. two sequential serial dilutions) was sufficient to indicate an increase in titer of antibodies against the antigen in question. This was considered biologically significant as it signaled a four-fold increase in the level of antibodies. This mathematical formulation which is based upon optical density and titers has been studied very carefully in relation to herpes infections of the nervous system (encephalitis) where AI = Antibody Index, OD = Optical Density and DF = Dilution Factor [771].

$$\text{AI: OD} = \frac{\text{OD CSF} \times \text{DF CSF}}{\text{OD SER} \times \text{DF SER}} \times \text{QAlb} \quad [1101]$$

The next historical phase was the use of albumin in CSF and serum, and the quotient of these two values was used as a denominator for the quotient of viral-specific IgG in CSF and serum, directed against the HSV antigen.

$$\text{AI: Qalb} = \frac{\text{HSV antibody CSF}}{\text{HSV antibody SER}} \times \text{QAlb} \quad [580]$$

Although this was used successfully for many years, it was eventually supplanted by the more sophisticated model comparing like with like, namely specific IgG directed against the HSV antigen in CSF and serum as a quotient, divided by the quotient of total IgG in CSF and serum, once again expressed as a quotient in the denominator (~ RSA, Relative Specific Antibody) [286]. This obviated a need for the determination of albumin and has a very strong intuitive value, e.g. in the serum of a patient with herpes encephalitis, the antigen-specific IgG could represent, say, 2 per cent of the total serum antibody, whereas in the corresponding CSF it might represent 20 per cent of the CSF IgG. This is a dramatic ten-fold increase which could in no way be accounted for by any form of barrier dysfunction. Once again it is formally independent of the determination of albumin.

$$\text{AI: QIgG} = \frac{\text{HSV antibody CSF}}{\text{HSV antibody SER}} \times \text{QIgG} \quad [1201]$$

There is a notable difference in these formulae when applied to the data in Table 3 of this article, namely that the patient on line 6 was 'false positive' by the Reiber formula (which includes QLim, i.e. QAlb) but was then true negative by the Ukkonen formula, since it only included QIgG (i.e. not QAlb).

It must be reiterated yet again that qualitative analysis has been widely acknowledged as being much more discriminating in the study of total IgG as well as antigen-specific IgG. This can be seen in a simple example, as everyone can easily visualize differences of less than 1 per cent in the total population of IgG molecules in CSF and/or serum, whereas to reach the same level of diagnostic certainty using any qualitative analysis of CSF and/or serum could not produce credible results for the same diagnostic problem.

Differential diagnosis of serum bands

Particularly important is the superimposition of systemic disease into the differential diagnosis of neurological diseases, as detailed in Zeman's article [1316]. When there is a simple mirror pattern, namely the same systemic oligoclonal proteins are passively transferred into the CSF, then there is no real discrimination amongst the various categories in the differential diagnosis for the systemic immune response, namely infection/autoimmune or neoplastic/paraneoplastic or even a peripheral neuropathy/Guillain–Barré syndrome (see Table 10.9). A few patients with a vascular/degenerative disease have a mirror pattern. However there can be superimposition of antibodies within the central nervous system, giving a 'greater than' or 'mirror plus' pattern, because of the extra bands found only in CSF. This latter pattern is essentially only found in infectious diseases and multiple sclerosis and, to a lesser degree autoimmune diseases. It is not found in vascular/degenerative or peripheral neuropathy/Guillain–Barré syndrome and only very rarely in neoplastic/paraneoplastic disorders. It is therefore very important to be able to distinguish a systemic immune response as well as any superimposed intrathecal synthesis, and is therefore a useful guide in the differential diagnosis of neurological diseases.

TABLE 10.9 Differential diagnosis for systemic IgG bands

	Mirror (systemic) pattern %	Local + systemic synthesis %
Infections	14	29
'Inflammations'	18	9
Paraneoplastic	5	4
Neoplastic	16	2
Guillain–Barré syndrome	10	0
Other peripheral neuropathies	18	0
Multiple sclerosis	2	57
Vascular	5	0
Degenerative	5	0

Serum oligoclonal bands in MS

Serum oligoclonal bands can be detected in almost half of MS patients (44 per cent) [1317].

In Figure 10.11 there is a sequential two-fold dilution of an oligoclonal pattern in CSF as it becomes progressively 'lost' in the polyclonal background of the serum. It is visible in a 1 in 8 dilution but is less apparent in the 1 in 16, but no longer recognizable by the 1 in 64 dilution. As mentioned previously, the normal dilution from CSF into serum is $\times$ 10 000.

The presence of serum bands clearly does not represent the 'return' of CSF IgG into the serum, where total CSF IgG could constitute only 3 per cent of the total serum IgG, and thus would be totally swamped by the polyclonal background. Once again the area under the curve for the serum oligoclonal bands is disproportionately higher than the expected 3 per cent which could be calculated from the CSF contribution, since dilution experiments have shown that concentrations of CSF IgG which are < 9 per cent of serum IgG become 'lost' in the background. What is intriguing is that such serum bands are 10 times more common in females than in males, namely 70 per cent (19/27) versus 7 per cent (1/14), respectively. The patients were also significantly older (46 years versus 38 years) as well as having a later age of onset (39 years versus 33 years). They also had a two-fold increase in the IgG index (2.2 versus 1.3). Serum bands are also more commonly associated with other serum markers of autoantibody production and therefore may represent a different constitution for the individual concerned or at least a different phase during the evolution of the disease. Naturally, it would be interesting to know what the responses of these individuals would be regarding the rather different types of immunomodulation such as steroids, mitoxantrone, intravenous immunoglobulin (IVIG), polymerized lysine (COP1) and/or beta-interferon.

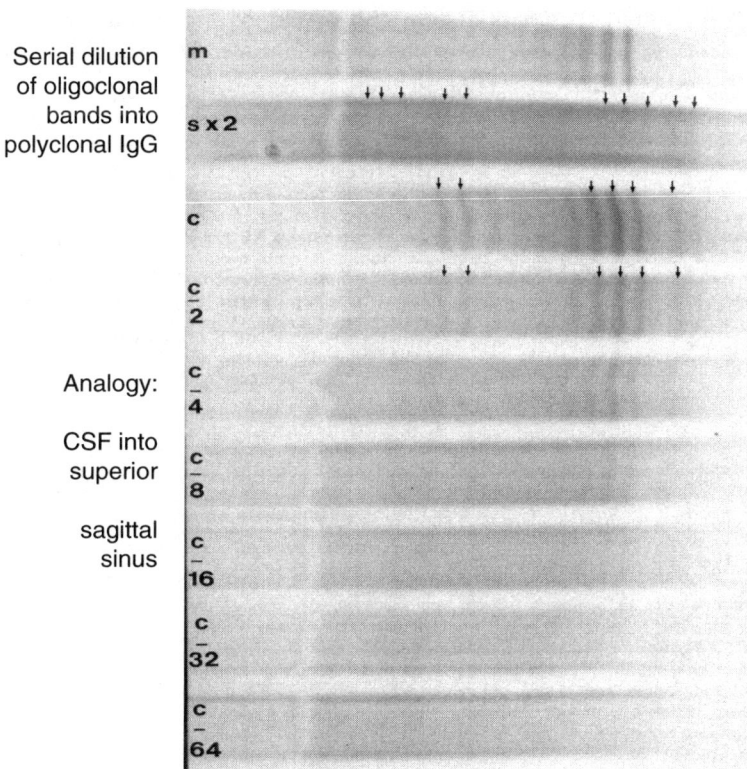

FIGURE 10.11 Experimental 10 000-fold dilution of CSF into serum. Note that the bands are 'lost' after 1 to 64 dilution.

Paraneoplastic serum and CSF antibodies

Affinity is also important in paraneoplastic syndromes, since a high titer of antibody may be the result of either a very small amount of very high-affinity antibody or a very large amount of a very low-affinity antibody. These can be distinguished only by an independent assay using NaSCN.

Paraneoplastic antibodies are assayed primarily in the blood (see Chapter 4), although their presence in the CSF can also be due to local synthesis as shown by the work of Dalakas [188], who found 93 per cent of patients (13/14) with Stiff-Man syndrome had oligoclonal bands in the CSF. McLean had previously shown this in 3/3 patients [733]. Antibodies against potassium channels can cause limbic encephalitis [1253] and four out of eight patients had local synthesis of these antibodies in their CSF. Local synthesis is crucial. Antibodies are not an epiphenomenon. Just as patients with Stiff-Man syndrome have anti-GAD antibodies, so do those with Sydenham's chorea, i.e. half (6/13) had local synthesis, as demonstrated by oligoclonal bands [148].

Conclusion

The measurement of relative specific antibody (following densitometry of immunofixed oligoclonal bands) again shows the great advantage of qualitative over quantitative analysis (see also Chapter 6) and the most sensitive and specific method is the performance of antigen-specific 'Eastern' (IEF) blots on parallel CSF and serum.

CHAPTER

11

Non-immunoglobulin proteins

Introduction
Serum proteins modified within the CNS
    Readily visualized proteins
        Haptoglobin polymers
        Tau protein (beta-2 transferrin)
        Group components
        Orosomucoid
        Paraprotein
        Alpha-1-antitrypsin
        Prealbumin (transthyretin)
        Alpha-2-macroglobulin
    Proteins better demonstrated by immunofixation
        C-Reactive protein
        Complement components
        Unbound (free) immunoglobulin light chains
        Fibrinogen degradation products
    'Brain-specific' proteins in the CSF
    Myelin basic protein
    Enolase (14-3-2, gamma isoenzyme: neuronal)
    Glial fibrillary acidic protein (GFAP, alpha albumin, astroprotein)
    D-2 antigen (N-CAM protein)
    I-CAM protein
    S-100 protein (a calcium-binding protein)
    Myelin associated glycoprotein (MAG)
    Major oligodendrocyte glycoprotein (MOG)
    Proteolipid protein (PLP)
    Ferritin (intracellular iron binding protein)
    Creatine kinase (CPK, BB isoenzyme)
    Tau protein
    Neurofilament protein (heavy, light and phosphorylated)
    The 14-3-3 protein
    Neopterin
General problems of any protein index or quotient
Conclusion

Introduction

CSF immuoglobulins have attracted a disproportionately large amount of research interest over many years, yet many fascinating insights can be obtained by paying attention to some of the other CSF proteins.

Those CSF proteins which are not principally derived from a response to either infections or other 'antigen-specific' stimuli can be roughly divided into those which appear to have their primary source in either serum or brain.

Serum proteins modified within the CNS

Some proteins are present in relatively large amounts and are thus visualized (Section "Readily visualized proteins") by general protein stains. Others, especially those with lower molecular weight fragments, are better visualized by specific immunofixation (Section "Proteins better demonstrated by immunofixation").

Readily visualized proteins

Multiple causes for elevation of the total protein level (destruction of blood–CSF barriers, see Figure 11.1) can be found in any standard textbook of neurology.

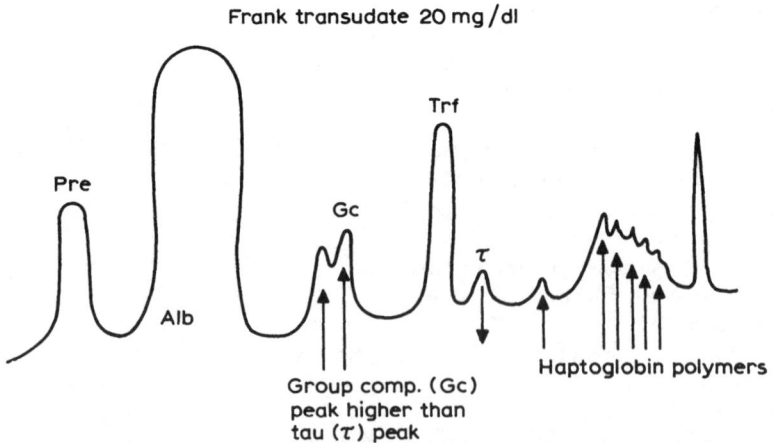

FIGURE 11.1 Frank transudate of serum proteins with normal CSF total protein level (20 mg/dl).

Haptoglobin polymers

Probably the most subtle of the pathological changes to be found in the CSF composition is the presence of the higher molecular weight haptoglobin polymers which were described by Blau [67] and Felgenhauer [268] (see earlier Figure 5.5).

Tau protein (beta-2 transferrin)

Tau protein may be elevated in the face of a 'normal' level of CSF protein, indicating the destruction of CNS parenchyma (Figure 11.2) due to the release of lysosomal hydrolases which cleave the NANA. Stasis of CSF flow, e.g. obstruction can produce the same elevation of tau protein levels. Decreased levels of tau protein are claimed in Rett's syndrome, in spite of the normal levels of CSF total protein [415].

Group components

In addition, attention should be drawn to the qualitative analysis of CSF proteins in those cases where the level of CSF total protein is ostensibly normal (e.g. 20 mg/dl

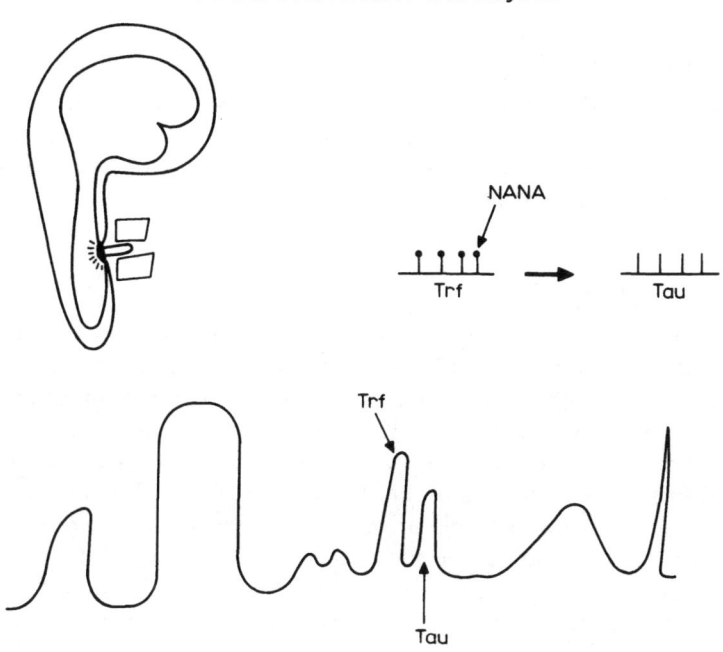

FIGURE 11.2 Elevated tau protein due to release of lysosomal hydrolases secondary to CNS parenchymal damage.

see Figure 11.1) and yet clear evidence for the destruction of blood–CSF barriers is seen as elevated amounts of group components proteins (Gc) compared to tau protein, hence Gc > tau.

Orosomucoid

An inflammatory transudate can even occur with 'normal' total protein levels as in metastatic, meningeal breast cancer (see earlier Figure 5.6).

Paraprotein

Myeloma proteins, being immunoglobulins, are normally selectively filtered out because of their large molecular size. However, in certain cases, the tumor may abut on the dura and with expanding necrosis may actually lyse the dura to release the paraprotein directly into the CSF. In this case, the level of CSF paraprotein exceeds that in the corresponding serum (Figure 11.3). Thus intrathecal cytotoxic therapy has direct access to the tumor site in question, rather than having to reach it via the serum.

Alpha-1-antitrypsin

There are several differences in the net charge on CSF alpha-1-antitrypsin (versus serum), some of which appear to be derived from more cathodic forms [1111]. Galvez found no change between the index for alpha-1-antitrypsin in CSF samples from tumors versus controls [334].

Prealbumin (transthyretin)

Prealbumin can be found in gross excess (almost as much as albumin) in cases of obstructive hydrocephalus, on the choroidal side of the obstruction. On the lumbar side of the obstruction, there are typically very low amounts (down to serum levels) of prealbumin. Glasner [375] classified patients with hydrocephalus into two categories, those who had back-diffusion of isotope into the ventricles (after injection into the cisterna magna), and those who did not. In the former category, he found elevated levels of prealbumin, provided there was still reabsorption of some of the isotope into the superior sagittal sinus. If there was no such reabsorption, he found a general increase in the level of all proteins. Others have found elevated prealbumin in those cases where the total protein level is usually not elevated [130, 1050].

Weisner suggested that the choroid plexus has the highest absolute level of prealbumin (56 µg/ml) [1273], but higher levels are actually found in serum (290 µg/ml)

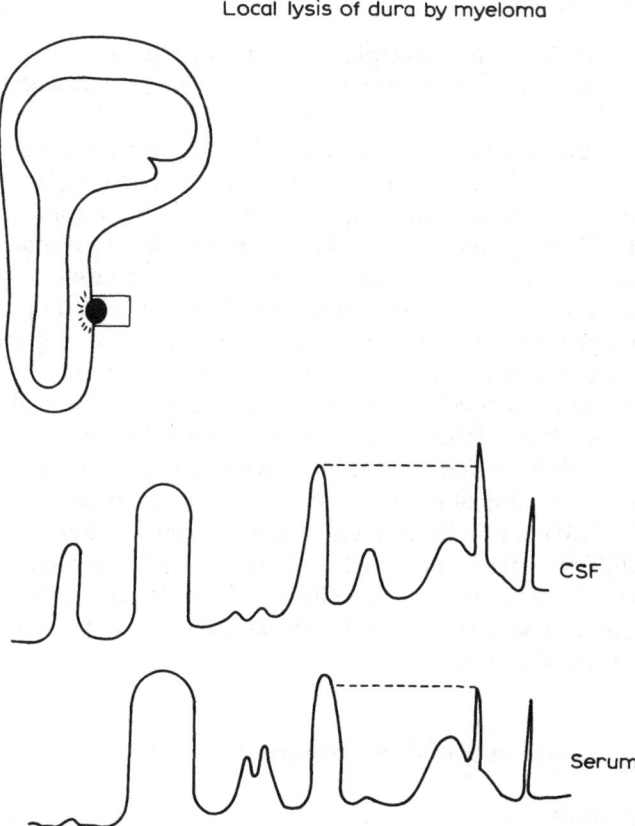

FIGURE 11.3 Myeloma paraprotein level (relative to transferrin as internal standard) is greater in CSF than serum.

and careful examination of the mathematics reveals a slight flaw of inversion. However, the more important conclusion is that, given the reasonable assumption that one gram of tissue contains essentially 1 ml of fluid, the highest relative specific concentration, which can be expressed as a 'percentage' of the albumin value, is to be found in the ventricular fluid (22.5 per cent) followed by the choroid plexus (4.3 per cent) and finally the serum (0.7 per cent). It is not yet known why the level of this protein remains relatively static with increased barrier damage [1274]. The finding of immunostaining with antibodies against prealbumin may also reflect the 'intermediate' status of the choroid plexus between the gradients (absolute and/or relative) for this protein in the serum versus the ventricular fluid [12], or indeed the likely local synthesis of prealbumin by the choroid plexus.

Alpha-2-macroglobulin

The main difference between alpha-2-macroglobulin in CSF versus serum is that there is a more cathodic form in CSF. There is also an extra anodic form in CSF [1040, 1041].

From the examples given earlier, it is clear that the 'normal' level of total protein can be quite deceptive if viewed in isolation from the protein pattern. Even when changes are allowed with increasing age, essentially all of the abnormal protein patterns (see Table 5.1 and Table 13.1) can be found with levels of total protein which would not be considered to be elevated. As a practical example, while it is known that in most cases of Guillain–Barré syndrome elevation of total protein is seen to be delayed with regard to the severity of clinical signs and/or symptoms, nevertheless, either high molecular weight transudate or even frank transudates can be seen, again with 'normal' levels of total protein. The data of Felgenhauer [989] also shows that the selectivity of the barriers (i.e. percentage transfer of the high molecular weight protein alpha-2-macroglobulin) can be damaged in Guillain–Barré when it is at the peak of its clinical severity [989]. The converse can also be seen, e.g. a patient aged 69 who eventually proved to have no demonstrable pathology had a total protein of 85 mg/dl but a normal protein pattern (decreasing gamma).

In conclusion, there are many examples in which the determination of the total protein level in the absence of qualitative fractionation can be incomplete or even misleading in its implication.

Proteins better demonstrated by immunofixation

C-Reactive protein

The level of C-reactive protein in serum is of more value in distinguishing viral from bacterial meningitis (since the latter has much higher levels) [1060]. In CSF, the CRP may complex with bacteria, since the levels are not as high in the CSF as the corresponding serum levels would lead one to expect [1060].

Complement components

Levels of complement proteins ($C'3$, $C'4$, $C'9$) may vary in CSF as well as in serum in relation to disease activity, but most immunochemical determinations do not distinguish between the converted versus the unconverted forms. Felgenhauer found no change in slope in a study of 300 patients [275], but Tourtellotte calculated an 'index' value (mathematically equivalent to his 'Formula', see Chapter 8) and claimed to find local synthesis of $C'3$ [1308]. A better method for the detection of complement conversion is the initial separation of precursor (Beta 1c) from product (Beta 1a) followed by immunochemical quantification of the two forms [27]. Using this method, changes have been found in cerebral Behcets in relation to clinical exacerbation [27] and later in several other diseases, but these were statistically significant only in association with those who had free light chains in their CSF (data not shown).

Unbound (free) immunoglobulin light chains

Why are free light chains seen in the CSF? Detection of this well-documented phenomenon can be complicated (see the Cross Index of References, Chapter 16). Free light chains occur in the initial stages of a new antigenic challenge either in germ-free animals or bone marrow transplants [111, 920]. Free light chains may reflect the higher rate of synthesis of light chains (relative to heavy chains) or perhaps to heavy chain 'switch' from IgM to IgG; or perhaps even the earlier switch from IgD to IgM or possibly the later switch from IgG to IgA. Light chains in serum are normally removed by the kidney to be found in the urine as Bence-Jones protein. The profusion of light chains in MS is also evident on SDS gels (Figure 11.4). In other inflammatory diseases, e.g. neurosyphilis, large numbers of free light chains are also associated with high white cell counts and the numbers of free light chains decrease coincident with treatment [1203]. In the case of myeloma, only a single light chain is found to proliferate (see Figure 11.5).

In any case, there is a clear statistical association between elevated numbers of free light chains in recent exacerbation of MS and conversely low numbers of free light chains with long duration of the disease. Perhaps this is a reflection of the clinical aphorism of the disease 'burning out'?

Fibrinogen degradation products

Fibrinogen degradation products typically show 20 per cent to 50 per cent transfer from serum to CSF [107, 469] and therefore are mainly produced within the thecal confines.

'Brain-specific' proteins in the CSF

Before listing individual proteins, some general points should be considered. Many, as yet incompletely characterized CSF proteins have been separated by two-dimensional electrophoresis under denaturing conditions using SDS detergent (see Cross Index of References, Chapter 16). Individual clonal fragments of heavy as well as light chains have been found in MS (see Figure 11.4). In myeloma, a single clone of heavy and light chains is predominant (see Figure 11.5). Many of the differences between serum and CSF (e.g. tau or beta-transferrin in CSF) are seen in the two-dimensional gels, although changes in other proteins tend to be

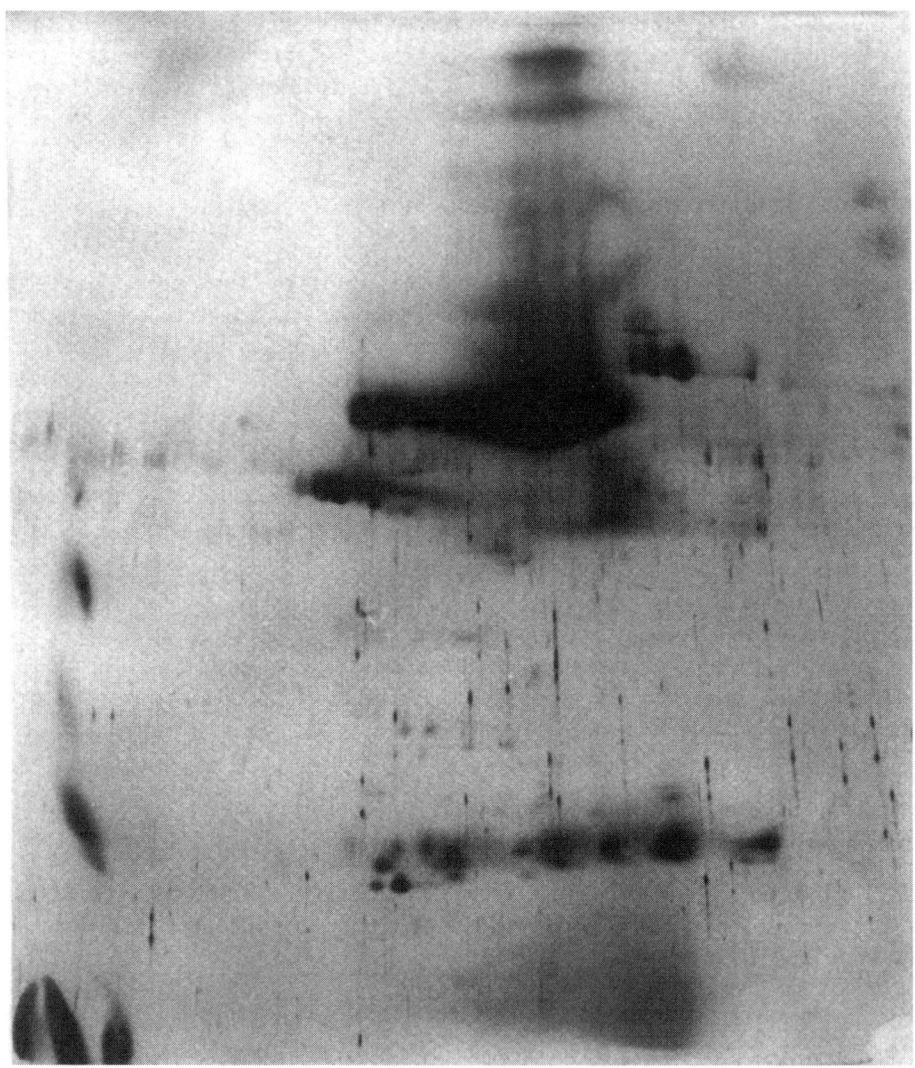

FIGURE 11.4 Two-dimensional electrophoresis of CSF from patient with MS. Near the bottom are multiple light chain bands.

rather small (about two-fold) [431]. Using the two-dimensional technique on brain tissue, in comparison with liver or other tissues, is another approach which has also been used extensively and various brain-specific proteins have been found [480, 481].

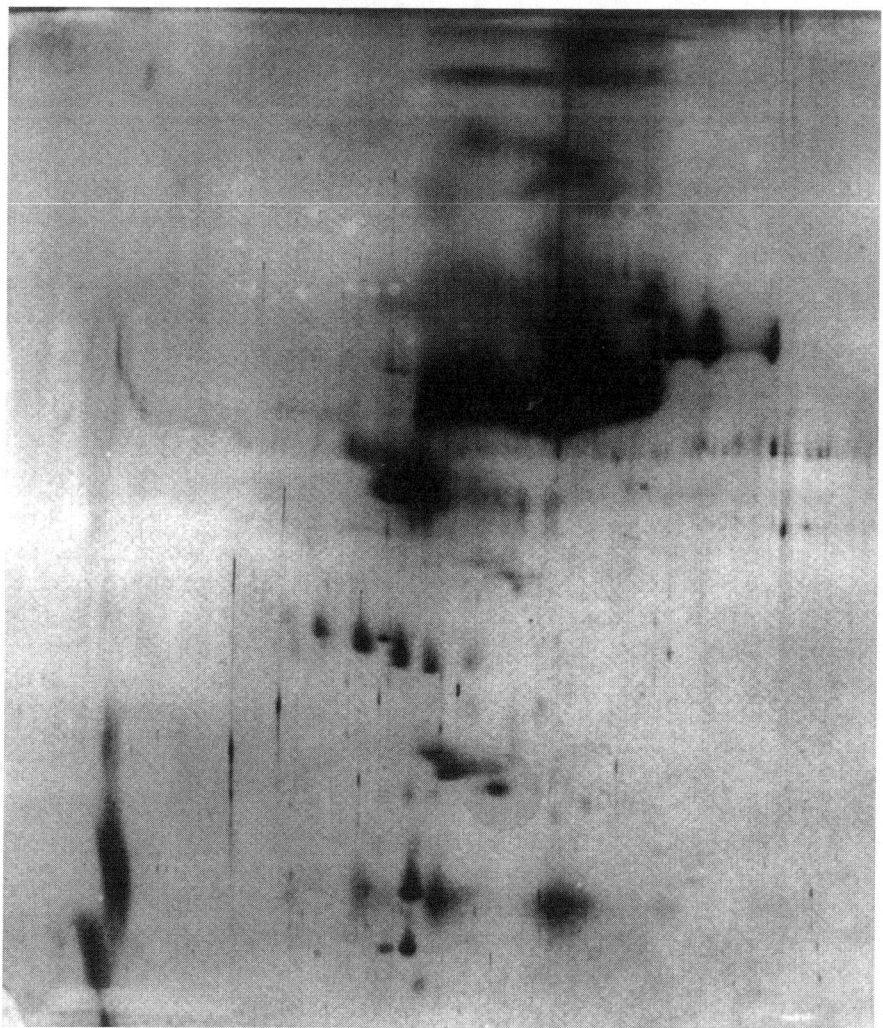

FIGURE 11.5 Two-dimensional electrophoresis of CSF from patient with myeloma. Near bottom is a major band of a single light chain.

It is worth reiterating the same three questions about the source of any of the CSF proteins.

1. Are there elevated serum levels?
2. Are there damaged barriers?
3. Are there intrathecal cells and/or fragments?

It is known that fragments of oligodendrocytes (i.e. pieces of myelin) are found in the CSF sediment [1187] and hence there would easily be proteins (and/or fragments) released into the CSF. The specific questions of the storage conditions of CSF always need attention.

1. Was there a protracted period at room temperature? (This can induce hydrolysis.)
2. Was the specimen frozen before removing cellular sediment? (Intracellular lysosomal hydrolases may be released into the supernatant CSF by the freeze–thaw action.) Even if the cellular material was spun out of the CSF, at what temperature, for how long, and how many times gravity was the centrifugation?
3. Were there elevated cell numbers in the CSF? (Lysozyme goes up in proportion to cell number [1160].) Even lysed red cells (e.g. traumatic tap) can show release of enolase into the serum [32].
4. Does the protein 'dilute out' in parallel to the standard? (There may be fragments having differing immunoreactivity to the intact protein [977].)

Considering the broader aspects of 'brain-specific' proteins, there is a clear historical precedent for comparing CSF with serum. The original work of Blake Moore was quite similar in principle to the two-dimensional gels, i.e. chromatography to separate proteins on the basis of charge, then molecular sieving on starch gel electrophoresis to sieve by size, and thus, many of the proteins which he first discovered have been re-mapped with two-dimensional gels. S-100, one of his original proteins, still does not have a definite cellular source (probably glial) and changes in its CSF level have been found in many neurological diseases [760, 769, 1058]. Another of Moore's proteins which has enjoyed considerable notoriety is 14-3-2 or enolase [74] which has been found to be more specific as a marker for neurons. However, this is also elevated in the CSF samples from diverse neurological disorders (see Cross Index of References, Chapter 16), so it may be a better indicator of the *extent* of CNS damage, rather than of specific diagnostic value [196].

In the following section, 'brain-specific' proteins are discussed.

Myelin basic protein

Produced by oligodendrocytes, this is probably the best studied protein of the brain, and clearly one of the most specific in terms of local synthesis by the CNS. However, once again the levels in serum or CSF appear to be more useful as prognostic indicators in severe head injury than in differential diagnosis, since elevated levels are found in many different neurological diseases [162]. This point has been clearly made in a comparison of its use as an indicator of recent relapse (extent of CNS damage?) whereas its ability to distinguish MS from other diseases is of less value [69]. The idea that it may be a useful guide to the application of various therapies in MS requires further investigation. Myelin basic protein is also affected by the discrepancy problem, i.e. the difference in immunoreactivity between the fragments versus aggregates, and also versus the intact protein. Because of its strong basic charge (pI 9 to 10) it readily binds non-specifically to alpha-2-macroglobulin (which was the basis for an early assay of MBP in CSF [742]) and even to the Fc portion of IgG [1056] or beta lipoprotein [827]. These different binding characteristics are best examined by technique use of dual dilution curves [206] employing both serial antibody and antigen dilutions. A more simple technique is to look for non-parallel curves in the test material (CSF) versus the standard [842]; since it must be certain be to compare like with like, lest the discrepancy between the immunochemical versus physical–chemical differences will be encountered.

This non-specific or low-affinity binding to other proteins is partly responsible for its non-parallel behavior when titrated in CSF [844]. Although fragments of myelin basic protein were noted earlier [118], another study has shown that while fragments occur in MS, aggregates occur in patients with stroke and following surgery [540].

The amounts of myelin basic protein in CSF are probably much higher than in serum but a normal range is difficult to define due to the low levels of the protein found in normal subjects. Most investigators have found less than 4 mg/l [162] or even less than 2 mg/l [1285].

Enolase (14-3-2, gamma isoenzyme: neuronal)

This protein is said to be specific for neurons, but if so, there are high serum levels of an immunoreactive congener (see Table 11.1). Thus enolase levels have not been found to be diagnostic of neuronal damage, and they may again be better employed as indicators of the *extent* of the CNS damage.

TABLE 11.1 Enolase levels in CSF and serum.
Levels of enolase (14-3-2) in CSF and serum (μg/l) with percentage transfer (%T), number of cases (#) and type of isoenzyme listed by the authors; for comparison, levels of 14-3-3 are also included.

First author (Date)	Reference	CSF	SER	%T	#	Type
Brown (1980)	[104]	10	5	200	64	Neuronal
Boston (1982)	[86]	15	50	30	82	(14-3-3)
Parma (1981)	[849]	2	9		5	Gamma
Royds (1981)	[959]	10			35	Gamma
Scarna (1982)	[980]	21	9	200	16	Gamma
Dauberschmidt (1983)	[196]	1	4	25	9	Neuronal
Gerbitz (1984)	[343]	–	9	–	40	Gamma
Median		10	10	100		

Glial fibrillary acidic protein (GFAP, alpha albumin, astroprotein)

This protein is found as a normal constituent of the intermediate filaments of astrocytes. Elevated levels are found in many diverse diseases (see Cross Index of References), but serum levels are very low. It seems a rather useful protein in the differential diagnosis of normal pressure hydrocephalus (NPH) from primary dementia, since elevated levels have been reported in about 75 per cent of patients with NPH [10].

D-2 antigen (N-CAM protein)

This protein (N-CAM: Neural Cell Adhesion Molecule) also appears to be found in serum (or cross-reacts with a similar epitope). Nevertheless, it also appears to be of limited use in the differential diagnosis between NPH versus primary dementia [1076], since about 60 per cent of NPH patients have decreased levels.

I-CAM protein

Another example used by Felgenhauer was that of ICAM (Intercellular Cell Adhesion Molecule) [925] to illustrate his concept of the chemically 'silent' brain, namely that parts of (mainly) the cerebral cortex could release proteins into the interstitial fluid which could in theory then pass down into the lumbar sac. However, he reasoned that it would be very improbable as these proteins are much more likely to be carried forward up into the superior sagittal sinus and thence back into the venous circulation. An example of this is the study of tumor markers such as carcino-embryonic antigen (CEA) [483]. Thus tumors in the cerebral cortex are unlikely to produce any abnormalities in the lumbar CSF.

S-100 protein (a calcium-binding protein)

The highest levels of S-100 seem to be found in the more acute diseases, e.g. vascular insults or viral infections [1058]. The diagnosis of CJD, typically reaveals very high amounts of all brain proteins, especially S-100, although most people use 14-3-3 protein. S-100 protein also shows an example of how rapidly brain proteins can be cleared from the CSF. Figure 11.6 gives a model based on an in dwelling ventricular drainage catheter, in which levels can fall three-fold over the course of 12 hours, and subsequently rise in the serum over the same time-course [395]. Some have found that serum S-100 levels correlated with infarct volume as well as outcome, [767] but others have offered caveats [444].

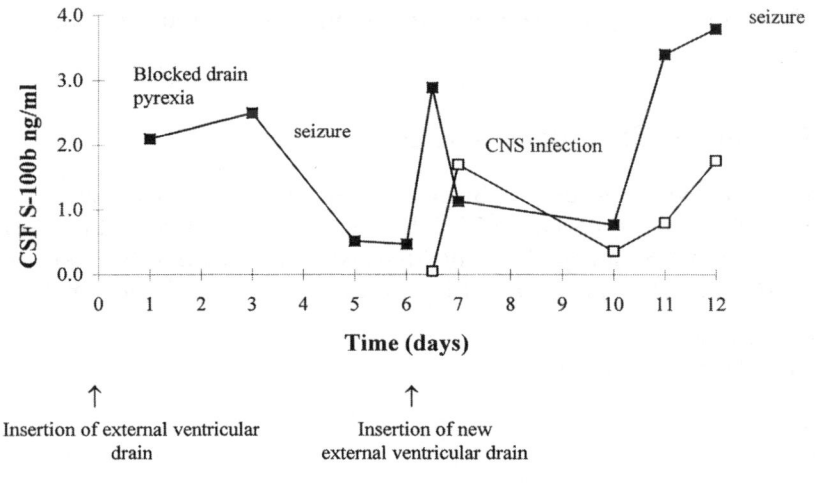

Serial CSF (■) *and serum S-100b* (□) *concentrations*

FIGURE 11.6 Ventricular drain showing the rapid clearance of S-100 protein from CSF into serum over 12 hours.

Myelin associated glycoprotein (MAG)

Although this protein does not seem to be found intact in the CSF, there appear to be fragments of this protein in normal subjects as well as in various neurological diseases [1311].

Major oligodendrocyte glycoprotein (MOG)

This protein has been much studied in experimental allergic encephalomyelitis, and recently was proposed as an early marker of those who would go on to develop MS [55]. However, our own work with coded samples being sent to the original investigators failed to find any such predictive value.

Proteolipid protein (PLP)

This is a major protein of the myelin sheath and yet, like MAG, its level in CSF does not alter greatly in various neurological diseases [211].

Ferritin (intracellular iron binding protein)

Since herpes encephalitis is characterized by hemorrhagic necrosis, high levels of ferritin have been found [1062], but this is not specific to the disease in question

[1059]. The molecule synthesized within the brain is different from that found in the serum [860].

Creatine kinase (CPK, BB isoenzyme)

Several enzymes are found in CSF and serum, but changes in their CSF levels are not very useful in diagnosis [38, 674]. They may have had a role in the estimation of the degree of CNS damage (see myelin basic protein, earlier) but too often there was not adequate allowance/correction for the two standard questions:

1. What is the serum level?
2. What is the barrier state?

Tau protein

Three proteins are usually measured to help discriminate between the different types of dementia, including Alzheimers, namely, beta-amyloid, tau and phospho-tau. Careful examination of the data reveals that the most useful of these is probably phospho-tau [155, 425, 476, 529, 1036]. With the exception of CJD [477] there is no dramatic improvement in the sensitivity/specificity when these three proteins are compared amongst vascular versus other forms of dementia, e.g. Lewy Body disease, Pick's disease or fronto-temporal dementia.

The ratio of phosphorylated to nonphosphorylated tau in CJD has been studied [477] and it was found that the level of phosphorylation, i.e. at serine-199 showed a disproportionately low percentage of the total tau, which was different from all the other forms of dementia. The other caveat is that the final diagnosis for the type of dementia is still best made at postmortem [155].

Neurofilament protein (heavy, light and phosphorylated)

Another group of proteins which can be measured from the neurons are of neurofilaments, either light chains or heavy chains. Rosengren [948–950] has performed many studies on the neurofilament light chains, and they appear to be less stable than heavy chains, in that they show over a 50 per cent loss in the 3–10 days following the lumbar puncture. The heavy chains are thankfully more resilient. The only problem has been that the antibodies used in Rosengren's laboratory were created 'in-house', specifically from immunized hen's eggs and therefore they are not commercially available to allow the rest of the international community to corroborate his work. We have used the heavy chains (NfH) not only because they are more stable, but because there is a whole series of commercially available antibodies, particularly the Sternberger monoclonals, and thus we have published a method for this [870]. We have found correlations in longitudinal studies between NfH and the clinical outcome in MS, namely the disability scale according to Kurtzke using the EDSS score.

The 14-3-3 protein

This has quite good sensitivity for classical (sporadic) CJD, but is less sensitive for the 'new' variant CJD (mad-cow disease) [396]. It is also found in many other diseases where there is a major destruction of the CNS tissue [707].

Neopterin

The bladder can naturally collect urine over many hours, especially during the eight hours of sleep. During the night, there is a six-fold increase in the production of CSF (see Figure 5.2) [361], and thus a sample of early morning urine will represent a composite integration of a large volume of CSF which has been passed from the brain through the superior sagittal sinus into the circulation where it is then filtered through the kidneys into the urine. A situation could be easily envisaged in which proteins might be 'missed', since they passed through the serum very quickly and were then filtered out by the kidneys, such that unless one sampled the serum continuously, the 'pulse' of protein would not be collected. We have seen from continuous monitoring of interstitial fluid of the brain, with in dwelling ventricular catheters, that there can be very rapid clearance of, say, S-100 protein, as demonstrated by the work of Green et al. [395] (see Figure 11.6). However, since all these proteins are being collected and 'summed' in the urinary bladder, the urine can of course be continuously monitored via an in dwelling catheter or, if the patient came into the emergency room, a sample of urine would represent a collection of all the proteins which had passed from the blood through to the kidneys up until the last time that the patient had emptied the bladder.

General problems of any protein index or quotient

One of the classical problems in interpreting a quotient for any protein in CSF and parallel serum is the caveat that an elevated quotient value is typically assumed which is consistent with local synthesis. However, a classical misinterpretation was shown in the article with Apo-E by Rifai [928] in which cases of multiple sclerosis were studied during relapse and remission and appeared to show a significant increase in quotient during the period of remission (and were normal during exacerbation). Nevertheless, it was astoundingly obvious that the CSF levels remained static, whereas the serum levels in fact decreased during relapse, and hence there was not an intrathecal increase but a systemic decrease, perhaps part of the repair process which patients may experience during their remission. They also had a slight decrease in CSF albumin, which could also serve to increase the index value of Apo-E. The same generic question applies to almost any therapy for multiple sclerosis, since it is invariably given systemically rather than intrathecally. This leads to the additional caveat that the albumin may also decrease in response to, say, a cachexia-like reaction (e.g. to TNF-alpha, which is also known to increase systemically during exacerbations). All this goes to reinforce the idea of qualitative analysis of CSF and parallel serum, rather than simple quantitative analysis, regardless of all the sophisticated mathematical formulations, including the determination of albumin, and not least the various 'correction factors' (namely the QLim). These have been introduced to compensate for the well-known idiosyncrasies introduced when attempts are made to measure quantitatively the amounts of 'abnormal' IgG compared to the standards, which are not always derived from the normal population. Yet again, it is not strictly like compared to like. The qualitative differences in CSF versus serum Apo-E are manifested with a much more acidic charge in CSF, due in part to more NANA [1310].

The differences between the CSF-derived proteins (versus serum) can be readily seen, not only amongst their amino acid sequences but also amongst their specific sugars and individual sequences, although differences have been described previously in the sugars of alpha-1-antitrypsin [1111] as well as orosomucoid [825]. More recent data from the HPLC determination of individual sugars and their sequences has revealed clear differences in the so-called 'biantenniary' sequence found not only in the beta-trace (prostaglandin synthase) but also in beta-2 transferrin (tau). These proteins have therefore been termed 'brain-type' N-glycosylation [457, 458]. The brain-derived CSF sugars have substantial amounts of bisecting N-acetyl glucosamine, proximal fucose and absent galacto chains compared with their serum counterparts which lack the bisecting N-acetyl glucosamine as well as the proximal fucose, and do contain galacto chains.

Physiological/pharmacological effects on the beta-transferrin glycoforms in Parkinson's disease can reveal significant changes in the serum transferrin isoforms which are not reflected in the parallel CSF from the same patient. Specifically, abnormalities (i.e. more acidic transferrin) were found among untreated patients who are notoriously 'rigid' and can have serious limitations in their normal bodily movements. The patients had more sialation of transferrin than controls, however upon subsequent treatment, when patients had improved movement and a return to more normal bodily functions, the serum transferrin reverted to the normal level of sialation [1213]. This reiterates the relative independence of the two types of transferrin, one derived systemically (serum) versus one derived intrathecally (CSF).

Conclusion

Determination of 'brain-specific' proteins in CSF has not yet proved to be the diagnostic panacea which some had hoped for. The various proteins appear to be more of prognostic value or for assisting differential diagnosis between two specific diseases (see glial fibrillary acidic protein, and D-2 antigen earlier). Further clinical–chemical correlations are needed rather than open-ended searches for 'one protein – one disease'. The biochemists should team up with the clinicians to investigate areas where a marker would make an important differential point in diagnosis, treatment, even prognosis.

CHAPTER

12

Monitoring therapy

Multiple sclerosis
Neurosyphilis
Tuberculous meningitis
CNS leukemia
Hydrocephalus
Herpes simplex encephalitis

A controlled trial of immunosuppression in MS is one of the few situations which involves monitoring with repeated lumbar punctures. It has also been suggested by a European Consensus that herpes simplex encephalitis should be so monitored.

Multiple sclerosis

Changes in the individual immunoglobulin bands have been followed during MS treatment with azathioprin and anti-lymphocyte serum. There were equally dramatic changes in band patterns in both immunosuppressed patients and placebo-treated controls. We found no changes either in IgG as a percentage of total protein or in the peak height of the most prominent oligoclonal band when compared with double-blind controls [757]. Measurement of free light chains may prove more helpful since they are associated with recent exacerbations [1203], and in a study of ciclosporin we found that there was indeed a statistically significant change in kappa and lambda free light chains, as is shown in Table 12.1.

Although changes in the levels of total IgG have been claimed in response to therapy [1169], the problem remains that the changes may reflect the presence of the fragments and/or aggregates of IgG, in addition to the molecules of intact (unbound) IgG, as has been discussed in Chapter 7. There may also be changes in the levels of myelin basic protein, but once again adequately controlled double-blind trials are required to clarify this issue.

TABLE 12.1 Change in free light chain bands after treatment

	Increased kappa		Increased lambda	
	Positive	Negative	Positive	Negative
Ciclosporin	2	7	2	7
Placebo	7	3	8	2
Fisher exact	$p < 0.05$		$p < 0.02$	

There have been several studies of beta-interferon, but perhaps the most interesting has been that published by Jacobs et al. [484]. Looking at their treatment table (Table 12.2), a number of points stand out:

1. The two halves of the table show essentially the same results after 12 months of treatment (top) or 18 months of treatment (bottom).

TABLE 12.2 Change in EDSS scores after beta interferon

		Placebo		Beta-interferon	
		#	%	#	%
Unsustained change		$N = 87$		$N = 83$	
Better	> 1.0	10	11.5	16	19.3
	0.5	10	11.5	13	15.7
Same	0.0	18	20.7	20	24.1
Worse	0.5	18	20.7	14	16.9
	1.0	3	3.4	5	6.0
	1.5	9	10.3	7	8.4
	2.0	4	4.6	4	4.8
	> 2.5	15	17.2	4	4.8[a]
Sustained change		$N = 56$		$N = 55$	
Better	> 1.0	5	8.9	10	18.2
	0.5	9	16.1	14	25.5
Same	0.0	14	25.0	13	23.6
Worse	0.5	11	19.6	8	14.5
	1.0	4	7.1	3	5.5
	1.5	2	3.6	3	5.5
	2.0	1	1.8	2	3.6
	> 2.5	10	17.9	2	3.6[a]

[a] Significant $p < 0.05$

2. Looking down the rows, the patients were subfractionated according to increasing changes in EDSS which progress from negative (improvement) to stronger levels of positive (deterioration). It is obvious that the only row which shows any statistical significance is the bottom row, namely those who have had the highest change in EDSS score over the period of time (both top and bottom halves show essentially the same results).
3. Only in the minority of the population (about 18 per cent), does the drug have any significant effect.
4. In conclusion, the drug only works on those who have the most 'malignant' form of the disease, namely the highest change in EDSS over the course of either 12 or 18 months.

Many different types of drugs are used in the treatment of MS, as we shall see in the five examples set out below, which raises the important question, what is the 'molecular basis' for their action? Further, what does this tell us about their separate roles in the pathogenesis of the disease? Should we be changing therapy at different stages of the disease? Beyond any longitudinal changes, what about the different subtypes of multiple sclerosis such as relapsing–remitting versus primary progressive and the benign form at the one end of the spectrum and the burnt-out form at the other? Clearly there is much to learn from further studies, not only of different therapies, but also of their appropriate controls, which again can be matched according to the different subtypes. The following is a list of five rather different types of therapy:

1. *Steroids* – which cause a generalized depression of the immune response.
2. *Mitoxantrone* – which can induce quite strong immunosuppression.
3. *Intravenous immunoglobulin* (IVIG) – which may have an anti-idiotype effect.
4. *Polymerized lysine* (COP1) – which exerts a net negative charge by the gross excess of primary amines on the side chains of lysine.
5. *Interferon* – which, among its other effects, blocks the replication of viruses.

In vitro synthesis of peptides and/or proteins is typically performed in cell factories and starts with *E. coli*, but such proteins are not glycosylated, a process which is much more important when synthesizing beta interferon, where the glycosylated versions are produced in the eukaryotic cells rather than the prokaryotic cells.

Neurosyphilis

The other indication for monitoring therapy is in the measurement of IgM oligoclonal bands in neurosyphilis, since these patients tend to get re-infected on a continuing basis. The question is often raised: was the original infection treated adequately, or is this a new infection? The problem with IgG levels (and oligoclonal bands) is that they are present for years following successful treatment [1236]. However, while IgM rises more rapidly than IgG, fortunately it also has a more rapid fall with successful treatment.

Tuberculous meningitis

We had a typical, chronic case whose CSF reverted to a normal IgM pattern on isoelectric focusing within three months of successful therapy, while the IgG pattern remained oligoclonal. This patient with tuberculous meningitis showed that during the course of successful treatment the oligoclonal IgG response evolved over three months (see Figure 12.1 top), whereas that against IgM fell off (see Figure 12.1 bottom); indeed the responses moved in opposite directions as demonstrated by the plot over the interval of four months (see Figure 12.2). This is most

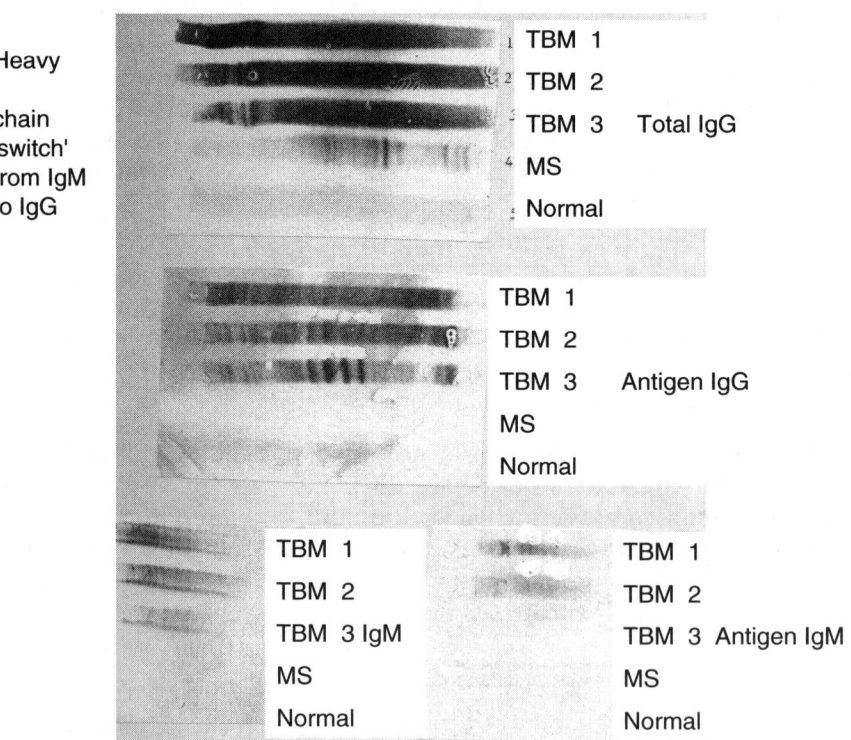

FIGURE 12.1 Time-course of tuberculous meningitis (TBM) showing total IgG and IgM plus antigen specific IgG and IgM.

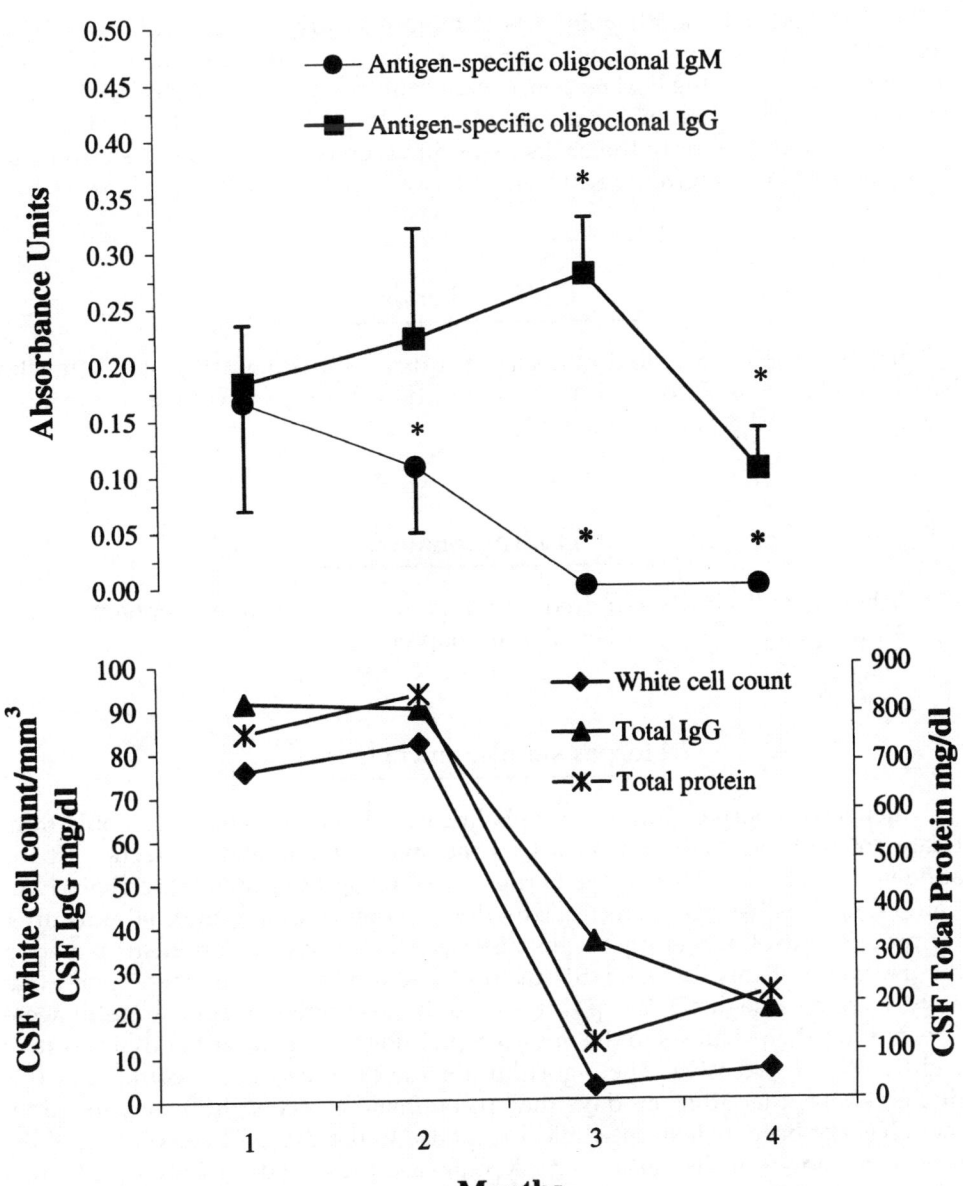

FIGURE 12.2 Densitometry scans of IgG and IgM along with cell counts plus total protein levels.

striking at the three-month point where there is clearly an adequate amount of total IgM, but the same volume of CSF shows no hint of specific anti-TB IgM. It is also noteworthy that the IgM response went from oligoclonal to polyclonal, which was the converse of the IgG response, and in both cases the total IgG or IgM was polyclonal, and thus only the antigen-specific (versus TB) response showed the 'Nossal switch' from IgM to IgG in the CSF.

CNS leukemia

One other use for repeated lumbar punctures is the monitoring of CSF myelin basic protein levels is to track CNS destruction during the course of cytotoxic therapy for intrathecal leukemia [162].

Hydrocephalus

Further work is clearly required with regard to chronic hydrocephalus, using e.g. D-2 antigen (N-CAM) or GFAP (see Chapter 11).

Herpes simplex encephalitis

Patients with herpes simplex encephalitis have benefited greatly not only from treatment with aciclovir, but also from advances in the early detection of the antigen, namely the DNA of the herpes virus, using the polymerase chain reaction (PCR). This has again emphasized the importance of longitudinal punctures for proper control of therapy in this disease. The European Consensus paper on the treatment of this disease [151] makes the important pathogenetic point that early in the disease the PCR is positive though the antibody response is still negative, but this then changes to the opposite situation where the antibody is positive and the PCR is negative. The algorithm for the diagnosis and treatment of this disease shows that after ten days they recommend a second lumbar puncture if there has not been a clear-cut clinical response to the drug. The evolution of the disease is shown in the somewhat atypical case presented by Coren *et al.* [168] where the longitudinal punctures showed clear evolution of the pattern over the course of one week. A similar case is shown in Figure 12.3 with representative scans shown in Figure 12.4, but this time over the course of a month. The IgG patterns in Figure 12.3 show very clearly the kind of 'molecular Darwinism', namely the evolution of the pattern with time (i.e. production of new clones) as well as

Serial Blots against Herpes Simplex; same amounts of Total IgG

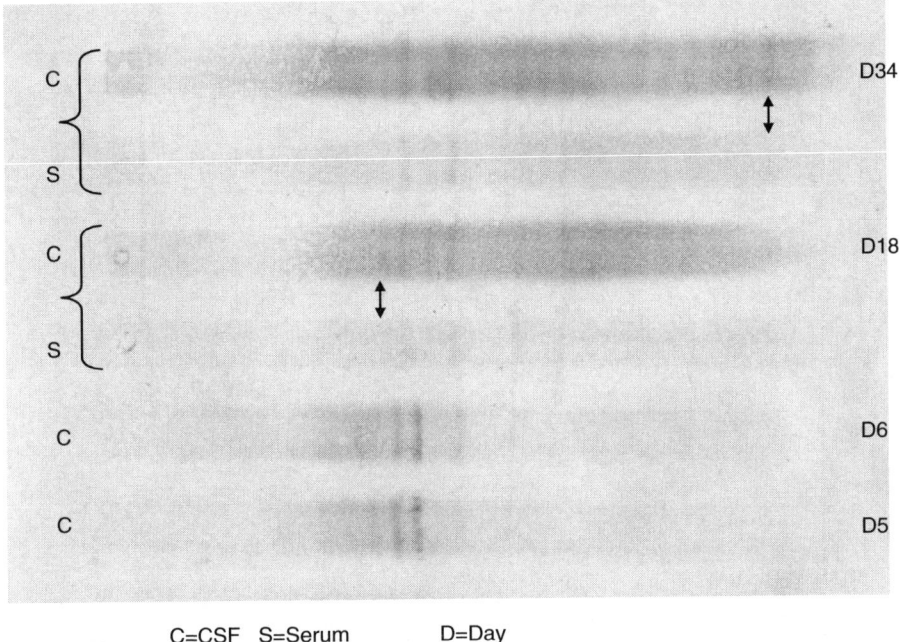

C=CSF S=Serum   D=Day

FIGURE 12.3 Antigen immuno-blots of herpes during the course of treatment. Note difference of CSF versus serum (double headed arrows).

the clear differential between the serum response (which was even seen on day three!) and the intrathecal pattern (see Figure 12.4, which presents three CSF scans from Figure 12.3). As mentioned previously, there is also a statistically significant difference in patterns from the patients whose samples (from the CSF bank, taken in the days before aciclovir therapy) were studied retrospectively. The current illustration reinforces the same point made about pathogenesis, namely that patients who survived had an early serum oligoclonal response against herpes, as opposed to those who did not, i.e. who had just a simple polyclonal response (see Chapter 10, Table 10.7) [53].

Antibody can be detected either by the ELISA technique (quantitative) or by isoelectric focusing (qualitative) against the specific Herpes antigen [771] with the important caveat that there is low-affinity binding in the case of multiple sclerosis as opposed to high-affinity binding in a chronic infection of the brain [686, 687],

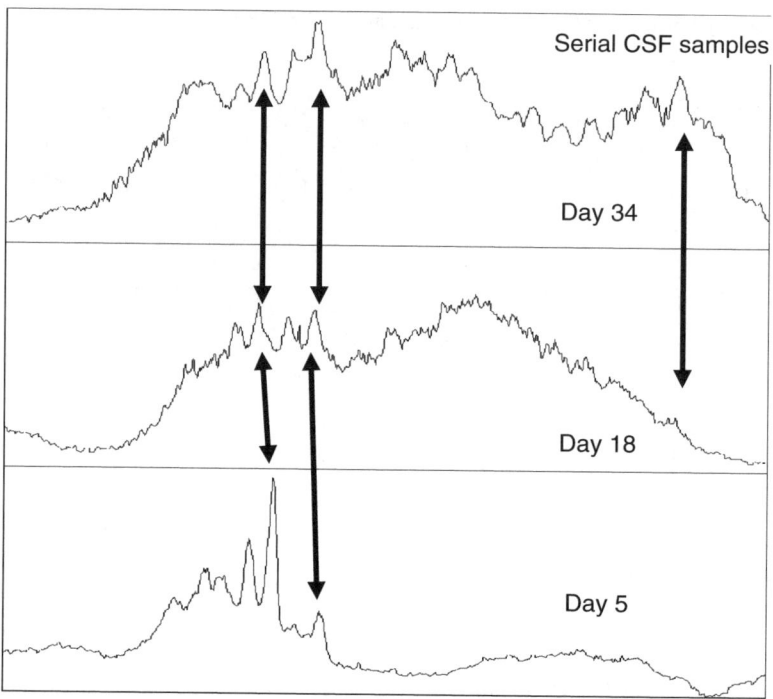

FIGURE 12.4 Densitometry scans of 3 CSF samples (from Figure 12.3) showing the evolution of the banding pattern.

see also Chapter 7, Figure 7.5. Once again, the comparison of quantitative (ELISA or AI) with qualitative (focusing or 'Eastern blots') revealed a statistically significant difference ($p < 0.005$) for the improved sensitivity (only six false negatives) and specificity (only two false positives) of qualitative over quantitative analysis (67 patients studied) [771].

CHAPTER

13

Prognostic protein levels

Introduction
Multiple sclerosis
Optic neuritis
Meningitis
Guillain–Barré syndrome
Head injury
Other diseases with possibly treatable and/or infectious origins

Introduction

Some CSF proteins may have prognostic value in individual diseases, as described here.

Multiple sclerosis

Changes in barrier function, specifically the abnormal presence of high molecular weight haptoglobin oligomers, have been found to be associated with the progressive forms of MS. The mechanism is not clear but the statistical association is strong ($p < 0.002$). Type 2-1 haptoglobin polymer in CSF is the only phenotype which can reliably be analyzed, and patients with the progressive from of MS have much higher incidence of haptoglobin polymers in their CSF [1262]. Whether this represents increased leakage and/or part of acute phase response, remains to be clarified.

Elevated amounts (relative to the internal standard of transferrin) of basically charged immunoglobulins clearly correlate with elevated levels of the Kurtzke disability score. This is not a simple reflection of the duration of the disease, since we subdivided the patients according to this criterion. In mechanistic terms, it may be that the total plaque 'volume' is reflected in the amount of immunoglobulin found in the CSF. Once again, although the precise role in pathogenesis is uncertain, the statistical associations are clear. Our earlier reporting of the association between a high Kurtzke score and elevated amounts of slow gamma on electrophoresis [1262] has also been found by the Schuller formula with an analogous measure of local synthesis of IgG [1003].

The occurrence of 'free' light chains, unbound (i.e. not linked with any heavy chains to form normal immunoglobulins) is typically seen early in the course of antigenic stimulation, and elevated numbers of 'free' light chains i.e. those outside the gamma region (and which are negative for immunofixation with heavy chains of IgG, IgA or IgM) is a phenomenon associated with a recent exacerbation in MS [1203]. Thus, there is probably increased antigenic stimulation at this time during the course of the disease. We have also noted that a decrease in free light chains is associated with longer duration of MS [1203]. Immunochemically, this may reflect the 'burning out' of the disease.

It has also been argued that oligoclonal bands are associated with 'malignancy' in MS [1092]. The problem was, however, that one group consisted of patients with both short duration and low disability (Kurtzke), whereas the other group had long duration and higher disability scores. Thus one could not distinguish which of the two features (duration or disability) or perhaps some combined effect was responsible. A slightly better index of malignancy was introduced by Poser [879], namely to divide the disability scores by the duration. In any case we examined all three of these variables, and found that only disability was correlated with the quantitative amounts of basically-charged immunoglobulins in our patient population [1262].

Patients with suspected MS and the presence of oligoclonal bands appear to be more likely to develop clinically-definite MS, but this is not a sufficiently strong association to allow prediction for individual patients, an obvious limitation which has been clearly stated [779].

Davies [198] followed up a total of 27 patients over an interval of 6 years, and of those who had only one band on isoelectric focusing, he found that 9 developed a full-blown oligoclonal pattern (3 developed MS versus 0/18 who were not oligoclonal positive) and therefore a single band should still remain as negative in terms of qualifying for an oligoclonal pattern, while the standard definition remains two or more bands. An example of the differential staining of kappa versus lambda bands is shown Figure 13.1B illustrating one lambda band in CSF which is absent from the corresponding serum, as well as absent from the stain for kappa light chains. A clearly different situation is shown in Figure 13.1A, where a single band is stained in CSF for both kappa and lambda light chains. This may represent an immune complex which is not found in the corresponding serum.

In a 3-year follow-up study, there was a correlation between CSF neurofilament levels and the Kurtzke EDSS score ($r = 0.54$, $p < 0.01$), an ambulation index measuring gait ($r = 0.42$, $p < 0.05$), and the 9-hole PEG test of upper limb function ($r = 0.59$, $p < 0.01$), [241]. In addition, there were higher levels of CSF neurofilaments in patients with primary and/or secondary progressive MS (0–0.12 pg/l) than in the relapsing/remitting form of the disease (0–0.29 pg/l), suggesting accumulation of axonal injury in the progressive phase of MS.

In summary, there is probably a consensus for a clear correlation between quantitative amounts of IgG as well as (possibly) neurofilament heavy chains and disability. Likewise, free light chains appear to be associated with recent exacerbation.

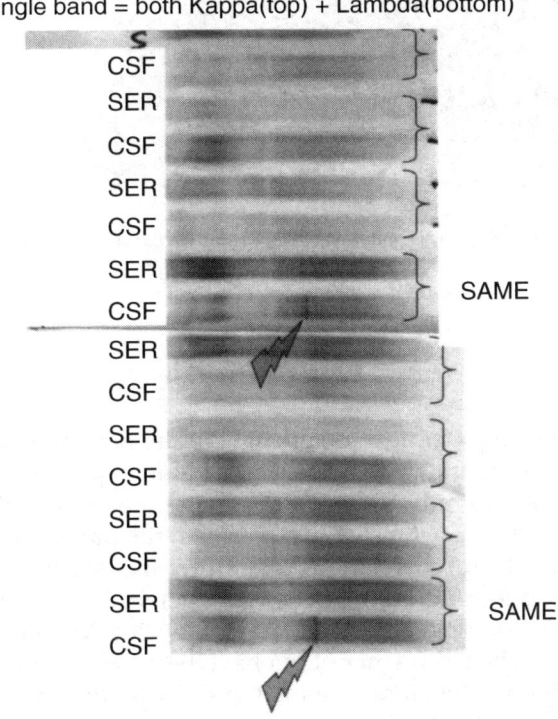

FIGURE 13.1A Kappa/Lambda staining to show the same band with both light chains.

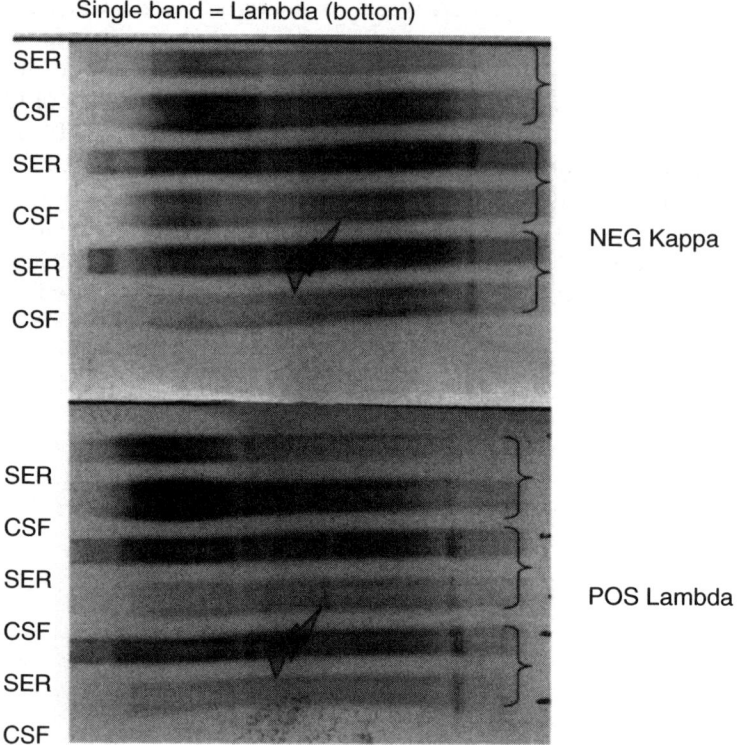

FIGURE 13.1B Kappa/Lambda staining to show a band with only lambda (bottom).

Optic neuritis

What is the meaning of oligoclonal bands in the patient with optic neuritis? It has been suggested that they have predictive value in patients with optic neuritis in that they can help to indicate those who will go on to develop MS. However, performing the Fischer exact test (Chi square) on the data of Sandberg–Wollheim shows no such significant correlation [977]. In another study [1093] the use of two groups with different durations of disease (3.6 versus 5.8 years) meant a probable selection against those who might eventually develop MS, in the group with the shorter follow-up. The results may thus have been quite tautological. However, quantitative analysis of IgG index seemed to show predictive value in at least one study of optic neuritis [986] if the patients were age-matched and if the disease was for a similar duration.

Meningitis

In patients with meningitis, the percentage transfer of albumin appears to have better predictive value for survival than that of alpha-2-macroglobulin percentage transfer [989]. Thus a high level of albumin transfer carries a worse prognosis than a proportionally high level of alpha-2-macroglobulin transfer.

Guillain–Barré syndrome

In patients with Guillain–Barré syndrome, it is known that the rise in total protein may be quite delayed, relative to the severity of symptoms. In this case the more dramatic changes in the percentage transfer of alpha-2-macroglobulin as opposed to albumin may be linked with clinical severity [989].

Head injury

In head injury, higher levels of serum myelin basic protein are associated with poor prognosis [844]. Likewise, in coma high levels of gamma enolase (14-3-2) are also associated with poor prognosis [980].

The point about prognosis was also made in Petzold's work [868] where elevated serum S-100 levels predicted an increase in intracranial pressure (ICP), almost a week before it actually occurred. This is therefore a significant predictor of patients who will require more attention as their hospital stay progresses. Specifically, there was a fourteen-fold increased risk of developing the morbidity associated with very high levels of intracranial pressure. Missler also found a similar predictive value of high levels of serum S-100 (but not NSE) for a bad prognosis after vascular insult and/or trauma [767].

Other diseases with possibly treatable and/or infectious origins

Those laboratories who still employ polyacrylamide gel electrophoresis should always be alert for possibly treatable or even completely curable diseases. With the help of computer 'prompts', one can enumerate various possibilities which should be considered when a particular chemical–pathologic protein pattern is encountered. The Table lists some of the diagnoses with particular emphasis on those which have a better prognosis. Based on some 21 possible patterns on polyacrylamide gel, one should consider at least the following differential diagnostic possibilities (see Table 13.1).

TABLE 13.1 Differential diagnosis for polyacrylamide gel electrophoresis

#	Pathological diagnosis	Peak ratios	Molecular basis	Anatomic basis	Pathophysiology
1.	G4/G5	G5/G2,G5/Tau	IgG, IgA, a few clones	'B' cells	Subacute stimulus
2.	Diffuse IEF'+'; K,L'+'	G5/G2, G5/Tau	IgG, IgA, many clones	'B' cells	Chronic stimulus
3.	Paraprotein (local)	'M'/Trf Urine, B-J/Trf	IgG/M/A, single clone	'B' cell clone	Lysis through dura
4.	Paraprotein (leakage)	'M'/Trf Urine, B-J/Trf	IgG/M/A, single clone	'B' cell clone, meninges	Non-CNS source
5.	Inflammatory	Oro/Trf	Orosomucoid	Meninges	'Hot' areas
6.	Tau increase	Tau/Trf	Lysosomal hydrolysis Trf	CNS cell death	Central disc pressure
7.	Cauda equina	Hp2/Trf	High amount haptoglobin polymers	Roots	'Hot' near LP site
8.	Obstruction (ventricle)	Pre/Trf	High amount prealbumin	Choroidal fluid	Stopped flow
9.	Frank transudate	GoTau	Group Comp protein > Tau	CNS parenchyma	Major leakage
10.	High molecular weight leak	Hp2/Trf	High amount haptoglobin polymers	CNS parenchyma	'Hot' spot
11.	CRP extra	Gl/Trf	C-Reactive protein dimer?	Macrophage liver	Bacteria not virus
12.	Diffuse IEF'−'; K,L'−'	G5/G2. High Hp2/Trf	Non-CNS many clones	'B' cells	Chronic stimulus
13.	Diffuse Hpl-1 frank	G5/G2. Yet N1 Hp2/Trf	Group Comp protein > Tau	CNS parenchyma	Major leakage
14.	Obstruction below	Pre/Trf	Low amount prealbumin	Froin's syndrome	Stopped flow
15.	Previous bleed		No free haptoglobin	Blood vessels	CSF circulation/8 hours
16.	Slightly traumatic		Hb–Hp complex	'No' RBC	Lysed RBC
17.	Traumatic decrease RBC				Normal
18.	Flat gamma				Normal
19.	Decrease gamma				Normal
20.	Normal ventricle				Normal
21.	Normal cistern				Normal

#	Assoc. path	Rule out	Suggested investigation	Consider therapy	Prognosis
1.	Circulating 'B' cells	SSPE	Measles titres	ACTH if free light chains	High G5 with high Kurtzke
2.	Pleocytosis	Abscess	CT with enhance/EEG	Antibiosis if bacterial	Good if syphilis
3.	Skull/vertebrae	Myeloma or CNS lymphoma	Marrow Bence-Jones	Intrathecal cytotoxicity	Good if local
4.	Marrow/Peyer	Benign, with old age	Marrow Bence-Jones	Systemic cytotoxicity	Good if benign
5.	Breast	Metastatic	CSF cytology	Intrathecal cytotoxicity	Good if local
6.	Alc. Cerebral degeneration	Remote cancer CJD	Myelography	Decompression	Good if early
7.	Vertical collapse	Compression	Area X-ray	Orthopedic support pin	Good if local
8.	Arachnoid block	Parasites	Radio-albumin scan	Shunt to drop pressure	Linked with shunt
9.	Barrier lysed	Organic disease	Multiple causes	Diagnose first	If MS progressive
10.	Not hysteric	Organic disease	Psychometry	Good if inflamed	If MS progressive
11.	Necrosis	Bacterial infection	Blood culture	Antibiosis if bacterial	Good if bacterial
12.	Normal WBC count	Remote abscess	Body scan	Antibiosis if bacterial	Good if bacterial
13.	Barrier lysed	Meningioma	Myelography	Decompression	Good if early
14.					
15.					
16.					
17.					
18.					
19.					
20.					
21.					

CHAPTER

14

Future growth areas

Introduction
Clinical
   Herpes encephalitis
   Guillain–Barré
   Multiple sclerosis
Biochemical
   Brain-specific proteins
   Prealbumin (transthyretin)
   Diffuse gamma 1 (acidic)
   Gamma trace
Technology
   Proteomics
   Chemiluminescence
   Point of care testing
Conclusion

Introduction

There are still several areas which require further work as noted briefly in earlier chapters. Some particularly relevant growth areas will probably increase our basic knowledge considerably as well as leading to improve patient care. These can be considered under three sections:

1. Clinical
2. Biochemical
3. Technological

Clinical

Herpes encephalitis

This is a potentially curable disease. Like meningitis, early diagnosis is crucial to allow early institution of therapy in order to obtain the best prognosis. Once coma sets in, the outlook is often poor [1068].

Although PCR is the best test for the first seven days of illness, we still require a rapid (less than one day to complete) and reliable antibody test, but the detection of antigenic fragments also in reliable and rapid fashion would obviate the need to wait 10 to 14 days for the antibody response to be mounted by the host [134].

Guillain–Barré

The claim that serum antibody binds to peripheral nerve tissue is encouraging [1240] (re: choice for plasmapheresis) though we need to know more about the nature of the antigen and whether this particular response is local and/or systemic.

Multiple sclerosis

Our own (unpublished) studies have shown that only about 50 per cent of patients from the Far East have oligoclonal bands in the CSF, versus 95 per cent in Europe and the United States. This may be related to their different haplotypes and thus may bear on the pathogenesis [327].

Biochemical

Brain-specific proteins

Although initially slightly discouraging results were found due to the lack of differential diagnostic value, nevertheless the work should be pursued, probably using more refined antigens as well as monoclonal antibodies. Specific questions of differential diagnosis and/or treatment should be addressed.

Although we have made significant advances in the study of S-100 protein, this can certainly be extended, as can work on changes in neurofilament proteins. We need to pursue the molecular basis for the different degrees of phosphorylation, not just for neurofilaments but also for the tau proteins.

Prealbumin (transthyretin)

What makes prealbumin such a good marker for ventricular fluid? Why are there different immunochemical forms? Why is it so avidly transferred into the CSF? Why is retinol binding protein not bound in CSF (unlike plasma)?

Most of the it questions have been answered by the finding of messenger RNA (mRNA) in the epithelial cells of the choroid plexus, which thus act as an intrathecal source of synthesis of this particular brain protein. It can be synthesized by astrocytes as well as macrophages.

Diffuse gamma 1 (acidic)

What is its nature? Why is so much present in CSF? Is it a brain protein? Is its presence in CSF the result of some local modification?

Gamma 1 is probably Apo-E lipoprotein. It is present in the CSF as the result of local synthesis, since the intrathecal portion is almost 20 per cent of the serum levels. I have also included the sessile macrophages (microglia) as there is probably some local modification which is achieved by binding to local phospholipids and cholesterol, since both are major constituents of the myelin sheath. Although gamma 1 could also be a lipid complexed with Apo-A rather than Apo-E, the latter is more likely. First, on the basis of higher molecular weights (range 150–600 kD, similar to IgG 155 kD), Apo-E is also more acidic since it has a higher level of N-acetyl neuraminic acid and on agarose gel electrophoresis it shows a much broader range of charge than does Apo-A, whose molecular weight is only 35 kD [1310].

Gamma trace

Gamma trace has been identified as cystatin C by use of the appropriate antibody, as the so-called highly alkaline fraction, which moves very close to the cathode on isoelectric focusing of CSF [939].

Technology

Proteomics

Proteomics is the new frontier. We now live in a postgenomic era but unfortunately many of medicine's problems remain to be solved in spite of the massive amount of data which has been produced by sequencing the entire genome. A new frontier has certainly been opened to the drug companies, who now see the sequencing of all human proteins as the future of 'big business' (www.Ciphergen.com).

Chemiluminescence

Regarding chemiluminescence we can say that there have been major advances in technology, including its ready availability at very affordable prices for almost all laboratories [48]. Traditionally it has been expensive to buy scanning densitometers sufficiently sensitive to measure output from chemiluminescence, as those command prices in excess of $40 000. Currently one can buy a video camera with a very low sensitivity to light (< 0.007 lux) for about $700. This is then connected to a video board (also costing about $700) which is inserted either into a Macintosh- or IBM-compatible PC. This allows one to integrate weak signals over time using memory installed on the PC card. This is analogous to either chemiluminescence exposed to photographic film, or to autoradiography exposed to X-ray film. In a matter of a few minutes one can capture an image which is then saved on the hard disk in JPEG format, and can subsequently be analyzed using public domain software freely downloadable from the NIH website (www.rsb.info.nih.gov/IJ). This includes an international standard for optical density which not only allows SI units to be compared around the world, but also, through the use of internal standards, one can in effect calibrate the system in terms of nanogram sensitivity for the protein(s) in question. Having previously mentioned the different degrees of phosphorylation for either the neurofilament proteins or the tau proteins, these can be separated on SDS gels and then transferred to nitrocellulose blotting paper which is then exposed to the appropriate antisera to determine the relative amounts of the different isoforms related to the different molecular weights. This technology also allows rapid bedside measurement of brain-specific proteins such as the previous example of S-100 for the prediction of elevated intracranial pressure and consequent mortality/morbidity and thus the attendant need for more aggressive therapy. In the laboratory this would also allow the rapid detection of a number of different proteins based upon their individual kits, which had previously been optimized using a grouped 'checkerboard' screen of the specific antigens and antibodies relevant to the particular protein(s).

The way forward is to have an output of 'quantitative exposure' i.e. chemifluorescence of the different subtypes of tau (e.g. phosphorylated). We can therefore use an internal control of increased A/B in disease 'X' versus increased B/A in disease 'Y'. The latter also applies to neurofilaments with differing numbers of phosphate groups.

Point of care testing

We can now study on-line sampling of brain interstitial fluid, since some neurosurgeons routinely insert a catheter containing a small dialysis membrane through a burr-hole. Thus a relatively low molecular weight protein such as the S-100 molecule can be monitored by continually washing through with an artificial CSF which equilibrates with the brain interstitial fluid. Dramatic alterations over several orders of magnitude have been seen over the course of a few hours. These are illustrated in Figure 14.1, where one sees that a patient with a clipped aneurysm had 2 successive bouts of high intracranial pressure (ICP), each followed by 2-log increase (and subsequent decrease to normal) of S-100 protein in his extra cellular fluid (ECF), before his eventual recovery. One of the technological spin-offs from the increase in international terrorism has been a 'field laboratory', which is essentially a portable 'black box' which can be set up in the most awkward circumstances, (including the bedside of a patient in, e.g. a surgical intensive care unit), and can perform immunoassays within 15 minutes (www.bioveris.com). This could prove exceedingly helpful since we have previously shown that increases in S-100 can predict a lethal rise in intracranial pressure 5 days later,

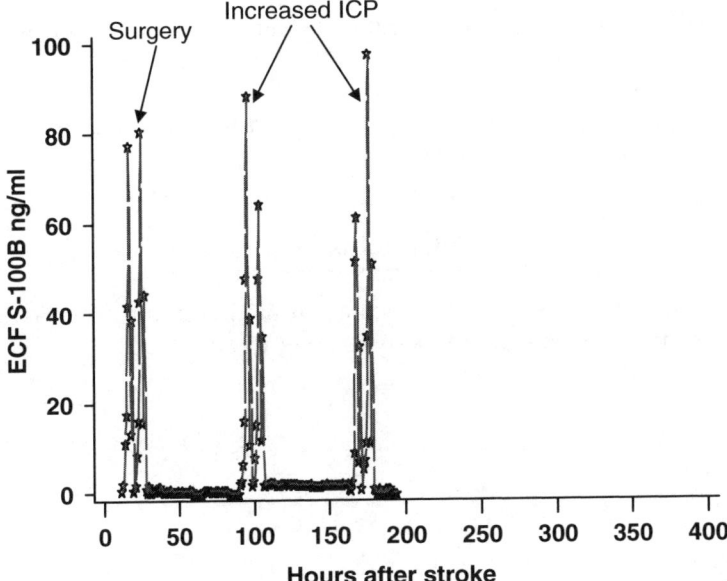

FIGURE 14.1 Interstitial fluid (ISF) drawn from indwelling dialysis catheter showing 2 orders of magnitude change over 1 hour in a patient who survived an aneurismal clipping.

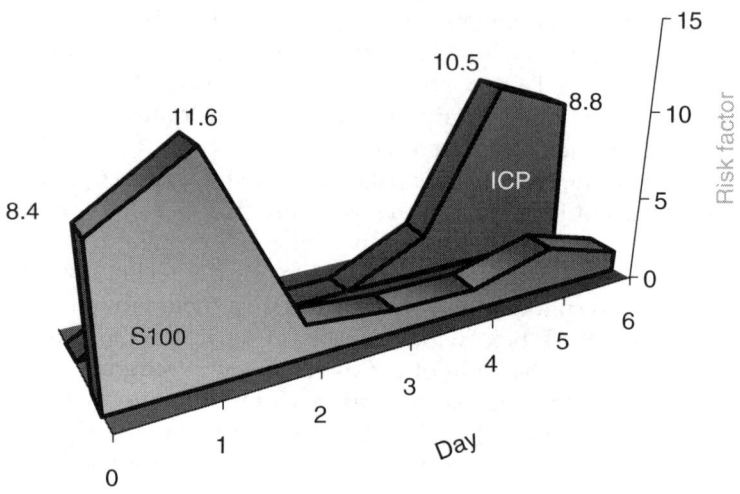

Initial high S-100b => Eleven-fold increased risk for high ICP 4–5 days later

FIGURE 14.2 Risk of mortality is predicted by almost 1 week using S-100 in serum.

and therefore one may be forewarned of the need for more dramatic therapeutic interventions (see Figure 14.2) [868].

Conclusion

Perhaps it is better to stop the speculation at this point. The fact should be clear: in the future, CSF is still not likely to be ignored.

PART 4

Appendices

CHAPTER

15

Appendices of methods

Methodology notes
Appendix 1: Separation
Appendix 2: Estimation
Appendix 3: Specific methods
    Polyacrylamide disc electrophoresis for CSF proteins
    Agarose electrophoresis
    Total protein
    Isoelectric focusing
    Analytical log template

Methodology notes

Most of this chapter is devoted to specific notes on methods, but towards the end there is a ready reference section on normal values for 'rapid' consultation.

It is worth making a few observations about the relative merits of different specific techniques regarding (1) separation and (2) estimation before the final (3) specific details.

Appendix 1: Separation

The addition of sodium dodecyl sulfate (SDS) to proteins will unfold all the tertiary structure (hydrogen bonding). In order for proteins with disulfide bridges to be completely unfolded, a reducing agent must also be added, typically beta mercaptoethanol or dithiothreoitol, in order to unfold the secondary structure [874]. This leaves only the primary structure i.e. differing chain lengths. These can be separated by chromatography [299] on the basis of their different chain lengths or on polyacrylamide gels [609]. The most widely used method of molecular weight determination is the modification of the Ornstein and Davis technique [200], with SDS added to maintain the chains in an unfolded state [609].

We have studied several hundred CSF specimens with this technique and noted that the changes are mainly associated with barrier damage, i.e. CSF becomes more 'serum-like' in its pattern. However, Ivanian and colleagues have found 'oligoclonal bands' in MS patients [146] using a slightly modified unorthodox technique without a reducing agent. They also found 'bands' in amyotrophic lateral sclerosis and glioma. It is therefore possible that the bands represent different degrees of disulfide cross-linkage. The technique of isoelectric focusing has been widely applied to CSF and some typical examples are given in Figure 15.1. Combining the two techniques, i.e. by molecular charge in the first dimension and then SDS separation by molecular size in the second, is shown in an earlier Chapter (Figure 11.4). Separation by size only (SDS) is shown in Figure 15.2.

Electrophoresis using polyacrylamide gel (molecular sieving), but in the absence of SDS, allows the native charge (which would be completely swamped by the sulfate of SDS) to play a role in the separation. Hence in a routine lab the use

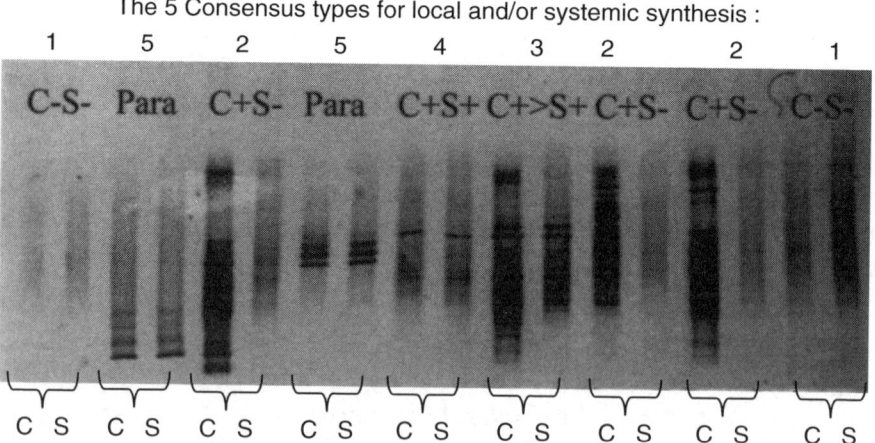

FIGURE 15.1 Isoelectric focusing of CSF followed by specific immunofixation against IgG, foot region (Fc).

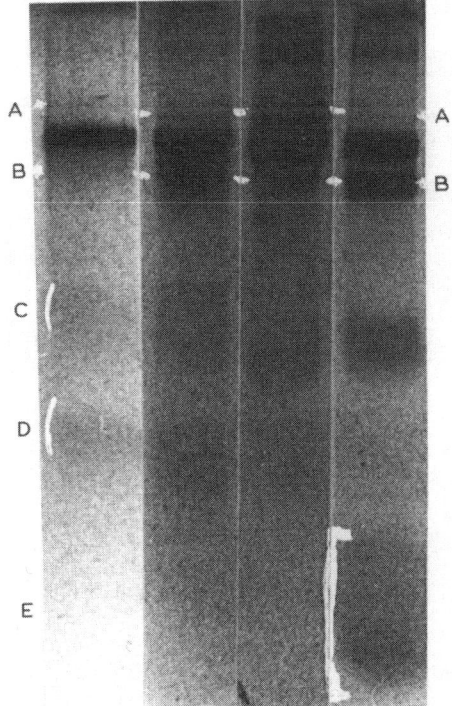

FIGURE 15.2 SDS gel electrophoresis of CSF (after reducing agent) to show elevated amounts of heavy chains (white lines, B) compared to amounts of transferrin (white lines, A). The upper arc (C) is light chains; the lower arc (D) is prealbumin. Samples are (left to right): normal, MS, brain tumor (serum leak), and paraprotein (note lower molecular weight fragments delimited by squared white lines, E).

of acrylamide without SDS offers more useful diagnostic information. Although in principle two-dimensional gels offer much more information, in practice they provide no major additional diagnostic information. However, further work is still required with this technique. It does involve considerably more technical effort, i.e. two gels and the variability is notoriously much greater than with one-dimensional acrylamide gels.

However, isoelectric focusing is very attractive since it 'spreads' the gamma region out to some 10 cm in length (typically 2 cm on acrylamide). In practice, however, the ampholytes are not always homogeneous and therefore artefactual bands which have been produced due to slight inhomogeneities or 'steps' in the gradient must be looked out for (Figure 15.3).

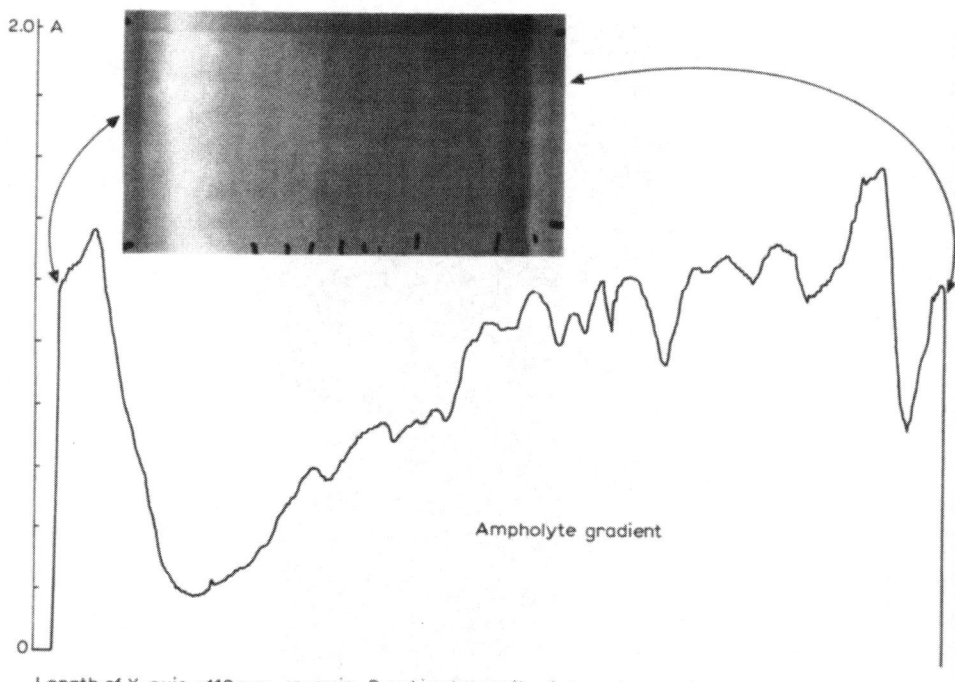

FIGURE 15.3 Multiple 'step functions' seen as white lines due to inhomogenous gradient of agarose and ampholytes (no protein present).

A single band might be 'split' by being forced to settle on two steps (see Figure 15.4). It is worth noting that the most homogeneous patterns are achieved by use of Pharmalytes [549].

A second problem is that the steps are zones of poor conductivity in some ampholytes (much less with Pharmalytes) and would therefore also be zones of increased electrical resistance and hence 'hot spots'. These may give rise to heat-denaturation of proteins during the course of exposure to the high voltages which are applied to effect the separations. Studies of kappa and lambda light chains have shown that both can occur in a given single band, thus arguing for overlap in spite of the presumptive separation power of the technique [1260].

There is a third problem with focusing, namely its variability from run to run, in spite of constant power output and volt-hour integrators. Even though it is claimed to be an 'equilibrium' technique, due to continual cathodic drift, it is in fact not as reproducible as standard acrylamide electrophoresis [141].

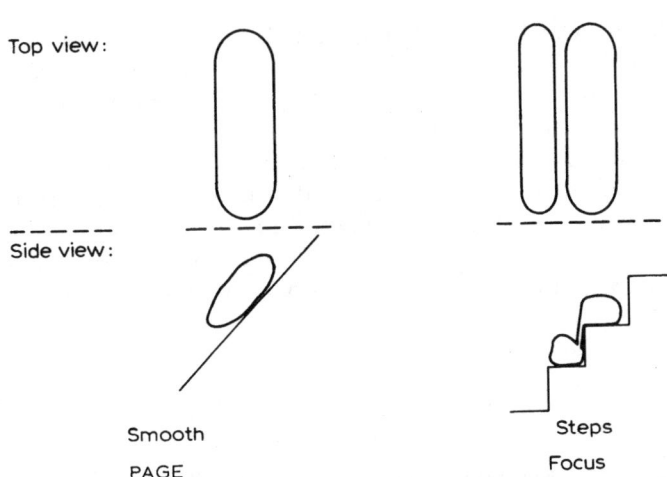

FIGURE 15.4 Probable effect of 'step function' (right) on the artificial dissection of a single band, compared with no artefactual steps (left).

Not only are the particular patterns of bands on IEF not pathognomonic for any given disease, but they easily vary from day to day in spite of constant volt-hours being applied. The heterogeneity of different commercial ampholytes has been clearly documented [16]. The other major issue is the number of bands which should be present in CSF but absent from the parallel serum from the same patient. If a large number of 'artefactual' bands are present as part of the 'background noise' then the number of bands may need to be four or five to be confident of local synthesis. Conversely, with a more homogenous mixture of ampholytes, two bands are sufficient.

In spite of this caveat about large numbers of 'common' bands due to inadequate ampholytes, [549] it does seem to be generally true that many locally synthesized bands seem to be more cathodic than the general pattern of serum IgG [625].

Paraproteins also tend to have a typical harmonic pattern (see Figure 15.1). The variability from plate to plate is such that strictly comparable results are only possible within the same plate, namely the same conditions throughout.

Overall, however, we have moved from separation on the basis of size (PAGE) to charge (IEF) since the latter offers higher discrimination and thus increased sensitivity/specificity outweigh all the problems discussed earlier. There is also International Consensus for this technique [316], plus the five types of pattern which should be used to classify the interpretation (see Figure 15.1).

There is one drawback to acrylamide gel electrophoresis which does not generally cause any limitation (except for a small percentage of myeloma cases) but nevertheless should be clearly recognized as an inherent problem. Because the gel buffer is fixed at pH 8.9, no protein which is more basic (pI 9.0 or greater) will enter the gel. However, even with focusing, care is required to prevent very cathodic proteins from running into the cathodic wick. The safest method in this regard is agarose gel electrophoresis [548]. The other uncommon, but nevertheless important limitation of acrylamide gel electrophoresis is the requirement for 100 μg of protein. Yet again agarose is important since it only requires 4 μl of CSF [548].

Appendix 2: Estimation

It was fashionable for some time to expect poor results from densitometry for electrophoretic separations of proteins. The preferred method was felt to be immunochemical. As we have seen, however, this is by no means ideal for CSF proteins. Indeed, setting out with a more positive inclination to densitometry, it is clearly feasible to obtain day-to-day coefficients of variation (CV) with typical values of 5 per cent, which are quite comparable to those from immunochemical techniques [502]. The most reliable method is the use of a specific protein as an internal control. We typically use transferrin (rather than albumin) since its level is of similar order to many other CSF proteins (see Table 15.1), while albumin is an order of magnitude higher and thus less directly comparable.

For specific identification, it is still best advised to use immunochemical techniques, and of these, the most recent nitrocellulose blotting seems to be the most suitable in several regards. Once again an internal standard (e.g. a myeloma paraprotein) can be used and so can obtain a run-to-run CV of < 10 per cent [845].

TABLE 15.1 Limits for normal ratios on PAGE

Protein	Cut-Off
Prealbumin	> 1.1 (< 0.3)
Orosomucoid	< 0.2
Alpha-1-antitrypsin	< 0.7
Group components	< 0.5
Tau	> 0.7 (< 0.3)
Pre gamma	< 0.4
Gamma 5	< 0.6
Gamma 5/Gamma 2	< 1.1
Group components/tau	< 1.0

Returning to acrylamide gel electrophoresis, this has a clear advantage since it does not have the artefactual 'steps' (Figure 15.3) seen with focusing. A second advantage is the separation based upon size plus charge. While focusing gives a wide (though artefact-prone) separation of the gamma region, it does less well for the acidic proteins. This is simply due to the fact that most proteins carry a negative charge while the gammaglobulins are unusual in that they carry a positive charge, as well as having the values for the positive charges spread over three orders of magnitude, thereby reflecting the broad spectrum of individual pI values.

For specificity of identification, acrylamide gels allow replicate blots (see Figure 15.5 top and bottom) hence kappa versus lambda patterns. Once again, unlike focusing, typically, overlap of the two light chains for an individual clone is not seen. Given the element of molecular sieving by differing pore sizes on acrylamide, it is clearly seen that bands outside the gamma region represent 'free' or unbound light chains. Since multiple blots can be prepared, it is possible to compare the replicate blots for IgG heavy chains and IgA heavy chains as well as complement C'3 conversion and C-reactive protein, for example.

In conclusion, we used to recommend polyacrylamide gel electrophoresis without SDS followed by immunofixation for kappa, lambda light chains, IgG and IgA. This was by far the most reliable routine method and all the steps were easily accomplished by a single technician in one working day. If this was not possible, our second preference was agarose electrophoresis followed by immunofixation for IgG. However, we now recommend routine focusing, in spite of the variable patterns causing some difficulties in interpretation. If restricted to measuring IgG and albumin in serum and CSF, this is by far the least desirable method, but as a simple minimum the result should be plotted in log–log fashion for the percentage transfer of IgG versus the percentage transfer of albumin. Notorious differences in the quality of commercial antibodies require that you examine your own 'normal' samples before deciding on a 'cut-off' (see Nomogram at end of this chapter).

Appendix 3: Specific methods

Polyacrylamide disc electrophoresis for CSF proteins

The method of PAGE used here is essentially that of Ornstein and Davis [200] but with the following modifications:

1. No stacking gels. This simply traps basic immunoglobulins (as noted by Felgenhauer [269]) and is quite unnecessary to achieve sharp bands.
2. A repipette to deliver the unpolymerized acrylamide to the glass rods quickly and in the exact volume (if not available, use a Pasteur pipette).

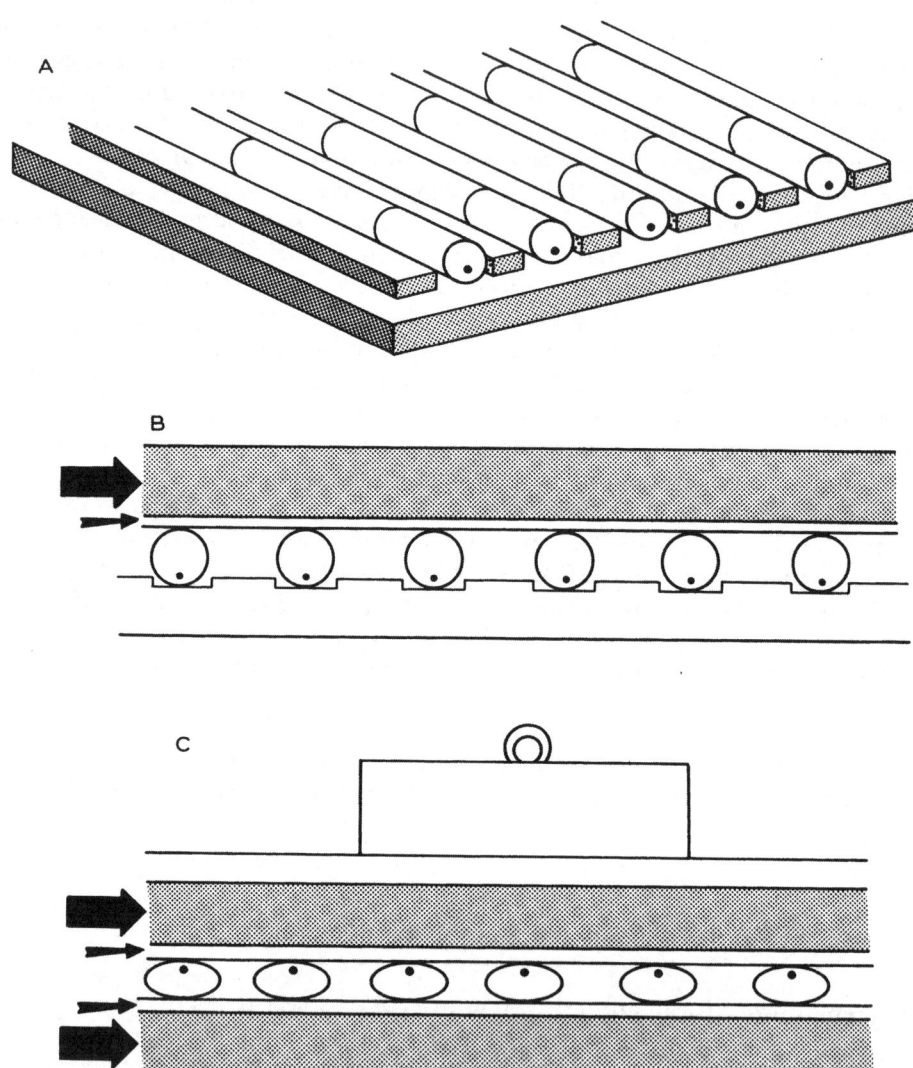

FIGURE 15.5 Gel rod immunoblotting, (A) gel holder, (B) first nitrocellulose paper (thin arrow), (C) second nitrocellulose paper, following inversion.

3. Overlaid with water-saturated isobutanol during polymerization then replaced with gel buffer.
4. Gels can be made in advance and stored at 4°C for a few weeks.
5. Naphthalene black stained gels are NOT required if immunofixation is performed.

Reagents required

1. Bromophenol blue for use as a marker dye (Kodak).
2. Tris (Trizma Base, Sigma).
3. Acrylamide. This can be obtained from a number of sources (Kodak, etc.), but we use Serva as supplied by Uniscience. This product has been recrystallized twice.
4. Bis-acrylamide. Again, this is like acrylamide, and we use Serva which has been recrystallized twice.
5. Glycine, ammonia-free (Kodak).
6. TEMED (Serva or Kodak); tetramethylenediamine.
7. Ammonium persulphate (A.R. quality).
8. Hydrochloric acid (A.R. quality).
9. Sodium chloride (A.R. quality).
10. Coomassie brilliant blue G-250 (Sigma).
11. Naphthalene black (Gurr–Searle).
12. Acetic acid (A.R. quality).
13. Perchloric acid 72 per cent (A.R. quality).
14. Photoflo 600 (Kodak).
15. Water-saturated butanol.

Apparatus

1. Electrophoresis power pack.
2. Glass tubing, Pyrex. The length of the tube depends on the final length of gel required and the amount of material to be electrophoresed. A tube of approximately 5 mm internal diameter and 7 mm external diameter, 20 cm length, is used with the ends ground on emery cloth, since flame polishing (i.e. rounded in the flame) may narrow the orifice.
3. Tanks. These are made out of round polystyrene food containers of about 10 cm diameter. Twelve holes are drilled in the base of one container, equidistant from the center, and rubber electrical grommets (7 mm internal diameter) fitted so that the glass tubing can be inserted easily but with a good seal. An electrode is fitted in the lid of the tank; this becomes the negative electrode. A similar electrode is fitted into the base of a 1-liter plastic beaker of about 10 cm diameter; this then becomes the positive electrode. These electrodes are best made from platinum wire.

4. Rubber base for standing tubes during polymerization of gels: this consists of two pieces of sheet rubber 6 mm thick. In one piece, a number of holes are cut, to hold the glass tubes which will contain the gel. The holes must be cut exactly and then stuck together to make a base for the tubes.
5. Pasteur pipettes. These can be made from any glass tubing, but they must be wide, long enough to touch the bottom of the tube, and strong enough so as to be filled and emptied of acrylamide quickly. One Pasteur pipette, used for overlaying butanol, must not be too broad at the base, but again not too narrow as the overlaying must be done with minimum mixing. The best internal diameter is approximately 2 mm.
6. A 10 ml or 20 ml syringe and a long fine needle (7 cm, size 22 is suitable).

Solutions

1. 1 N HCl   48 ml
   Tris   36.6 g
   TEMED   0.23 ml
   Water to pH 8.9   100 ml
2. Acrylamide   28.0 g
   Bis-acrylamide   0.735 g
   Water to   100 mL
3. 0.01 per cent bromophenol blue in water
4. 0.14 g ammonium persulphate
   Water to   100 ml
   (This solution will keep a few days under refrigeration)
5. Stock buffer
   Tris   6.0 g
   Glycine   8.8 g
   Water to pH 8.3   1.0 l
   For tank buffer (5A), dilute 1 to 9 of water is diluted plus 3 drops of solution (3)
6. Naphthalene black 1 g
7. 7 per cent acetic acid 100 ml 7 per cent acetic acid
8. Photo-flo   1 ml
9. Water   600 ml

All solutions can be kept at 4°C for a month

Preparation of gels

Glass tubes are washed well in soap and water to remove all greasy substances, and rinsed well in distilled water. The tubes are then soaked in 10 per cent (w/v) nitric acid overnight followed by distilled water wash. Any loose pieces of gel are removed with a cotton wool-tipped stick prior to the nitric acid wash. Chromic acid is less successful. The tubes are not siliconized since the flow of water down the side walls is too sporadic. After washing, the tubes are coated with Photoflo solution (8) by standing them in a measuring cylinder with sufficient solution to cover the tubes. The tubes are then dried in a hot-air oven (60°C).

Before making the gels, the tubes are marked with the length of gel required. Gels, 12 cm long, are satisfactory; tubes are placed on a rubber rack or any other type of holder; they are then ready to be filled.

Each tube will require about 2.5 ml of acrylamide solution. This is prepared as follows:

To make sufficient gel solution for 10 tubes, 30 ml of gel solution is required. In order to prepare this, all solutions should be at room temperature (22°C), 4 ml of buffer solution (1) is placed in a small Erlenmeyer 100 ml flask, 8 ml of acrylamide solution (2) and 4 ml of distilled water are added. The flask is then gently swirled to ensure even mixing. This is solution (2A). Into another flask, about 20 ml of the ammonium persulphate solution (4) is poured. Into a third flask, 15 ml of the buffer acrylamide–water solution (2A) is pipetted and then 15 ml of the ammonium persulphate solution (4) is added to this, with a gentle flow rate down the side of the flask to avoid any absorption of air. Thorough mixing is then achieved by gently swirling the flask. Using a large Pasteur pipette each tube is then filled to the mark with this solution. When all tubes have been filled, they are gently tapped to allow any trapped air bubbles to rise to the surface, and then overlaid in order with water-saturated isobutanol. This is done slowly and gently by using a Pasteur pipette with finger control; a 4 to 5 mm layer of butanol is all that is necessary to exclude air and to give a flat surface to the gel. No mixing must occur as this will result in a weaker top for the gel.

The tubes are then left for the acrylamide to polymerize. It is to be noted that about 10 to 15 minutes after over-layering with isobutanol, the demarcation line between the two solutions disappears, but after a few minutes, reappears a few millimeters lower. After the acrylamide polymerizes, the isobutanol should be removed and replaced with gel buffer. The gels are ready for use the following day. Before placing the tubes containing the gel into the appropriate tanks, it is necessary to rinse the top of the gel with the diluted tank buffer (5A) to remove the watery solution left behind after polymerization. The tubes can be labeled for sample with any marker pen. Generally, gels are prepared the day before use.

The protein solution or CSF (a volume containing 100 pg of total protein for Coomassie, 200 pg for naphthalene black) for electrophoresis is then pipetted on top of the gel in the tubes, and then carefully overlaid with tank buffer (5A). No sucrose is added to the sample (see later). 450 ml and 1 l of buffer solution are then added to the top and bottom tanks respectively. The upper tank contains the gel tubes and rests on a retort ring. The gels are immersed in the lower tank buffer, so that the bottom 12 cm of the tubes dip into the bottom tank. The top tank already has two or three drops of bromophenol blue solution (3) added and has been gently stirred.

Bubbles from the bottom of the tubes are removed with a Pasteur pipette having a bent 'U' tip. The lid containing the negative electrode is then placed on the upper tank – wires for both electrodes connected to the power pack, and the gels are ready to run. A maximum of 2.5 mA per gel is all that is required to obtain good separation. The power is switched off when the marker dye (bromophenol blue) is seen to be about 5 mm from the bottom of the tube (time approximately 4 to 5 hours). The gels are now ready for staining. To remove the gels from the tubes, the 20 ml (or 10 ml) syringe are filled with distilled water, the needle is fixed on the syringe, then the needle is gently inserted into the bottom end of the tube, injecting water at the same time. This must be between the gel and tube wall. By gently pushing the needle further into the tube, and rotating the tube so that the needle and water separate the gel from the tube all round, it will be seen that the gel suddenly moves; on removing the needle, continuing to inject water, the gel should then gently slide out and is ready for staining either in Coomassie or naphthalene black solution.

Staining with Coomassie blue solution: The gel is then transferred to the staining solution which is made by dissolving 100 mg of Coomassie brilliant blue G 250 (Not: R 250) in about 200 ml of water. To this solution 10 ml of perchloric acid (72%) is slowly added (color turns brown) with vigorous mechanical stirring. The volume of the solution is then brought to 250 ml with water. Proper preparation is essential for achieving small micelles of stain in the solution. This solution is stirred for 1 hour, and then filtered thorough small pore filter paper Watman No. 42. Before use, the filtered solution is warmed to 60°C. The gel is then incubated in this solution for 2 hours. During this process, the protein bands became visible. However, the background will have a pale brown color. It is then placed into a tube of 7 per cent acetic acid (7) for final destaining (overnight). The gel can be stored in a small volume of the aforementioned solution (7) in test tubes or plastic bags sealed with tape or heat, and will keep for years.

Staining with naphthalene black takes 1 hour only (6); then destain (7). If it is not convenient to stain over only 1 hour in strong stain, the gel is placed in a 1 in 10 dilution (6A) of the stain with (6) and left overnight. Destaining for both dyes may be carried out electrophoretically using 7 per cent acetic acid (7), or by placing the gel in a 150 cm × 2.5 cm tube and continually changing the solvent (7) until the gel is properly destained (3 to 4 days). To destain electrophoretically, a slightly wider tube (6 mm i.d.) is required for the gel, with one end of the tube partially closed (fire polished) so that the gel will not pass through. When destained (current 5 mA/gel) the gels are stored in narrow tubes using 7 per cent acetic acid (7) for naphthalene black stained gels.

Preparation of sample

As the protein sample is no longer incorporated in a large pore gel, it has been suggested that the addition of solid sucrose to make a 25 per cent solution in the protein mixture is necessary in order to be able to overlay the buffer on the sample. Some workers advocate glycerol. However, with care, it is better to use neat CSF without the addition of sucrose or glycerol. Needless to say, a rather distorted protein pattern is obtained, using sucrose and the pattern can be altered according to the method of mixing, whether vigorous or not [517].

It is therefore essential that the same method be used throughout in order to produce comparable results. Now, neither sucrose nor sample agitation is used. The gels stained with naphthalene black were scanned using a Joyce Lobel chromoscan, as were the Coomassie gels, each using a combination of two Wratten filters Nos 9 and 15. Both gels are also examined on a viewing screen, again using two Kodak Wratten filters, Nos 9 and 15, with the gels placed in the test tubes (12 mm i.d.) on top of the filters. It has been found that viewing in this way allowed clearer visualization of the protein bands, especially in the IgG region on the Coomassie-stained gel.

PAGE immunoblotting

The first blots (Figure 15.5), top and bottom (3-second application) are discarded, to remove surface detritus. The second blots (7 minutes) are stained for kappa and lambda. The third blots (20 minutes) are stained for IgA and complement C'3. The fourth blots (10 minutes) are stained for IgG (Fc) and CRP.

Upon completion of the acrylamide run (about 3 hours), the gels (Figure 15.5A) are temporarily placed in rows in a plastic mold and one sheet of nitrocellulose paper is applied to the top (Figure 15.5B) followed by five sheets of filter paper. The mold with gels is then carefully inverted, the mold removed, and a second sheet of nitrocellulose paper applied (Figure 15.5C), again followed by five sheets of filter paper. After placing a glass plate on top, a 5 kg weight is applied for 15 minutes (Figure 15.5C). Each piece of nitrocellulose paper is then removed and placed in a solution of 1 g each of bovine serum albumin and gelatin per liter of phosphate buffered saline (PBS-A) for 5 minutes to block unoccupied binding sites. If required, a second and third press could be performed on the same gels to give a total of six immunoblots per run. The acrylamide gel is stained with Coomassie blue as described earlier. After blocking, the nitrocellulose paper is incubated in 30 /A of mouse anti-human kappa or lambda polyclonal antibody (Atlantic, USA) in 40 ml of PBS-A added to each tray, followed by rocking at 22°C for 1.5 hours. The nitrocellulose papers are then rinsed with four changes of PBS-A over 10 minutes. This is followed by incubation in 25 pg of horseradish peroxidase conjugated anti-mouse immunoglobulin (Miles, U.K., was the most satisfactory) in 40 ml of PBS-A, and this is rocked for another 1.5 hours at 22°C. A final four washes with PBS-A are followed by a wash with 0.02 mol/l sodium acetate buffer pH 5.1. The color is then developed (about 10 minutes) using ethylaminocarbazole, per details mentioned earlier.

Agarose electrophoresis

The following points are important:

1. The degree of endosmosis must be such that the origin does not overlap the gamma region.
2. Cooling must be sufficient to keep the temperature < 5°C (e.g. a copper platten for maximal heat transfer).

We routinely only perform electrophoresis on the serum (24 samples per day).

A 1.5 per cent w/v agarose gel comprising 1 per cent w/v high electroendosmotic agarose (Miles Laboratories, Slough) plus 0.5 per cent w/v medium electroendosmotic agarose (Miles) dissolved in 0.06 M barbitone buffer according to Jeppsson et al. [489], is poured between two glass plates separated by a U-frame 210 × 105 × 1 mm cut from silicon rubber (Esco Rubber Co., U.K.), and left to cure for at least 1 hour at 4°C.

After equilibrating at room temperature the assembly is dismantled and excess surface fluid removed from the gel using a sheet of hardened filter paper (Schleicher and Schuell type 1575, available through Anderman and Co., East Molesey, U.K.), cut to the dimensions of the gel. This is best effected by carefully placing the short edge of the filter paper on the extreme edge of the gel and positioning the remainder of the paper on the gel surface by smoothly sweeping or rolling, ensuring even wetting as the absorbent paper contacts the surface. The gel surface is then smoothed by repeating the exercise using a sheet of dry nitrocellulose (Schleicher and Schuell Type BA85, available through Anderman and Co.) which is cut to the gel dimensions. The nitrocellulose sheet is carefully removed from the gel surface and placed in distilled water for future use.

A sample application mask (LKB, Croydon, U.K.) is positioned so that the sample slots are 1 to 1.5 cm from one long edge of the gel, being careful to exclude air bubbles. A volume of CSF containing 0.02 to 0.05 µg of IgG is loaded. This is the equivalent of 1 µl of CSF having a normal total protein concentration. Serum samples were diluted 1 in 200 in PBS and 1 µl used. Immediately after the samples enter the gel the sample application foil is removed and the gel positioned on the electrophoresis tank. Electrophoresis is carried out at 200 V (approximately 30 mA)/gel for 2 hours. Cooling water at 5°C is circulated. Migration of the proteins of interest is towards the negative electrode.

On completion of the run the gel surface is blotted with damp cellulose acetate (Sartorius, Sutton, U.K.) to remove buffer and non-specific interfering material (the precise nature of which is unknown). The cellulose acetate may be reused after washing in water. The sheet of nitrocellulose used previously is then blotted to remove excess water and placed on the gel surface, followed by a damp sheet of hardened filter paper, and six sheets of dry blotter (e.g. 3 MM, Whatman, Maidstone, U.K., or 'Postlip' paper, Postlip Mills, Gloucester, U.K.), a glass plate to cover the whole plate, and a 1 kg weight. After 15 minutes the nitrocellulose is removed from the gel surface and immersed in a solution of 1 per cent w/v bovine serum albumin (Sigma, London, U.K.), in phosphate-buffered saline (PBS, 0.15 M, pH 7.2). After 10 minutes the albumin solution is replaced by a mixture of 45 ml PBS containing 5 ml of the 1 per cent albumin solution and 25 µl sheep anti-human IgG Fc (Seward Immunodiagnostics, London, U.K.), and left rocking for 45 minutes at room temperature. The membrane is then washed in five changes of PBS over 10 minutes to remove unreacted antiserum and incubated for 45 minutes in a mixture of 45 ml PBS plus 5 ml 1 per cent albumin containing 25 µl horseradish peroxidase conjugated rabbit anti-sheep immunoglobulins (Dako, London, U.K.). Finally the membrane is again washed in five changes of PBS over 10 minutes and then rinsed in 50 ml 0.02 M acetate buffer (pH 5.1).

Substrate mix is prepared by dissolving 10 mg 3-amino-9-ethylcarbazole (Sigma) in 6 ml dimethylsulfoxide (BDH, Poole, U.K.) and adding 50 ml 0.02 M acetate buffer containing 50 µl 100 volume per cent hydrogen peroxide (BDH). The membrane is immersed in this and incubated at room temperature with gentle mixing until the pattern develops fully (10 to 20 minutes). The membrane is then washed in two changes of tap water, blotted and dried under a stream of warm air.

Total protein

To 2 ml of 30 per cent (v/v) polyethylene glycol is added 0.2 ml CSF while continuous and vigorous mixing is performed. This is heated to 100°C for 2 minutes, cooled and read in a nephelometer [538].

Normal values for protein patterns are given in Table 15.1.

Computer program

This program recognizes each individual peak and valley, then assigns names based on A-values (Figure 15.6 and Table 15.2). These are printed and then finally vetted to allow peak height ratios to be calculated by the program (see Figure 15.7).

Normal/abnormal values for common proteins are given in Table 15.3. A Nomogram for expression of quantitative IgG/Albumin results in given in Figure 15.8.

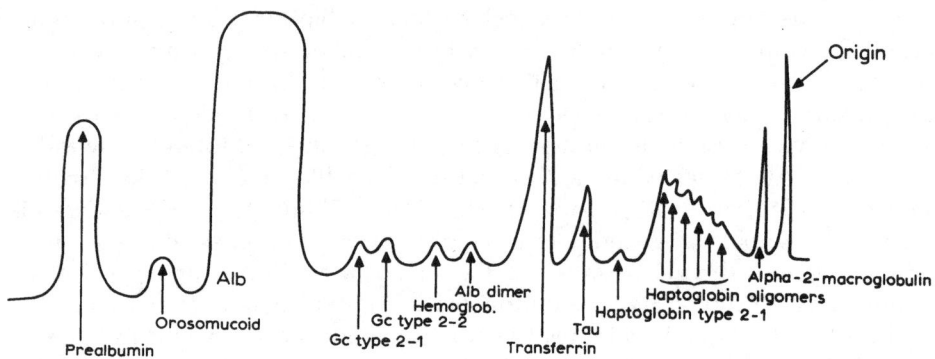

FIGURE 15.6 Identification of the major CSF proteins on polyacrylamide electrophoresis.

TABLE 15.2 R_f values assigned by computer program

Protein	R_f	Full name
Pre	1.40	Prealbumin
Oro	1.14	Orosomucoid
Gc1	0.82	Group components, type 1 allele
Gc2	0.76	Group components, type 2 allele
HbF	0.70	Hemoglobin free (unbound)
Alb2	0.64	Albumin dimer
Alb3	0.60	Albumin trimer
Trf	0.53	Transferrin
Tau	0.43	Tau (transferrin minus sialic acid)
Hp2	0.32	Haptoglobin, type 2 allele (monomer)
Gm12	0.21	Gamma region (first 2, anodic fifths)
Gm35	0.12	Gamma region (last 3, cathodic fifths)
A2M	0.07	Alpha-2-macroglobulin
BLP	0.02	Beta lipoprotein

Isoelectric focusing

Introduction to IEF

The United States Food & Drug Administration (FDA) has now given its approval for the use of the Helena kit based upon the method of isoelectric focusing followed by immunofixation against IgG. This has recently been approved by an International Consensus of neurologists and clinical pathologists [316] as the gold standard for the diagnosis of multiple sclerosis, being 98 per cent sensitive

APPENDIX 3: SPECIFIC METHODS

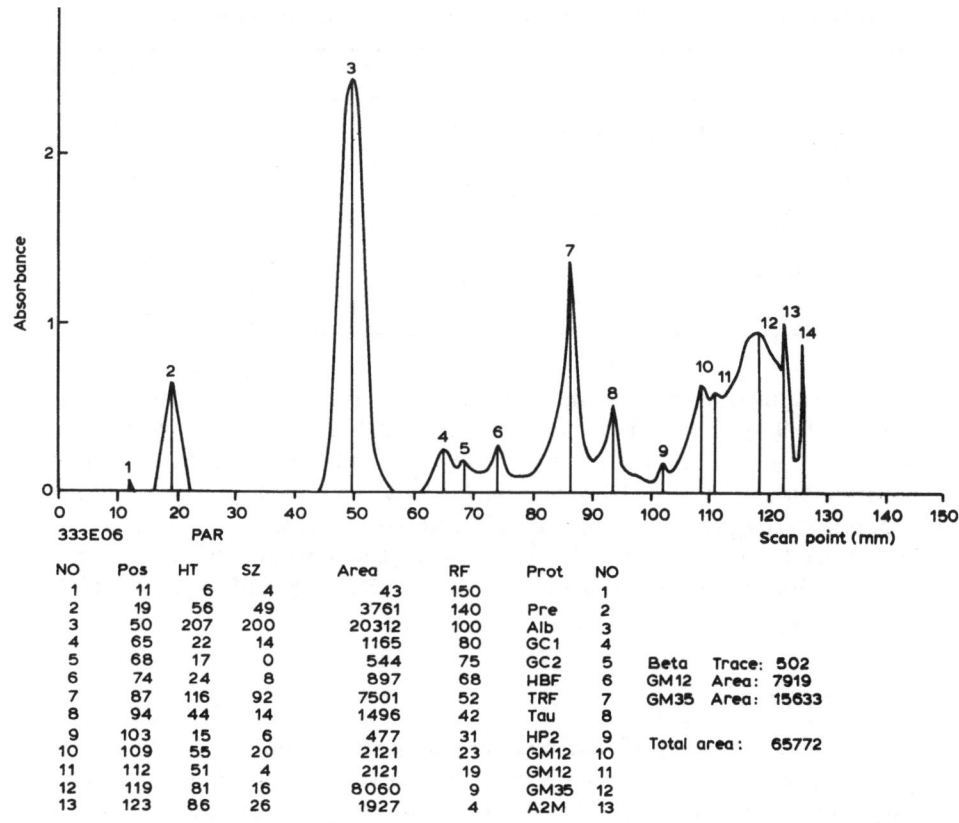

FIGURE 15.7 Screen dump from high resolution graphics display for densitometer scan of Coomassie-stained acrylamide gel from a patient with MS. Data is given regarding position (Pos), peak height from baseline (HT), peak height from previous valley (SZ) and protein names based upon Rf values.

and 87 per cent specific, which is higher than MRI, not least because it is not age-dependent and it measures IgG rather than H_2O.

Analytical principle

Proteins in CSF or diluted serum are separated according to their isoelectric points within a pH gradient that is generated by electrophoresis and stabilized using zwitterionic ampholytes ('Pharmalytes'). Once separated the proteins are passively blotted on to nitrocellulose membrane and unoccupied protein binding sites blocked with non-fat milk. IgG is identified using a double antibody

TABLE 15.3 Normal and abnormal levels for commonly determined CSF proteins

	Normal	Abnormal
Total protein	35	> 55 mg/dl
Alb % transfer	0.40	> 0.70
IgG % transfer	0.20	> 0.35
IgA % transfer	0.20	> 0.40
A2M % transfer	0.07	> 0.13
IgG % total	12	> 18
IgG index	0.50	> 0.70
A2M index	0.23	> 0.33

immunodetection system involving the sequential addition of goat antihuman IgG Fc antibody, followed by peroxidase-conjugated rabbit antigoat immunoglobulin. Bound peroxidase activity is then detected by a chromogenic reaction using ethylaminocarbazole and hydrogen peroxide. This gives a pink-brown precipitate that corresponds to the distribution of the IgG in the original sample.

Oligoclonal bands are seen as bands superimposed upon a diffuse background of polyclonal IgG. Oligoclonal bands may be seen in either CSF or in both CSF and serum, but never in serum alone.

Purpose of test

Isoelectric focusing is used to detect oligoclonal bands of IgG in serum and CSF. Of specific interest are oligoclonal bands that are present in CSF only, with no corresponding bands observed in paired serum. In this instance the oligoclonal bands are local to the central nervous system (CNS), and so are termed locally (or intrathecally) synthesized. Oligoclonal IgG is a sensitive and specific marker for intrathecal synthesis and is present in a large proportion of patients with demyelinating disease, such as multiple sclerosis (MS). Oligoclonal bands may also be found in other neurological diseases, including autoimmune disorders, paraneoplastic disorders and infections of the CNS.

In addition to being synthesized locally within the CNS in neurological conditions, oligoclonal IgG may also be found in the blood in a variety of systemic pathologies. Under these circumstances the oligoclonal band pattern of the blood is reflected in the CSF (a 'mirror' pattern).

Sample requirements

CSF samples should be preferably taken in a sterile plain tube, but as anticoagulants do not interfere, it is also possible to analyze samples that have been put into fluoride, EDTA or heparin tubes. On being received in the laboratory CSF

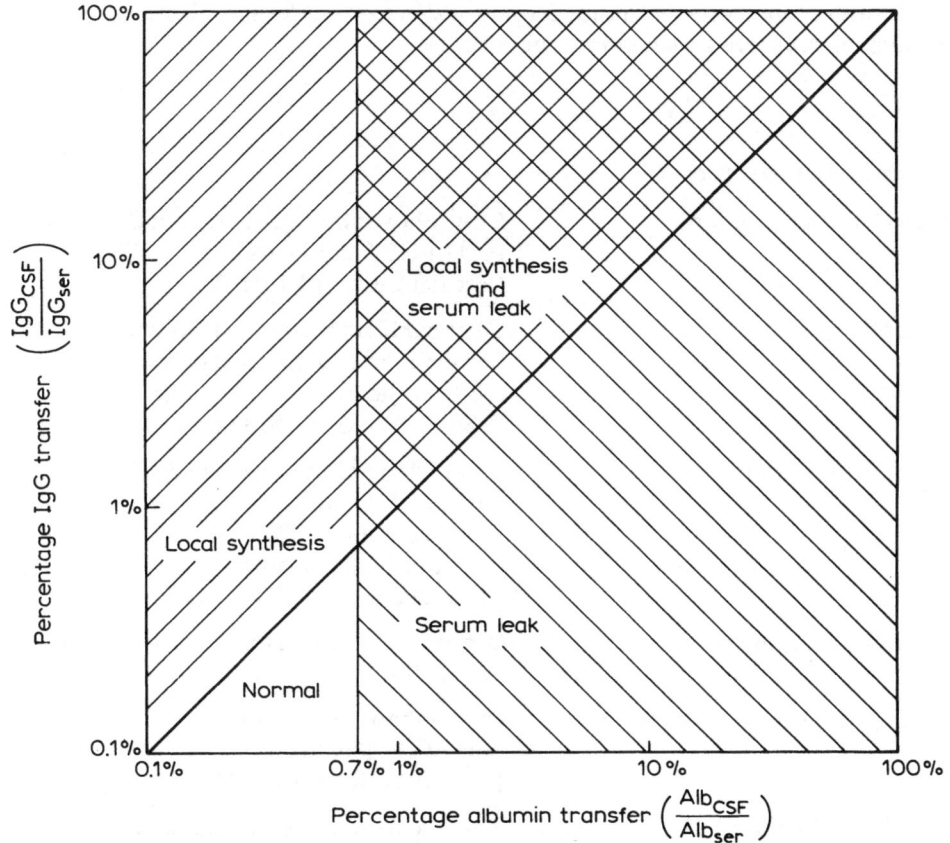

FIGURE 15.8 Use of the logarithmic plot to distinguish IgG barrier abnormalities (serum leak) from local synthesis.

samples should be stabilized by the addition of 20 μl of 2 per cent (w/v) aqueous sodium azide (NaN_3) per ml of sample.

IMPORTANT: adding sodium azide will interfere with bioassays and cell culture (including microbiological culture).

Serum is preferable, but plasma (with heparin, EDTA or fluoride anticoagulant) is also suitable. All samples are stable at room temperature for several days, as long as they are not contaminated and have been stabilized with azide. Samples can be sent to the laboratory by first class post at ambient temperature. Following receipt by the laboratory, samples are stable at 4°C for up to 2 weeks, after which they a frozen at −25°C for long term storage. All CSF and serum samples are retained the laboratory specimen bank for 5 years. Where possible, samples on interesting or unusual cases are retained indefinitely.

Equipment required

Isoelectric focusing tank	Multiphor II tank (Amersham Biosciences, part of GE Healthcare) Product Code: 18-1018-06
Power supply unit	ECPS 500/1000 power supply unit (Amersham Biosciences, part of GE Healthcare) Product Code 18-1018-06
Volt-hour integrator	VH-1 (Amersham Biosciences, part of GE Healthcare) Product code 19-7300-02 (also requires adaptor, product code: 18-1751-01)
Circulating water cooler	Grant LTD-6 (available through VWR). Set at 10°C
Rectangular IEF casting frame	(Amersham Biosciences, part of GE Healthcare) Product code 18-1016-82
Sample application foil	These are no longer available from Amersham Biosciences (part of GE Healthcare). A suitable substitute is available from Helena, UK
Micropipette and tips	To pipette volumes between 1 and 5 µl
Leveling table with bubble levels:	(Amersham Biosciences, part of GE Healthcare) Product code 18-1016-88
Thermostatted water bath	any suitable model, set to 65°C.
Boiling water bath	any suitable model
Rocker	either tilting or orbital models
Glass plates	255 × 130 × 5 mm with ground glass edges Available from local glazier (Preedy Glass Ltd.)

50 ml glass bottles with caps	(Aimer Ltd.) product code BR120-20 (bottles) product code BR230-12 (caps)
Miscellaneous	Rubber roller (photographic), hair dryer, retort stand and clamps, spring clips ('Bulldog' variety)

Materials required

Nitrocellulose membrane:	Trans-Blot Nitrocellulose. (Bio-Rad Ltd.) Product code: 162-0094
Postlip (Phoprinto) paper	(Hollingsworth and Vose Company Ltd.) Grade: thick. Fibre Free 203 × 254 mm.
Fine grain hardened filter paper	(Whatman Labsales) product code 1450917 Whatman Grade 50 46 × 57 cm.
Electrolyte wicks	254 × 10 mm long strips cut from the Postlip paper Each wick consists of two thicknesses of paper. Each plate requires 2 wicks.
GelBond	Gel-Fix for Agarose, sheets 265 × 125 mm (Serva product 42981) Available through AMS Biotechnology

Chemicals required

Glacial acetic acid (CH_3COOH)*	VWR (Merck), product code 10001 6X
Glycerol	VWR (Merck), product code 10118 4K
Hydrogen peroxide (H_2O_2; 30 volume per cent)	VWR (Merck), product code 10128 4N
Sodium acetate trihydrate ($CH_3COONa. 3H_2O$)	VWR (Merck), product code 10235
Sodium chloride (NaCl)	VWR (Merck), product code 10241
Sodium hydroxide (NaOH)*	VWR (Merck), product code 10438
D-Sorbitol	VWR (Merck), product code 10464
1M Sulfuric acid (H_2SO_4)*	VWR (Merck), product code 191687E
Methanol (CH_3OH)	Hayman Ltd., product code MB/2C (SIN 1230)
Marvel Dried Skimmed Milk	Supermarket grade
3-amino-9-ethyl carbazole*	Sigma-Aldrich, product code A575H

Pharmalyte 3-10	Amersham Biosciences, product code 17-0456-01
Pharmalyte 8-10·5	Amersham Biosciences, product code 17-0455-01
Agarose IEF	Amersham Biosciences, product code 17-0468-01
Goat anti-human IgG Fc	Diasorin, product code 80261
Peroxidase conjugated rabbit Anti-goat immunglobulins**	Dako Cytomation, product code P130

*NOTES: Refer to a hazard warning information on these chemicals. Read this carefully before undertaking the procedure.

**The peroxidase-conjugated antiserum is particularly prone to contamination which leads to a loss of activity. For this reason the peroxidase-conjugated antiserum is stored in 1 ml aliquots at 4°C.

Reagent preparation

1. Catholyte: 1 M Sodium Hydroxide
   It is prepared fresh each month.
   40 g sodium hydroxide is dissolved in water and made up to 1 l. It is stable at room temperature.
2. Anolyte: 0.05 M Sulfuric Acid
   It is prepared fresh each month.
   50 ml of 1 M sulfuric acid is added to water and made up to 1 l. It is stable at room temperature.
3. Stock: (10× concentrate) acetate buffer 0.2 M, pH 5.1
   27.2 g sodium acetate trihydrate is dissolved in water, 4.5 ml glacial acetic acid is added and made up to 1 l. That is stable at room temperature.
   The working strength buffer is prepared by diluting 1 volume of stock buffer with 9 volumes of water.
4. Saline:
   8.5 g of sodium chloride is dissolved in water and made up to 1 l. It is stable at room temperature.
5. Blocking Solution:
   2.5 g of Marvel is dissolved in 150 ml saline. This is prepared fresh each day.
6. Color Developer:
   This is prepared just before use (takes 10–15 minutes)
   50 mg of ethyl amino-carbazole is dissolved in 20 ml ethanol, then 100 ml of working strength acetate buffer and 100 µl of 30 volume per cent hydrogen peroxide are added.

Technique

Stage 1: Preparing the plate 5 ml of glycerol is diluted with 50 ml of water. 30 ml of this solution is added to a 50 ml glass bottle.

Then 3.6 g D-Sorbitol is added and dissolved.

0.3 g of IEF agarose is added and mixed well. This mixture is placed in a boiling water bath for 15 minutes, mixing occasionally, to dissolve the agarose.

The molten gel is placed in the 65°C water bath and allowed of the temperature to equilibrate.

A glass plate and gel casting frame are cleaned. A sheet of Gel-Fix for Agarose is cut to the size of the casting frame. About 1 ml of 50 per cent aqueous methanol is sprayed on to the glass plate and then the hydrophobic surface of the Gel-Fix is placed down on the plate (the hydrophobic surface is easily identified as the one on which water 'beads'). A paper towel is placed over the gel and surplus methanol is removed using the photographic print roller. The casting frame is positioned on the Gel-Fix and clamped to the glass plate using several spring clamps.

The casting assembly is placed on the leveling table and confirmed to be horizontal using bubble levels. The glass plate is warmed using the hair dryer.

Using 2.5 ml and 1.0 ml syringes and needles, the rubber cap is pierced and 2.0 ml Pharmalyte 3–10 and 0.5 ml Pharmalyte 8–10.5 are added to the gel solution. The contents are constantly mixed using a swirling action. The gel is replaced in the water bath for a further 5 minutes to bring the solution back to temperature.

After switching off the hair dryer, the bottle of gel is then removed and dried. The gel is poured into the casting frame and the gel solution is spread throughout the framed area using the neck of the bottle. This should be completed quickly to get the gel to spread before it starts to set.

The gel is allowed to set at room temperature for 10 minutes, then the clamps are removed from around the frame. The gel, still on the glass plate, is placed on two damp paper towels in a plastic box with a lid. The lid is closed and the gel is stored at 4°C until use.

The Analytical Log Sheet is filled in.

Stage 2: Preparing the run

Sample preparation CSF samples are generally used undiluted. To simplify the method a fixed volume of 3 µl is applied, with no adjustment for variation in total protein. If the CSF total protein is particularly high, then the sample should be diluted in water to a protein concentration of approximately 500 mg/l.

Serum or plasma samples are diluted 1/400 in water prior to analysis (10 µl + 4 ml water).

Using marker proteins Colored markers are not routinely run to monitor the IEF run. These are available commercially. Alternatively, a strong solution of either hemoglobin (red cell hemolysate) or myoglobin (crystalline proteins available from Sigma-Aldrich) can be used. To use, protein solution are spotted at both anode and cathode regions. On applying power, the proteins will migrate in opposite directions and when the spots fuse into a common band, then focusing is complete.

Loading the gel The gel is allowed to equilibrate to room temperature.

The samples are prepared.

The plate is removed from the box and placed on the bench. Using a scalpel blade the outermost 2 mm from all four edges of the gel are trimmed. This removes the slight meniscus created by casting using an open frame. If this meniscus is not removed it can prevent the electrodes of the IEF lid from sitting flat on the surface of the electrolyte wicks.

The Gel-Fix around the edge of the gel is carefully wiped with a paper tissue to remove any flecks of gel.

1 ml of 50 per cent aqueous methanol is sprayed on to the cooling surface of the IEF tank and the gel is carefully positioned on the tank, taking care to avoid trapping air bubbles. A paper tissue is used to remove surplus fluid from around the edge of the Gel-Fix.

A sheet of nitrocellulose is cut to the same size as the gel. This is used to dry the surface of the gel. Starting at one end of the gel, the edge of the membrane on the surface is carefully positioned and allowed to wet evenly. The rest of the nitrocellulose membrane is then gently folded on to the gel surface. Once it has become evenly wet the membrane is then removed from the gel, being careful not to tear the gel. The membrane can then be washed in water, dried and re-used. The sample application foil is then placed on the gel, about 2 cm in from the anodic (+) edge.

3 µl of sample is applied to the application slot. A separate pipette tip for each individual lane is used to avoid carry-over.

Two 1 cm wide electrode strips are soaked in 0.05 M sulfuric acid. They are lightly blotted to remove surplus electrolyte and positioned at the anodic (+) side of the gel, about 1 cm in from the outermost edge.

Another two strips are soaked in 1 M sodium hydroxide. These are lightly blotted to remove surplus electrolyte and are applied at the cathodic (−) side of the gel, about 1 cm in from the edge.

Paper hand towels are placed on the outermost long edges of the gel, to soak up surplus fluid which accumulates at the electrodes during the IEF run. It is to be made sure that the towels are on the gel, but are not actually touching the paper electrode wicks.

The electrodes in the IEF lid are adjusted to make sure that they lie in the middle of the electrode wicks and then the lid is placed on the tank.

Running the plate The gel is run at 15 W (constant) for an integrated time of 850 Vh.

The ECPS 3000/150 power supply unit is set up as follows:

The SET button is pushed (the LED comes on)
The V button is pushed (the LED comes on), adjusting the maximum voltage to 1000 V
The mA button is pushed (the LED comes on), adjusting the maximum current to 150 mA
The W button is pushed (the LED comes on) and the power adjusted to 15 W (if two gels are being run set this to 30 W)
The SET button is pushed again (the LED goes off)

Remember that the power supply is disabled whilst the SET red LED is ON. The SET button must be switched OFF before starting the run.

The volt–hour integrator is set-up as follows:

Alarm is pressed off (the red LED is off)
The fast/slow set is adjusted to read 850
The alarm button is pressed so that the LED is ON and the alarm will count down
When the power is switched on the volt–hour integrator will steadily count down to 0 and trigger the alarm to signify that the run has finished.
The power supply is switched ON using the ON/OFF toggle switch.

The initial readings are recorded on the Analytical Log Sheet. The actual values will vary from gel to gel. Typically the voltage will start at around 350 V, this will dip slightly in the first few minutes and then rise steadily throughout the run, reaching the limiting value (1000 V) in the last 10 minutes or so. An abnormally high or low voltage at the start of the run is bad news. The current will be around 45 mA and this will fall during the run. The power should remain steady at 15 W, falling slightly to 12 W during the last 10 or so minutes. Runs will typically take 1 hour to 1 hour 15 minutes.

Approximately 10 minutes into the run, when the volt-hour integrator shows 550 V-h remaining, the power supply is switched off and the lid is taken off. The sample application foil is carefully peeled off from the surface of the gel and the paper towels are replaced with dry ones. The lid is replaced and the power supply is switched back on.

Protein blotting and immunodetection   Two sheets of nitrocellulose membrane, a single sheet of hardened filter paper and several sheets of Phoprinto filter paper are cut to the same dimensions as the agarose gel.

At the end of the run the power supply is switched off and the IEF lid is removed. Using the paper towels, the surplus fluid is removed from the electrode wicks, then the wicks are removed and discarded.

The surface of the gel is preblotted using a single sheet of dry nitrocellulose. Starting at one end of the gel, the edge of the membrane is positioned on the gel surface and allowed to wet evenly. The rest of the nitrocellulose membrane is then gently folded on to the gel surface. It may be necessary to use a finger to get the membrane into the valleys on the gel surface. Once it has become evenly wet, the membrane is then removed from the gel, being careful not to tear the gel. This membrane has to be discarded. The same procedure is carried out with a second sheet of dry nitrocellulose, followed by a sheet of damp Whatman 50 paper and 12 sheets of Phoprinto blotter. Finally, a glass plate is placed on top with a 3 kg weight.

The proteins are allowed to transfer from the gel to the membrane for 20 minutes.

The filter papers are removed and discarded. The nitrocellulose is carefully peeled off from the gel and placed 'protein side up' in 100 ml of 2 per cent Marvel in saline. This is placed on the rocker for 30 minutes to block any unoccupied protein-binding sites.

The protein diluent is prepared by diluting 50 ml block solution to 500 ml with saline.

The blocking solution is discarded and the membrane rinsed once with saline. It is replaed with 50 ml protein diluent solution and 50 µl primary antibody (Goat anti-human IgGFc) is added. This is incubated at room temperature with rocking for 1 hour.

The membrane is washed in several changes of tap water over 10 minutes followed by a 5-minute wash in 100 ml protein diluent solution.

The wash solution is decanted and replaced with 50 ml protein diluent. 50 µl of a second antibody (peroxidase conjugated rabbit anti-goat immunoglobulin) is added, and incubated for 1 hour at room temperature with rocking.

The membrane is washed in several changes of tap water over 10 minutes, followed by a 5-minute wash in saline.

The final saline wash is decanted and a color developer is added. The patterns are allowed to develop sufficiently, then the reaction is stopped by washing the membrane in several changes of de-ionized or distilled water.

The immunoblot is then dried using the hair dryer and stored. The developed membranes are light sensitive and slowly darken when exposed to bright light. They are nevertheless stable for several years, if stored protected from light. Critical membranes should be photographed or digitally scanned.

Troubleshooting guide

The following is a list of potential problems and possible solutions.

Unusual running conditions

1. Marker proteins do not focus.
   Power supply is not connected.

SET button is switched on (LED is lit). When the set button is on, the power supply is disabled.

These problems are generally recoverable.

2. The volt x hour integral is decreasing slower than usual.
  The polarity of the electrodes is wrong.
  This problem is usually not recoverable.
3. The current (mA) at the start of the run is much higher than usual.
  The gel has been prepared using saline rather than water.
  This problem is not recoverable.
4. Poor patterns.

Unfortunately these faults are found only after the procedure has been completed. Most problems are thus not recoverable and require that the whole procedure be repeated.

1. *Generally bad patterns, poor resolution, distortion of patterns.* Gel is too hot when ampholytes are added. The recommended temperature for handling Pharmalytes is 75°C. A temperature at which the ampholytes are more stable, i.e. 65°C, is generally used. Pharmalytes are rapidly destroyed at temperatures higher than 85°C.
2. *Distortion of several adjacent lanes.* This is usually due to an uneven gel. Thin areas of gel have a lower conductivity than thicker regions. Fluid loss during the run has a greater effect on thin versus thick regions, and will emphasize any distortion. It is important that the gel is kept cold during the run, and ideally should be cooled to 10°C or less. If tap water (generally 15–20°C) is being used to cool the gel then a lower power ($\sim$ 10 W/gel) should be used. This will increase the length of time for the run to complete. When placing the gel on the cooling platter, care should be taken to not trap air bubbles under the gel as this reduces heat transfer efficiency and causes local hot spots.

    If the outermost sample lanes bend out towards the edge of the gel, then water is added on the outermost lanes to reduce this effect. The use of volumes of sample (20–30 µl) will also cause the lanes to balloon and displace the neighboring tracks. This can be minimized by flanking such samples with water.

Lesions on immunoblot Pattern in one lane shows streaks of enhanced staining.

This is due to pipette tip gouging a hole in the surface of the gel into which the sample pools.

Using small sample volumes can lead to a meniscus effect at the edge. This causes the lane edge to stain more intensely than the center. This can be corrected by applying a larger volume of diluted sample.

Unstained spots on immunoblot These are undissolved flecks of agarose. These do not take up the ampholytes and thus do not contribute to the formation of the pH gradient. Proteins do not enter these gel deposits so there is a blank spot. There may be streaking of the proteins as they move around and beyond the knot. Such knots also show themselves at the end of the IEF run as nodules rising out of the surface of the gel.

Mottling of the immunoblot Failure to preblot the gel (the preblot is carried out at the end of the run, but before the proper protein blot) gives rise to extensive non-specific staining. The nature of this material is unknown but probably represents lipophilic compounds such as high MW lipoproteins, cell membrane remnants, fibrin fragments etc.

Generalized mottling can still occur after effective preblotting. Sometimes for no apparent reason air bubbles appear between the NCM and the gel and are probably released from the gel under the pressure of the weight. Using a lighter weight may help.

Some nitrocellulose membranes are 'sided'. For example, the membrane may have one side that is shinier than the other. It is important when blotting that the shiniest of the sides is placed in contact with the gel, otherwise small air pockets will develop between the membrane and gel surface. These probably arise because the dull side is slightly more hydrophobic. So long as the 'sidedness rule' is obeyed there is no difference in the performance of the various membranes.

No staining of sample lanes

(a) Where the background nitrocellulose is unstained

   (i) No samples are loaded.

   (ii) Primary or secondary antibody is omitted.

   (iii) Incorrect primary or secondary antiserum is used.

   (iv) No hydrogen peroxide is added to the color developer.

   (v) Inhibition of peroxidase by sodium azide (often used to preserve reagents)

(b) Where the background nitrocellulose is stained
   Background staining heavy and even

   (i) blocking stage omitted (or inadequate)

   (ii) inadequate washing before addition of either second antibody or color developer.

   Background staining uneven

   (i) preblot omitted (or inadequate)

Sample lanes stained but faint
   (i) inadequate sample
   (ii) insufficient time for development of color
   (iii) outdated or contaminated hydrogen peroxide
   (iv) outdated or contaminated peroxidase-conjugated antibody
   (v) incorrect pH of acetate buffer
   (vi) ethylaminocarbazole is oxidized

Sample tracks overstained
   (i) excessive sample
   (ii) overdevelopment

Sample tracks distorted
   (i) inadequate cooling during IEF
   (ii) condensation on gel surface during IEF (usually due to circulating excessively cold water, or high environmental humidity)
   (iii) short circuit due to fluid accumulation under the sample application foil
   (iv) bacterial degradation of samples (can cause ballooning)
   (v) saline used instead of water to prepare gel (current high, gel distortion during IEF)
   (vi) too large a sample volume applied (may cause ballooning)

Inadequate separation
  (i) No power:
      (a) power supply not connected
      (b) lid electrodes not connected or broken
  (ii) No ampholytes in gel (characterized by low current at beginning of IEF)

Bacterial contamination of samples Bacterial contamination of either CSF or serum can cause problems in the interpretation of the IEF patterns since IgG can undergo a partial proteolysis and/or deglycosylation. This leads to the formation of several bands with acidic isoelectric points (pI). Although the resulting bands resemble those seen with a paraprotein they are generally not as regularly spaced and one or more satellite bands usually accompany each major band.

Differentiating between a true paraprotein and the consequences of bacterial contamination is relatively easy: Usually only the CSF or the plasma, rarely both, is contaminated. This means that the bands are seen in only one fluid. The normal polyclonal background immunoglobulin is missing, because it has been partially digested by the bacteria. There will be a typical polyclonal background pattern in either CSF or plasma.

The fragments produced by the partial digestion have lower molecular weights so they tend to have a lower affinity for the nitrocellulose membrane. Consequently, they are often found to be present on both the front surface and on the back surface of the blot. Often the fragments will stain a slightly different color from the intact molecules, being slightly more bluish-purple when compared to the typical pink-brown of the normal pattern; looking at the sample, shows that it will usually be cloudy.

Addresses of suppliers

Amersham Biosciences
Part of GE Healthcare
Nightingales Lane
Chalfont St.Giles
Bucks HP8 4SP
Tel: +44 (0) 800 515313

Bio-Rad Laboratories Ltd.
Bio-Rad House
Maylands Avenue
Hemel Hempstead
Herts HP2 7TD
Tel: +44 (0) 800 181134

Dako Cytomation Ltd.
Denmark House
Angel Drove
Cambridgeshire
CB7 4ET
Tel: +44 (0) 1353 669911

VWR International Ltd.
Hunter Boulevard
Magna Park
Lutterworth
Leics LE17 4XN
Tel: +44 (0) 800 223344

AMS Biotechnology
63b Milton Park
Abingdon
Oxon
OX14 4RX
Tel: +44 (0) 1235 828200

Hollingsworth & Vose Co. Ltd.
Postlip Mills
Winchcombe
Cheltenham
Glos
Tel: +44 (0) 1242 602227

The Labsales Company
(Whatman)
Over Industrial Park
Norman Way,
Cambridge CB4 5GR
Tel: +44 (0) 1954 233190

Sigma-Aldrich Co. Ltd.
Fancy Road
Poole
Dorset
BH12 4QH
Tel: +44 (0) 800 717181

Preedy Glass
33 Crawford Street
Hughenden Avenue
High Wycombe
Bucks
Tel: +44 (0) 1494 452016

Aimer Products Ltd.
56/58 Rochester Place
Camden Town
London
NW1 9JY
Tel: +44 (0) 20 7485 3618

Diasorin Ltd.
Charles House
Toutley Road
Wokingham
Berks RG41 1QN
Tel: +44 (0) 1189 364200

Analytical log template

Analytical log for IgG IEF

Date: /   /   Analyst:
Details of Plate
Gel Prepared On: /   /   Gel Prepared By:
Glycerol Batch Number:
D-Sorbitol Batch Number:
Agarose IEF Batch Number:
Pharmalyte 3–10 Batch Reference:
Pharmalyte 8–10·5 Batch Reference:
Details of Run:

Time IEF Started: Time IEF Finished:
Duration of Run:
Initial Voltage: Current:
Volt-Hours:
Details of Immunodetection:

Nitrocellulose Batch Number:
Anti-human IgG Fc Batch Number: Expiry date:
HRP-anti goat Batch Number: Expiry date:
Ethyl Amino Carbazole Batch Number:
Solvent – Ethanol/Methanol Batch Number:
Acetate Buffer Batch Number:
Hydrogen Peroxide Batch Number:
Assessment of Technical Quality:

CHAPTER

16

Cross-index of references

The numbers refer to those given to the alphabetic list of references. For specific page numbers within the text, consult the index to text which follows the references The numbers below are not page numbers but reference numbers.

14-3-2, *see* Enolase

2-Dimensional, *see* Two-dimensional electrophoresis

ACTH, *see* Adrenocorticotrophin

Acute phase proteins [142], [296]

Adrenocorticotrophin (ACTH) [994], [1169]

Adrenoleukodystrophy [60]

Albumin [8], [9], [59], [73], [185], [186], [242], [294], [335], [336], [513], [514], [528], [537], [583], [634], [653], [654], [676], [783], [836], [864], [921], [934], [1163], [1176], [1248]

Aldolase C4 [1161], [1296]

Alpha-1-acid glycoprotein, *see* Orosomucoid

Alpha-1-antitrypsin [334], [1111]

Alpha-l-microglobulin [243], [478]

Alpha-2-macroglobulin [41], [73], [178], [944], [987], [1079]

Alpha lipoprotein, *see* Lipoproteins

Alzheimer's, *see also* Dementia

Alzheimer's disease [133], [246], [254], [634], [1075], [1197], [392], [393], [425], [437], [447], [446], [476], [477], [494], [529], [869], [947], [1036], [1065]

Amyloid proteins [231], [254], [376], [377], [883], [1067], [155], [156], [437], [447], [529], [543], [544], [1065]

Amyotrophic lateral sclerosis [4], [409], [456], [633], [937]

Apolipoproteins, *see* Lipoproteins

16. CROSS-INDEX OF REFERENCES

Aseptic meningitis [13], [113], [307], [324], [325], [326]

Astroprotein, *see* Glial fibrillary acidic protein

Ataxia telangiectasia [673]

Behcet's disease [11], [27], [112], [350], [778], [821], [956], [1009]

Bence-Jones protein, *see* Unbound light chains

Bence-Jones proteins [111], [1151]

Beta-2-microglobulin [57], [669], [1118], [1119]

Beta lipoprotein, *see* Lipoproteins

Beta trace protein [626], [642], [652], [669], [831], [859], [351], [710], [744], [458]

Blood CSF barriers [8], [9], [90], [91], [98], [99], [101], [218], [236], [240], [273], [284], [290], [291], [372], [375], [408], [463], [634], [665], [902], [909], [987], [989], [1083], [1114], [1172], [1179], [1277], [282], [552], [586], [1023], [1019], [1014], [1137]

Brucellosis, *see* Neurobrucellosis

C-reactive protein [169], [429], [799], [816], [858], [996], [1060]

Cerebral infarction [9], [463], [634]

Cerebral lupus [60], [113], [414], [433], [610], [914], [1035], [734], [1193]

Cerebrovascular diseases [407], [841], [953], [955]

Child's CSF, *see* Pediatric values

Complement proteins [27], [64], [166], [207], [324], [333], [414], [440], [774], [885], [962], [1079], [1280], [1287], [1308]

Creatine kinase [54], [873]

Cryptococcus [877]

D-2 antigen [510], [1076]

Dementia [40], [75], [245], [415], [456], [634], [1197], [1292], [1294], [155], [710], [927], [210], [392], [393], [1016], [390], [396], [869], [487], [156], [425], [529], [811], [1065], [1066], [948], [947]

Diagnostic criteria from CSF [43], [68], [69], [83], [100], [120], [123], [122], [136], [162], [179], [181], [215], [295], [337], [374], [452], [464], [470], [486], [496], [499], [579], [582], [588], [593], [599], [677], [680], [681], [701], [731], [740], [742], [793], [795], [856], [878], [888], [889], [934], [958], [982], [984], [1038], [1062], [1120], [1124], [1168], [1214], [1239], [1278], [1286]

Down's syndrome [245]

Electro convulsive therapy [14]

Encephalitis [100], [117], [164], [235], [256], [345], [399], [400], [486], [512], [579], [580], [593], [595], [635], [691], [781], [795], [978], [1055], [1062], [1068], [1069], [1070], [1115], [1214], [1230], [1232], [1286], [477], [1109], [927], [1253], [686], [687], [210], [722], [168], [390], [465], [53], [353], [803], [151], [191], [208], [1162], [592], [134]

Encephalitogenic protein, *see* Myelin basic protein
Enolase (14-3-2) [74], [85], [86], [104], [196], [343], [468], [704], [769], [849], [957], [958], [959], [980], [1088], [1089]
Eosinophil cationic protein [422]
Epilepsy [311]
Epstein-Barr virus [93], [895]
Ferritin [1059], [1062], [175], [210], [437], [550], [860], [867], [865]
Fibrinogen [22], [107], [108], [159], [193], [469], [1184], [1185], [1270]
Free light chains, *see also* Bence-Jones protein
Free light chains, *see* Unbound light chains
Gamma trace protein [42], [404], [405], [626], [642], [668], [669], [859], [939], [411], [865], [812], [698], [1193], [818], [1195], [97], [460], [949], [950], [946], [871]
Glial fibrillary acidic protein [10], [139], [436], [536], [676]
Group components proteins [247]
Guillain-Barré syndrome [262], [383], [466], [601], [644], [1045], [1240], [1024], [209], [1021]
Haptoglobin [67], [398], [1206], [132]
Harada's, *see* Uveo-meningitis
Head injury [89], [760], [844], [1120], [1121]
Herpes simplex encephalitis [100], [117], [136], [164], [286], [508], [524], [525], [526], [558], [579], [593], [595], [594], [605], [635], [691], [781], [795], [809], [863], [1068], [1070], [1115], [1214], [1227], [1230], [1232], [1286], [1299], [771], [726], [210], [722], [168], [53], [803], [151], [221], [1162], [1267], [592], [134]
HIV [149], [150], [191], [210], [365], [391], [390], [764], [807], [1015]
Hydrocephalus [10], [128], [130], [1050], [1076]
IgA [211], [212], [450], [583], [630], [631], [750], [766], [802], [1001], [1061], [1071], [1102], [1232], [1238], [1061], [521], [523], [1142], [819], [1148], [1152], [629], [910], [1267]
IgD [801], [884]
IgE [802], [1063]
IgG [8], [81], [116], [117], [142], [146], [218], [228], [239], [261], [286], [309], [310], [312], [325], [335], [336], [346], [348], [402], [408], [445], [450], [471], [526], [527], [534], [566], [583], [605], [628], [630], [631], [634], [648], [653], [654], [664], [672], [701], [711], [713], [715], [716], [718], [719], [720], [741], [746], [748], [752], [766], [782], [781], [783], [787], [802], [804], [813], [815], [836], [837], [840], [862], [864], [876], [882], [952], [985], [986], [1003], [1004], [1005], [1057], [1056], [1071], [1091], [1095], [1106], [1156], [1163], [1167], [1169], [1172], [1173], [1174], [1177], [1178], [1179], [1180], [1181], [1183], [1201],

IgG *(continued)* [1205], [1207], [1217], [1218], [1221], [1222], [1225], [1228], [1232], [1233], [1235], [1238], [1240], [1245], [1259], [1263], [1265], [1282], [380], [308], [800], [771], [426], [708], [895], [354], [521], [522], [970], [598], [1196], [369], [552], [1298], [684], [737], [549], [686], [683], [728], [633], [727], [735], [733], [1152], [724], [1018], [1029], [726], [725], [149], [434], [473], [687], [734], [209], [1317], [1318], [723], [241], [910], [829], [597], [928], [663], [803], [148], [221], [796], [911], [1144], [1297], [1267], [906], [1193], [417], [1241]

IgM [95], [103], [307], [315], [397], [403], [450], [451], [526], [566], [583], [605], [630], [631], [670], [741], [766], [785], [792], [802], [999], [1001], [1054], [1071], [1201], [1224], [1232], [1238], [1295], [1322], [380], [308], [800], [771], [426], [708], [895], [354], [521], [522], [970], [598], [1196], [369], [552], [1298], [684], [737], [549], [686], [683], [728], [727], [735], [733], [1152], [724], [1018], [1029], [726], [725], [149], [434], [473], [687], [734], [209], [1317], [1318], [723], [241], [910], [829], [597], [928], [663], [803], [148], [221], [796], [521], [523], [737], [1026], [1022], [1027], [1023], [724], [1018], [1029], [1028], [1030], [784], [929], [923], [1267]

Immune complexes [28], [170], [172], [173], [194], [378], [438], [440], [509], [682], [696], [697], [855], [975], [1055], [1056]

Immunoglobulin allotypes [971]

Immunoglobulin idiotypes [31], [35], [233], [794]

Jakob-Creutzfeldt disease [105], [333], [712], [830], [1077]

Lactoferrin [1119]

Light chains, *see* Free light chains

Lipoproteins [448], [449], [467], [827], [940], [115], [137], [542], [545], [544], [543], [1310], [928], [447], [1065]

Lupus, *see* Cerebral lupus

Lymphoma [588], [1200], [1263]

Lysozyme [295], [427], [1119]

Malignant tumor, *see* Tumors

Measles encephalitis [29], [58], [79], [81], [80], [119], [176], [256], [315], [418], [419], [420], [507], [525], [562], [568], [651], [746], [752], [782], [805], [809], [815], [1069], [1074], [1173], [1174], [1187], [1218], [1220], [1221], [1225], [1228], [1227], [1235], [1299], [800], [221], [911], [417]

Meningitis [108], [113], [159], [169], [283], [324], [325], [326], [336], [410], [527], [565], [567], [599], [648], [702], [775], [864], [877], [918], [969], [1055], [1201], [1223], [1229], [1287], [1109], [500], [1152], [1014], [1319], [210], [390], [1101], [151], [1162], [592], [258]

Metastatic tumor, *see* Tumors

Motor neurone disease, *see* Amyotrophic lateral sclerosis

Multiple sclerosis [2], [7], [26], [28], [29], [30], [31], [35], [43], [44], [45], [47], [60], [63], [69], [77], [78], [79], [81], [80], [87], [92], [93], [101], [116], [118], [120],

16. CROSS-INDEX OF REFERENCES

[121], [124], [138], [142], [157], [170], [172], [174], [194], [195], [214], [216], [218], [233], [238], [239], [240], [244], [250], [251], [252], [261], [263], [306], [305], [307], [315], [322], [323], [346], [348], [378], [379], [389], [412], [418], [419], [431], [440], [441], [452], [461], [464], [470], [479], [496], [497], [498], [499], [513], [555], [557], [560], [568], [572], [589], [590], [596], [602], [613], [617], [628], [641], [643], [647], [649], [650], [654], [655], [657], [661], [664], [682], [690], [701], [703], [716], [717], [718], [719], [720], [731], [741], [745], [748], [750], [751], [752], [757], [759], [763], [774], [778], [779], [794], [804], [809], [814], [815], [832], [833], [835], [840], [852], [855], [862], [864], [876], [878], [879], [886], [887], [897], [938], [954], [962], [963], [964], [965], [966], [967], [968], [971], [973], [974], [975], [976], [993], [998], [1007], [1013], [1037], [1044], [1057], [1086], [1092], [1093], [1095], [1097], [1106], [1114], [1124], [1149], [1150], [1165], [1166], [1167], [1168], [1169], [1172], [1174], [1175], [1176], [1177], [1178], [1179], [1181], [1189], [1190], [1205], [1217], [1221], [1222], [1226], [1228], [1231], [1235], [1237], [1245], [1255], [1261], [1262], [1269], [1283], [1293], [1295], [1302], [1307], [1308], [1313], [1321], [1322], [820], [708], [327], [756], [1204], [738], [736], [1026], [686], [727], [735], [733], [1018], [1029], [1028], [1031], [1158], [739], [1025], [1032], [1017], [434], [687], [371], [1317], [1318], [362], [395], [359], [358], [612], [367], [366], [629], [943], [364], [465], [867], [1052], [484], [55], [928], [663], [221], [730], [911], [915], [812], [698], [417], [688], [949], [303], [961], [1234]

Mumps [326], [512], [648], [775], [809], [918], [1201], [1223], [1227], [1229], [1299]

Myasthenia gravis [3], [232], [884]

Mycobacteria, *see* Tuberculous

Mycoplasma [383], [695]

Myelin associated glycoprotein [732], [1311]

Myelin basic protein [19], [39], [44], [63], [118], [140], [162], [163], [203], [204], [206], [205], [322], [346], [539], [540], [632], [658], [659], [827], [841], [843], [844], [846], [851], [1056], [1120], [1234], [1284], [1285], [820], [357], [465], [55]

N-CAM, *see* D2 Antigen

Neurobrucellosis [1049]

Neurofilaments [34], [97], [241], [460], [465], [688], [698], [811], [810], [812], [818], [870], [865], [945], [948], [947], [1066], [1065], [1077], [1193], [1195], [1314]

Neuropeptides, *see* Table 3.3

Neurosarcoid [566], [735], [734], [1315]

Neurosyphilis [126], [229], [406], [680], [681], [768], [789], [790], [791], [792], [793], [839], [857], [882], [1236], [503], [1101]

Nigrosin stain [922], [1034]

Optic neuritis [805], [977], [986], [1091], [1093], [1094], [1227], [1196], [728], [735], [434], [639]

Orosomucoid [84], [329], [825]
Parkinson's disease [456], [709], [1213], [191], [460]
Pediatric values [111], [128], [130], [169], [355], [418], [1074]
Peptides, see Table 3.3
Plasminogen [603]
Polio [253]
Prealbumin [12], [66], [73], [231], [265], [293], [511], [583], [602], [883], [1067], [1273], [1274], [995]
Proteolipid protein [1192]
Retinol binding protein [56], [1073]
Rheumatoid factor, see Immune complexes
Rubella [29], [173], [809], [1187], [1227], [1233]
S-100 protein [759], [760], [769], [1033], [1058], [1062], [392], [475], [545], [698], [868], [867], [869], [872], [865]
SDS, see Sodium dodecyl sulfate electrophoresis
Sarcoid, see Neurosarcoid
Schizophrenia [73], [177], [432], [1199], [1212], [1276]
Silver stain [171], [430], [678], [747], [749], [755], [776], [824]
Sjogren's syndrome [13]
Sodium dodecyl sulfate electrophoresis [388], [453]
Spasmodic torticollis [575]
Stroke, see Cerebrovascular diseases
Subacute sclerosing panencephalitis [58], [78], [119], [187], [256], [261], [420], [534], [535], [562], [640], [679], [716], [717], [718], [719], [720], [782], [815], [920], [938], [1037], [1047], [1048], [1057], [1074], [1078], [1104], [1218], [1219], [1221], [1225], [1228], [1261], [1268]
Syphilis, see Neurosyphilis
Systemic lupus, see Cerebral lupus
Tau (filament)[494], [1320], [1036], [155], [425], [477], [1109], [437], [927], [393], [390], [396], [156], [476], [529], [447], [1065], [446], [1314], [1066]
Tau protein, beta-2 transferrin [330], [1244], [1188], [1213], [744], [556]
Total protein [312], [313], [513], [656], [783]
Transferrin [73], [197], [317], [321], [330], [692], [1209], [1210], [1211], [1249], [457], [556], [744], [1213]
Transthyretin, see Prealbumin
Tuberculous infections [36], [410], [527], [565], [567], [599], [702], [969]
Tumors [199], [334], [436], [588], [831], [1043], [1256], [1266]

16. CROSS-INDEX OF REFERENCES   247

Twin studies in MS [1293], [1302], [1306]

Two dimensional electrophoresis [21], [23], [33], [222], [223], [227], [274], [272], [280], [382], [430], [431], [480], [620], [699], [755], [822], [1090], [1264], [1288], [1303]

Unbound (free) light chains [76], [78], [82], [628], [679], [1203], [1217], [1224], [1233], [380], [416], [1164], [598], [1157], [1204], [736], [1026], [1028], [612], [241], [1116], [796], [961]

Uveo meningitis (Harada's) [1275]

Vaccinia infections [557]

Varicella [345], [1235], [1238], [110], [353], [803], [911], [1267], [592]

Ventricular fluid [1264], [1271]

Viral infections [114], [174], [244], [277], [306], [305], [399], [498], [580], [648], [713], [780], [782], [888], [935], [954], [972], [978], [1037], [1226], [1237], [1302], [1305], [110], [737], [686], [1027], [726], [725], [784], [465], [289], [353], [48], [191], [208], [1162], [592], [417]

Vogt-Kornadi-Harada, *see* Uveo meningitis

Web search for (selected) CSF proteins:
[1], [2], [3], [4], [5], [6], [7], [8], [9], [10], [11], [12], [13], [14], [15], [16], [17], [18], [19], [20], [21], [22], [23], [26], [27], [28], [29], [30], [31], [32], [33], [34], [35], [36], [37], [38], [39], [40], [41], [42], [43], [44], [45], [46], [47], [49], [51], [52], [54], [56], [57], [58], [59], [60], [61], [63], [64], [65], [66], [67], [68], [69], [70], [72], [71], [73], [74], [75], [76], [77], [78], [79], [81], [80], [82], [83], [84], [85], [86], [87], [88], [89], [90], [91], [92], [93], [96], [95], [98], [99], [100], [101], [102], [103], [104], [105], [107], [108], [111], [112], [113], [114], [116], [117], [118], [119], [120], [121], [125], [123], [124], [122], [126], [127], [128], [129], [130], [133], [135], [136], [138], [139], [140], [141], [142], [143], [144], [145], [146], [153], [154], [157], [158], [159], [160], [161], [162], [163], [164], [165], [166], [167], [169], [170], [171], [172], [173], [174], [176], [177], [178], [179], [180], [181], [182], [184], [185], [187], [186], [193], [194], [195], [196], [197], [199], [200], [201], [203], [204], [206], [205], [207], [211], [212], [213], [214], [215], [216], [217], [218], [219], [220], [222], [223], [224], [225], [226], [227], [228], [229], [230], [231], [232], [233], [234], [235], [236], [237], [238], [239], [240], [242], [243], [244], [245], [246], [247], [248], [249], [250], [251], [252], [253], [254], [256], [257], [259], [260], [261], [262], [263], [264], [265], [266], [267], [268], [269], [270], [271], [274], [272], [273], [275], [276], [277], [278], [280], [281], [283], [284], [285], [286], [287], [288], [290], [291], [292], [293], [294], [295], [296], [297], [298], [299], [300], [301], [302], [304], [306], [305], [307], [309], [310], [311], [312], [313], [314], [315], [317], [318], [319], [320], [321], [322], [323], [324], [325], [326], [329], [330], [331], [332], [334], [333], [335], [336], [337], [339], [340], [341], [343], [344], [345], [346], [347], [348], [349], [350], [352], [355], [372], [373], [374], [375], [376], [377], [378], [379],

Web search for (selected) CSF proteins: *(continued)*

[381], [382], [383], [385], [386], [387], [388], [389], [397], [398], [399], [400], [401], [402], [403], [404], [405], [406], [407], [408], [409], [410], [412], [413], [414], [415], [418], [419], [420], [421], [422], [423], [424], [427], [428], [429], [430], [431], [432], [433], [436], [438], [439], [440], [441], [443], [445], [448], [449], [450], [451], [452], [453], [454], [455], [456], [459], [461], [462], [463], [464], [466], [467], [468], [469], [470], [471], [478], [479], [480], [481], [485], [486], [489], [491], [492], [495], [496], [497], [498], [499], [502], [505], [507], [508], [509], [510], [511], [512], [513], [514], [515], [517], [518], [520], [524], [525], [526], [527], [528], [530], [531], [532], [533], [534], [535], [536], [537], [538], [539], [540], [546], [547], [548], [554], [555], [557], [558], [559], [560], [562], [563], [564], [565], [566], [567], [568], [570], [569], [571], [573], [572], [575], [574], [576], [577], [578], [579], [580], [581], [582], [583], [584], [585], [587], [588], [589], [590], [591], [593], [595], [594], [596], [599], [600], [601], [602], [603], [604], [605], [607], [608], [609], [610], [613], [615], [616], [617], [618], [619], [620], [622], [621], [623], [624], [626], [625], [627], [628], [630], [631], [632], [633], [634], [635], [636], [637], [638], [640], [642], [641], [643], [644], [645], [646], [647], [648], [649], [650], [651], [652], [653], [654], [655], [656], [657], [658], [659], [660], [661], [662], [664], [665], [666], [667], [668], [669], [670], [671], [672], [673], [674], [675], [676], [677], [678], [679], [680], [681], [682], [689], [690], [691], [692], [694], [695], [696], [697], [699], [700], [701], [702], [703], [704], [705], [706], [709], [711], [712], [713], [714], [715], [716], [717], [718], [719], [720], [721], [731], [732], [740], [741], [742], [743], [745], [746], [747], [748], [749], [750], [751], [752], [753], [754], [755], [757], [759], [760], [761], [762], [763], [765], [766], [768], [769], [770], [772], [773], [774], [775], [776], [777], [778], [779], [780], [782], [781], [783], [786], [785], [787], [788], [789], [790], [791], [792], [793], [794], [795], [797], [798], [799], [801], [802], [804], [805], [806], [808], [809], [813], [814], [815], [816], [817], [821], [822], [824], [825], [826], [827], [830], [831], [832], [833], [834], [835], [836], [837], [838], [839], [840], [841], [842], [843], [844], [845], [846], [848], [849], [850], [851], [852], [853], [854], [855], [856], [857], [858], [859], [861], [862], [863], [864], [873], [874], [875], [876], [877], [878], [879], [880], [881], [882], [883], [884], [885], [886], [887], [888], [889], [891], [892], [896], [897], [898], [900], [901], [902], [909], [912], [913], [914], [918], [919], [920], [921], [922], [924], [926], [930], [931], [932], [933], [934], [935], [936], [937], [938], [940], [941], [942], [944], [951], [952], [953], [954], [955], [956], [957], [958], [959], [960], [962], [963], [964], [965], [966], [967], [968], [969], [971], [972], [973], [974], [975], [976], [977], [978], [979], [980], [981], [982], [983], [984], [985], [986], [987], [989], [988], [990], [991], [992], [993], [994], [996], [997], [998], [999], [1000], [1001], [1002], [1003], [1004], [1005], [1006], [1007], [1008], [1009], [1010], [1012], [1013], [1033], [1034], [1035], [1037], [1038], [1039], [1040], [1041], [1042], [1043], [1044], [1045], [1046], [1047], [1048], [1049],

16. CROSS-INDEX OF REFERENCES

[1050], [1051], [1053], [1054], [1055], [1057], [1056], [1058], [1059], [1060], [1061], [1062], [1063], [1067], [1068], [1069], [1070], [1071], [1073], [1074], [1075], [1076], [1077], [1078], [1079], [1083], [1084], [1085], [1086], [1087], [1088], [1089], [1090], [1091], [1092], [1093], [1094], [1095], [1096], [1097], [1098], [1099], [1100], [1102], [1103], [1104], [1105], [1106], [1107], [1108], [1110], [1111], [1112], [1113], [1114], [1115], [1117], [1118], [1119], [1120], [1121], [1124], [1149], [1150], [1151], [1154], [1156], [1159], [1160], [1161], [1163], [1165], [1166], [1167], [1168], [1169], [1170], [1171], [1172], [1173], [1174], [1175], [1176], [1177], [1178], [1179], [1180], [1181], [1182], [1183], [1184], [1185], [1186], [1187], [1189], [1190], [1191], [1192], [1194], [1197], [1198], [1199], [1200], [1201], [1203], [1205], [1206], [1207], [1208], [1209], [1210], [1211], [1212], [1214], [1215], [1216], [1217], [1218], [1219], [1220], [1221], [1222], [1223], [1224], [1225], [1226], [1228], [1227], [1229], [1230], [1231], [1232], [1233], [1234], [1235], [1236], [1237], [1238], [1239], [1240], [1242], [1244], [1243], [1245], [1246], [1247], [1248], [1249], [1250], [1251], [1254], [1255], [1256], [1258], [1259], [1260], [1261], [1262], [1263], [1264], [1265], [1266], [1268], [1269], [1270], [1272], [1271], [1273], [1274], [1275], [1276], [1277], [1278], [1279], [1280], [1281], [1282], [1283], [1284], [1285], [1286], [1287], [1288], [1290], [1291], [1292], [1293], [1294], [1295], [1296], [1299], [1300], [1301], [1302], [1303], [1304], [1305], [1306], [1307], [1308], [1309], [1311], [1312], [1313], [1321], [1322], [380], [308], [110], [820], [800], [771], [411], [416], [1188], [426], [708], [1164], [351], [767], [494], [327], [1320], [895], [62], [354], [521], [1213], [523], [488], [115], [522], [970], [1036], [155], [425], [598], [1196], [24], [475], [444], [342], [995], [1202], [710], [744], [939], [860], [175], [477], [1109], [437], [927], [1253], [1123], [1126], [1125], [541], [758], [614], [1128], [1129], [1157], [551], [1132], [1137], [1138], [1140], [369], [1142], [1143], [1130], [1135], [1146], [183], [1145], [1155], [519], [890], [1127], [756], [1072], [500], [501], [504], [819], [1148], [1204], [503], [552], [1131], [1298], [553], [1133], [684], [685], [738], [737], [736], [1026], [1022], [549], [686], [683], [728], [727], [735], [733], [1027], [1023], [1152], [338], [724], [1018], [1029], [1028], [556], [726], [725], [807], [1015], [1014], [1031], [1030], [1019], [1020], [1136], [1158], [149], [550], [739], [1024], [1025], [1032], [1017], [1289], [1316], [1319], [25], [131], [150], [434], [152], [473], [490], [493], [687], [734], [784], [1139], [209], [356], [357], [371], [823], [1141], [1317], [1318], [210], [362], [363], [360], [361], [392], [395], [929], [109], [370], [359], [358], [365], [561], [586], [611], [612], [722], [1147], [168], [367], [394], [391], [393], [606], [723], [1021], [1016], [1315], [366], [390], [629], [764], [943], [132], [190], [368], [364], [396], [465], [1153], [137], [147], [192], [189], [868], [867], [53], [198], [241], [542], [545], [870], [872], [865], [869], [910], [917], [544], [487], [543], [328], [384], [1122], [905], [899], [255], [729], [904], [1081], [828], [474], [1257], [829], [597], [50], [847], [1080], [279], [188], [1052], [1082], [484], [435], [516], [1116], [893], [289], [55], [353], [1310], [483], [928], [663], [803], [1101], [48], [148], [151],

Web search for (selected) CSF proteins: *(continued)*
[156], [191], [221], [458], [457], [476], [482], [529], [707], [730], [796], [907], [911], [923], [915], [1144], [208], [1297], [1162], [1267], [906], [472], [592], [812], [945], [698], [1193], [811], [447], [810], [818], [1065], [446], [1314], [417], [1195], [1066], [97], [948], [688], [460], [947], [949], [950], [946], [925], [303], [94], [1134], [106], [134], [202], [258], [282], [316], [442], [639], [693], [871], [866], [894], [903], [908], [916], [961], [1011], [1064], [1241], [1252]

References

[1] Aarli, J. A. (1983). The immune system and the nervous system. *J. Neurol.*, **229**, 137–154.
[2] Abramsky, O., Lisak, R. P., Silberberg, D. H. *et al.* (1977). Antibodies to oligodendroglia in patients with multiple sclerosis. *N. Engl. J. Med.*, **297**, 1207–1211.
[3] Adornato, B. T., Houff, S. A., Engel, W. K. *et al.* (1978). Abnormal immunoglobulin bands in cerebrospinal fluid in myasthenia gravis. *Lancet*, **ii**, 367–368.
[4] Adornato, B. T., Houff, S. A., Engel, W. K. and Sever, J. L. (1979). Cerebrospinal fluid oligoclonal proteins in amyotrophic lateral sclerosis. *Arch. Neurol.*, **36**, 119.
[5] Ahonen, A., Myllyla, V. V. and Hokkanen, E. (1978). Measurement of reference value for certain proteins in cerebrospinal fluid. *Acta Neurol. Scand.*, **57**, 358–365.
[6] Ahonen, A., Myllyla, V. V. and Hokkanen, E. (1979). Cerebrospinal fluid protein findings in various lower back pain syndromes. *Acta Neurol. Scand.*, **60**, 93–99.
[7] Ahonen, A., Pyhtinen, J., Hokkanen, E. and Myllyla, V. (1982). Cerebrospinal fluid protein findings in cervical syndromes classified by myelography, and in multiple sclerosis. *Acta Neurol. Scand.*, **66**, 369–377.
[8] Al Kassab, S., Dittmann, L. and Olesen, H. (1979). IgG subclasses, barrier function for albumin and production of immunoglobulin G in the central nervous system. *Acta Neurol. Scand.*, **60**, 129–139.
[9] Al Kassab, S., Olsen, T. S. and Skriver, E. B. (1981). Blood–brain barrier integrity in patients with cerebral infarction investigated by computed tomography and serum–CSF albumin. *Acta Neurol. Scand.*, **64**, 438–445.
[10] Albrechtsen, M., Soelberg-Sorensen, P., Gjerris, F. and Bock, E. (1985). High cerebrospinal fluid concentration of glial fibrillary acidic protein (GFAP) in patients with normal pressure hydrocephalus. *J. Neurol. Sci.*, **70**, 269–274.
[11] Alema, G. (1970). Behcet's disease. In: P. J. Vinken and G. W. Bruyn (eds.), *Handbook of Clinical Neurology*, pp. 475–512. North-Holland Publishing Co.
[12] Aleshire, S. L., Bradley, C. A., Richardson, L. D. and Parl, F. F. (1983). Localization of human prealbumin in choroid plexus epithelium. *J. Histochem. Cytochem.*, **31**, 608–617.
[13] Alexander, E. L. and Alexander, G. E. (1983). Aseptic meningoencephalitis in primary Sjögren's syndrome. *Neurology*, **33**, 593–598.
[14] Alexopoulos, G. S., Kocsis, J. H. and Stokes, P. E. (1978). Increase in CSF protein in association with ECT. *J. Neurol. Neurosurg. Psychiatry*, **41**, 1145–1146.

[15] Allen, P. C., Hill, E. A. and Stokes, A. M. (1977). Isolation and properties of some main proteins. In: P. C. Allen, E. A. Hill and A. M. Stokes (eds.), *Plasma Proteins*, pp. 160–247. Blackwell.

[16] Allen, R. C. (1980). Rapid isoelectric focusing and detection of nanogram amounts of proteins from body tissues and fluids. *Electrophoresis*, **1**, 32–37.

[17] Allinquant, B., Musenger, C. and Schuller, E. (1984). Intrathecal origin of CSF ribonuclease. *Acta Neurol. Scand.*, **69**, 12–19.

[18] Alper, C. A. and Johnson, A. M. (1969). Immunofixation electrophoresis: a technique for the study of protein polymorphism. *Vox Sang.*, **17**, 445–452.

[19] Alvord, E. C., Hrudy, S. and Sires, L. R. (1979). Degradation of myelin basic protein by cerebrospinal fluid: preservation of antigenic determinants under physiological conditions. *Ann. Neurol.*, **6**, 474–482.

[20] Ames, G. F. L. (1974). Resolution of bacterial proteins by polyacrylamide gel electrophoresis on slabs. *J. Biol. Chem.*, **249**, 634–644.

[21] Anderson, L. and Anderson, N. G. (1977). High resolution two-dimensional electrophoresis of human plasma proteins. *Proc. Natl Acad. Sci.*, **74**, 5421–5425.

[22] Anderson, M., Matthews, K. B. and Stuart, J. (1978). Coagulation and fibrinolytic activity of cerebrospinal fluid. *J. Clin. Pathol.*, **31**, 488–492.

[23] Anderson, N. L., Nance, S. L., Pearson, T. W. and Anderson, N. G. (1982). Specific antiserum staining of two-dimensional electrophoretic patterns of human plasma proteins immobilized on nitrocellulose. *Electrophoresis*, **3**, 135–142.

[24] Anderson, R. E., Winnerkvist, A., Hansson, L. O. *et al.* (2003). Biochemical markers of cerebrospinal ischemia after repair of aneurysms of the descending and thoracoabdominal aorta. *J. Cardiotho. Vasc. Anesth.*, **17**, 598–603.

[25] Andersson, M., Alvarez-Cermeno, J., Cogato, I. *et al.* (1994). The role of cerebrospinal fluid analysis in the diagnosis of multiple sclerosis: a consensus report. *J. Neurol. Neurosurg. Psychiatry*, **57**, 897–903.

[26] Ansari, K. A., Wells, K. and Vatassery, G. T. (1975). Quantitative estimation of cerebrospinal fluid globulins in multiple sclerosis. *Neurology*, **25**, 688–692.

[27] Aoyama, J., Inaba, G. and Shimizu, T. (1979). Third complement component in cerebrospinal fluid in neuro-Behçets syndrome. *J. Neurol. Sci.*, **41**, 183–190.

[28] Arnadottir, T., Kekomaki, R., Lund, G. A. *et al.* (1982). Circulating immune complexes in patients with multiple sclerosis. *J. Neurol. Sci.*, **55**, 273–283.

[29] Arnadottir, T., Reunanen, M., Meurman, O. *et al.* (1979). Measles and rubella virus antibodies in patients with multiple sclerosis. *Arch. Neurol.*, **36**, 261–265.

[30] Arnon, R., Crisp, E., Kelley, R. *et al.* (1980). Anti-ganglioside antibodies in multiple sclerosis. *J. Neurol. Sci.*, **46**, 179–186.

[31] Arnon, R., Kelley, R., Schumaker, N. and Fahey, J. L. (1979). Idiotype anti-idiotype complexes in cerebrospinal fluids of multiple sclerosis patients. *J. Neurol. Sci.*, **43**, 149–156.

[32] Baark, J.-P., Berg, O. and Hemmingsen, L. (1983). Immunonephelometric determination of proteins in cerebrospinal fluid in various neurological disorders. *Clin. Chim. Acta*, **127**, 271–277.

[33] Baggio, G., Bertolotto, A., Moretti, M. G. and Palmucci, L. (1978). Two-dimensional immunoelectrophoresis of unconcentrated cerebrospinal fluid. *J. Neurol. Sci.*, **37**, 199–203.

[34] Bahmanyar, S., Moreau-Dubois, M. C., Brown, P. *et al.* (1983). Serum antibodies to neurofilament antigens in patients with neurological and other diseases and healthy controls. *J. Neuroimmunol.*, **5**, 191–196.
[35] Baird, L. C., Tachovsky, T. G., Sandberg-Wollheim, M. *et al.* (1980). Identification of a unique idiotype in cerebrospinal fluid and serum of a patient with multiple sclerosis. *J. Immunol.*, **124**, 2324–2328.
[36] Bal, V., Kamat, R. S., Kamat, J. and Kandoth, P. (1983). Enzyme-linked immunosorbent assay for mycobacterial antigens. *Indian, J. Med. Res.*, **78**, 477–483.
[37] Bammer, H. (1966). Liquorkomplement und multiple sklerose. *Deut Zeitsch Nerv.*, **188**, 271–288.
[38] Banik, N. L. and Hogan, E. H. (1983). Cerebrospinal fluid enzymes in neurological diseases. In: J. H. Wood (ed.), *Neurobiology of Cerebrospinal Fluid*, pp. 205–231. Plenum.
[39] Barbarese, E., Braun, P. E. and Carson, J. H. (1977). Identification of prelarge and presmall basic proteins in mouse myelin and their structural relationship to large and small basic proteins. *Proc. Natl. Acad. Sci.*, **74**, 3360–3364.
[40] Baringer, J. R., Gajdusek, D. C., Gibbs, C. J. *et al.* (1980). Transmissible dementias: current problems in tissue handling. *Neurology*, **30**, 302–303.
[41] Barrett, A. J., Brown, M. A. and Sayers, C. A. (1979). The electrophoretically 'slow' and 'fast' forms of the alpha 2-macroglobulin molecule. *Biochem. J.*, **181**, 401–418.
[42] Barrett, A. J., Davies, E. and Grubb, A. (1984). The place of human gamma-trace (cystatin C) amongst the cysteine proteinase inhibitors. *Biochem. Biophys. Res. Commun.*, **120**, 631–636.
[43] Bartel, D. R., Markand, O. N. and Kolar, O. J. (1983). The diagnosis and classification of multiple sclerosis: evoked responses and spinal fluid electrophoresis. *Neurology*, **33**, 611–617.
[44] Bashir, R. M. and Whitaker, J. N. (1980). Molecular features of immunoreactive myelin basic protein in cerebrospinal fluid of persons with multiple sclerosis. *Ann. Neurol.*, **7**, 50–57.
[45] Basten, A., Pollard, J. D., Steward, G. J. *et al.* (1980). Transfer factor in treatment of multiple sclerosis. *Lancet*, **ii**, 931–934.
[46] Bauer, H. (1953). Uber die bedeutung der papier-elektrophorese des liquors fur die klinische forschung. *Deut. Zeit. Nervenh.*, **170**, 381–401.
[47] Bauer, H. and Gottesleben, A. (1969). Quantitative immunochemical studies of cerebrospinal fluid proteins in relation to clinical activity of multiple sclerosis. *Pathol. Etiol. Demyel. Dis.*, **36**, 643–648.
[48] Baumgarten, A. (1987). Viral antigen detection in CSF. In: E. J., Thompson (ed.), *Advances in CSF Protein Research and Diagnosis*, pp. 129–149. MTP Press Ltd.
[49] Becker, K. L., Silva, O. L., Post, R. M. *et al.* (1980). Immunoreactive calcitonin in cerebrospinal fluid of man. *Brain Res.*, **194**, 598–602.
[50] Beer, S., Rosler, K. M. and Hess, C. W. (1995). Diagnostic value of paraclinical tests in multiple sclerosis: relative sensitivities and specificities for reclassification according to the Poser committee criteria. *J. Neurol. Neurosurg. Psychiatry*, **59**, 152–159.
[51] Behan, P. O., Geschwind, N., Lamarche, J. B. *et al.* (1968). Delayed hypersensitivity to encephalitogenic protein in disseminated encephalomyelitis. *Lancet*, **ii**, 1009–1011.

REFERENCES

[52] Behr, W., Sclimok, G., Firchau, V. and Paul, H. A. (1985). Determination of reference intervals for 10 serum proteins measured by rate nephelometry, taking into consideration different sample groups and different distribution functions. *J. Clin. Chem. Clin. Biochem.*, **23**, 157–166.

[53] Bell, J. B., Davies, R. A. and Thompson, E. J. (2003). Herpes simplex encephalitis. A study of seven patients and their immunological response prior to routine acyclovir treatment. *J. Infect.*, **47**, 161–163.

[54] Bell, R. D., Rosenberg, R. N., Ting, R. et al. (1978). Creatine kinase BB isoenzyme levels by radioimmunoassay in patients with neurological disease. *Ann. Neurol.*, **3**, 52–59.

[55] Berger, T., Rubner, P., Schautzer, F. et al. (2003). Antimyelin antibodies as a predictor of clinically definite multiple sclerosis after a first demyelinating event. *N. Engl. J. Med.*, **349**, 139–145.

[56] Bernard, A. M., Moreau, D. and Lauwerys, R. R. (1982). Latex immunoassay of retinol-binding protein. *Clin. Chem.*, **28**, 1167–1171.

[57] Bernard, A. M., Vyskocil, A. and Lauwerys, R. R. (1981). Determination of beta 2 microglobulin in human urine and serum by latex immunoassay. *Clin. Chem.*, **27**, 832–837.

[58] Bernard, D., Ripert, G., Haddad, A. et al. (1974). An immunological study of subacute sclerosing panencephalitis: measles and tissue antigens and immunoglobulins. *Ann. Immunol.*, **125 C**, 461–489.

[59] Bernhardt, W. and Weisner, B. (1978). Liquorbefund und lebensalter. Trivariate wertung der konzentrationen von albumin und immunglobulin G. *J. Clin. Chem. Chim. Biochem.*, **16**, 435–439.

[60] Bernheimer, H., Budka, H. and Muller, P. (1983). Brain tissue immunoglobulins in adrenoleukodystrophy: a comparison with multiple sclerosis and systemic lupus erythematosus. *Acta Neuropathol.*, **59**, 95–102.

[61] Bickerstaff, E. R. (1950). Changes in the cerebrospinal fluid during pneumoencephalography. *Lancet*, **ii**, 683–685.

[62] Bieschke, J., Giese, A., Schulz-Schaeffer, W. et al. (2000). Ultrasensitive detection of pathological prion protein aggregates by dual-colour scanning for intensely fluorescent targets. *Proc. Natl. Acad. Sci. USA*, **97**, 5468–5473.

[63] Biggins, J. A., Taylor, A. and Caspary, E. A. (1978). Cirulating antibody to myelin basic protein in relapsing-remitting multiple sclerosis. *J. Neurol. Neurosurg. Psychiatry*, **41**, 1131–1134.

[64] Bing, D. H., Almeda, S., Isliker, H. et al. (1982). Fibronectin binds to the C1q component of complement. *Proc. Natl. Acad. Sci.*, **79**, 4198–4201.

[65] Birry, A. and Kastiya, S. (1979). Evaluation of microhaemagglutination assay to determine treponemal antibodies in CSF. *Br. J. Vener. Dis.*, **55**, 239–244.

[66] Blake, C. C. F., Geisow, M. J. and Oatley, S. J. (1978). Structure of prealbumin: secondary, tertiary and quaternary interactions determined by Fourier refinement at 1.8 A. *J. Mol. Biol.*, **121**, 339–356.

[67] Blau, J. N., Harris, H. and Robson, E. B. (1963). Haptoglobins in cerebrospinal fluid. *Clin. Chim. Acta*, **8**, 202–206.

[68] Blomkeck, B. and Hanson, L. A. (1979). V. Diagnostic use of plasma protein analyses. In: B. Blomkeck and L. A. Hanson (eds.), *Plasma Proteins*, pp. 335–369. Wiley.

REFERENCES

[69] Bloomer, L. C. and Bray, P. F. (1981). Relative value of three laboratory methods in the diagnosis of multiple sclerosis. *Clin. Chem.*, **27**, 2011–2013.

[70] Blundell, T. L. and Humbel, R. E. (1980). Hormone families: pancreatic hormones and homologous growth factors. *Nature*, **287**, 781–787.

[71] Bock, E. (1973). Non-plasma proteins in cerebrospinal fluid. *Scand. J. Immunol.*, **2**, 119–124.

[72] Bock, E. (1973). Quantitation of plasma proteins in cerebrospinal fluid. *Scand. J. Immunol.*, **2**, 111–117.

[73] Bock, E. (1978). Immunoglobulins, prealbumin, transferrin, albumin, and alpha-2-macroglobulin in cerebrospinal fluid and serum in schizophrenic patients. In: D. Bergsma, A. L. Goldstein, B. Harber et al. (eds.), *Neurochemical and Immunologic Components in Schizophrenia*, pp. 283–295. A R Liss Inc.

[74] Bock, E. and Dissing, J. (1975). Demonstration of enolase activity connected to the brain-specific protein, 14-3-2. *Scand. J. Immunol.*, **4** (Suppl 2), 31–36.

[75] Bock, E., Kristensen, V. and Rafaelsen, O. (1974). Proteins in serum and cerebrospinal fluid in demented patients. *Acta Neurol. Scand.*, **50**, 91–102.

[76] Bollengier, F. (1979). Bound and free light chains in serum from patients affected with various neurological diseases. *J. Clin. Chem. Clin. Biochem.* **17**, 45–49.

[77] Bollengier, F., Delmotte, P. and Lowenthal, A. (1976). Biochemical Findings in Multiple Sclerosis, III. Immunoglobulins of restricted heterogeneity and light chain distribution in cerebrospinal spinal fluid of patients with multiple sclerosis. *J. Neurol.*, **212**, 151–158.

[78] Bollengier, F., Lowenthal, A. and Henrotin, W. (1975). Bound and free light chains in subacute sclerosing panencephalitis and multiple sclerosis serum and cerebrospinal fluid. *Z. Klin. Chem. Klin. Biochem.*, **13**, 305–310.

[79] Bollengier, F. and Mahler, A. (1979). Measles antibodies and kappa-lambda chain distribution in immunoglobulins of patients affected with multiple sclerosis. *J. Neurol.*, **220**, 105–112.

[80] Bollengier, F., Mahler, A. and Clinet, G. (1981). Measles antibodies, kappa-lambda light chain distribution and immunoglobulins in serum, cerebrospinal fluid and brain of a patient affected with multiple sclerosis. *J. Neurol.*, **225**, 135–143.

[81] Bollengier, F., Mahler, A., Clinet, G. and Lowenthal, A. (1978). Multiple sclerosis: oligoclonal IgG, kappa lambda light chain distribution and measles antibodies in brain extracts. *Brain Res.*, **152**, 133–144.

[82] Bollengier, F., Rabinovitch, N. and Lowenthal, A. (1978). Oligoclonal immunoglobulins, light chain ratios and free light chains in cerebrospinal fluid and serum from patients affected with various neurological diseases. *J. Clin. Chem. Clin. Biochem.*, **16**, 165–173.

[83] Booij, J. (1959). Agar-agar electrophoresis as an aid in cerebrospinal fluid diagnostics. *Folia Psychiatr. Neurol.*, **62**, 247–253.

[84] Bordas, M. C., Biou, D. R., Feger, J. M. et al. (1981). Involvement of sialic acid residues in electroimmunodiffusion of alpha 1 acid glycoprotein. A method for determining the degree of sialylation of serum glycoproteins. *Clin. Chim. Acta*, **116**, 17–24.

[85] Boston, P. F., Jackson, P., Kynoch, P. A. M. and Thompson, R. J. (1982). Purification, properties, and immunohistochemical localisation of human brain 14-3-3 protein. *J. Neurochem.*, **38**, 1466–1474.

[86] Boston, P. F., Jackson, P. and Thompson, R. J. (1982). Human 14-3-3 Protein: Radioimmunoassay, tissue distribution, and cerebrospinal fluid levels in patients with neurological disorders. *J. Neurochem.*, **38**, 1475–1482.

[87] Bottcher, J. and Trojaborg, W. (1982). Follow-up of patients with suspected multiple sclerosis: a clinical and electrophysiological study. *J. Neurol. Neurosurg. Psychiatry*, **45**, 809–814.

[88] Bouloukos, A., Lekakis, J., Michael, J. and Kalofoutis, A. (1980). Immunoglobulins in cerebrospinal fluid in various neurologic disorders. *Clin. Chem.*, **26**, 115–116.

[89] Bourguignat, A., Albert, A., Ferard, G. *et al.* (1907). Prognostic value of combined data on enzymes and inflammation markers in plasma in cases of severe head injury. *Clin. Chem.*, **29**, 1904–1907.

[90] Bradbury, M. (1979). *The Concept of a Blood-Brain Barrier*, pp. 1–516. Wiley, Chichester.

[91] Bradbury, M. W. B. (1984). The structure and function of blood-brain barrier. *Fed. Proc.*, **43**, 186–190.

[92] Bradshaw, P. (1964). The relation between clinical activity and level of gamma globulin in the cerebrospinal fluid in patients with multiple sclerosis. *J. Neurol. Sci.*, **1**, 374–379.

[93] Bray, P. F., Bloomer, L. C., Salmon, V. C. *et al.* (1983). Epstein-Barr virus infection and antibody synthesis in patients with multiple sclerosis. *Arch. Neurol.*, **40**, 406–408.

[94] Brendel, K., Meezan, E. and Nagle, R. B. (1978). The acellular perfused kidney: a model for basement membrane permeability. In: N. A. Kefalides (ed.), *Biology and Chemistry of Basement Membranes*, pp. 177–193. Academic Press.

[95] Brendel, S., Mulder, J. and Verhaar, M. A. T. (1974). Detection of IgM heterogeneity dependent on the support medium of immunoelectrophoresis. *Clin. Chim. Acta*, **53**, 29–33.

[96] Brendel, S., Mulder, J. and Verhaar, M. A. T. (1974). Heterogeneity of monoclonal immunoglobulin-G proteins studied by isoelectric focusing. *Clin. Chim. Acta*, **54**, 243–248.

[97] Brisby, H., Olmarker, K., Rosengren, L. *et al.* (1999). Markers of nerve tissue injury in the cerebrospinal fluid in patients with lumbar disc herniation and sciatica. *Spine*, **24**, 742–746.

[98] Broadwell, R. D., Balin, B. J., Salcman, M. and Kaplan, R. S. (1983). Brain-Blood barrier? Yes and no. *Proc. Natl. Acad. Sci.*, **80**, 7352–7356.

[99] Broadwell, R. D. and Salcman, M. (1981). Expanding the definition of the blood-brain barrier to protein. *Proc. Natl. Acad. Sci.*, **78**, 7820–7824.

[100] Brodtkorb, E., Lindqvist, M., Jonsson, M. and Gustafsson, A. (1982). Diagnosis of herpes simplex encephalitis – A comparison between electroencephalography and computed tomography findings. *Acta Neurol. Scand.*, **66**, 462–471.

[101] Broman, T. (1964). Blood-brain barrier damage in multiple sclerosis: supravital test observations. *Acta Neurol. Scand.*, **40**, 21–24.

[102] Broome, S. and Gilbert, W. (1978). Immunological screening method to detect specific translaton products. *Proc. Natl. Acad. Sci.*, **75**, 2746–2749.

[103] Brouwer, J. and Leeuwen-Herberts, T. V. (1983). Estimation of IgM in cerebrospinal fluid by enzyme immunoassay. *Clin. Chim. Acta*, **131**, 337–342.

[104] Brown, K. W., Kynoch, P. A. M. and Thompson, R. J. (1980). Immunoreactive nervous system specific enolase (14-3-2 protein) in human serum and cerebrospinal fluid. *Clin. Chim. Acta*, **101**, 257–264.

[105] Brown, P., Gibbs, C. J., Amyx, H. L. *et al.* (1982). Chemical disinfection of Creutzfeldt-Jakob disease virus. *N. Engl. J. Med.*, **306**, 1279–1282.
[106] Bruck, W., Bitsch, A., Bruck, Y. *et al.* (1997). Inflammatory central nervous system demyelination: correlation of magnetic resonance imaging findings with lesion pathology. *Ann. Neurol.*, **42**, 783–793.
[107] Brueton, M. J., Breeze, G. R. and Stuart, J. (1976). Fibrin-fibrinogen degradation products in cerebrospinal fluid. *J. Clin. Pathol.*, **29**, 341–344.
[108] Brueton, M. J., Tugwell, P., Whittle, H. C. and Greenwood, B. M. (1974). Fibrin degradation products in the serum and cerebrospinal fluid of patients with group A meningoccal meningitis. *J. Clin. Pathol.*, **27**, 402–404.
[109] Brydon, H. L., Keir, G., Thompson, E. J. *et al.* (1998). Protein adsorption to hydrocephalus shunt catheters: CSF protein adsorption. *J. Neurol. Neurosurg. Psychiatry*, **64**, 643–647.
[110] Burgoon, M. P., Hammack, B. N., Owens, G. P. *et al.* (2003). Oligoclonal immunoglobulins in cerebrospinal fluid during varicella zoster virus (VZV) vasculopathy are directed against VZV. *Ann. Neurol.*, **54**, 459–463.
[111] Bushell, A. C., Whicher, J. T. and Yuille, T. (1979). The progressive appearance of multiple urinary Bence-Jones proteins and serum paraproteins in a child with immune deficiency. *Clin. Exp. Immunol.*, **38**, 64–69.
[112] Bussone, G., La Mantia, L., Boiardi, A. and Giovannini, P. (1982). Chorea in Behcet's syndrome. *Neurology*, **227**, 89–92.
[113] Canoso, J. J. and Cohen, A. S. (1975). Aseptic meningitis in systemic lupus erythematosus. *Arthritis Rheum.*, **18**, 369–374.
[114] Cappel, R., Thiry, L. and Clinet, G. (1975). Viral Antibodies in the CSF after acute CNS infections. *Arch. Neurol.*, **32**, 629–631.
[115] Carlsson, J., Armstrong, V. W., Reiber, H. and Seidel, D. (1991). Clinical relevance of the quantification of apolipoprotein E in cerebrospinal fluid. *Clin. Chim. Acta*, **196**, 167–176.
[116] Caroscio, J. T., Kochwa, S., Sacks, H. *et al.* (1983). Quantitative CSF IgG measurements in multiple sclerosis and other neurologic diseases. *Arch. Neurol.*, **40**, 409–413.
[117] Carroll, J. F. and Booss, J. (1976). Cerebrospinal fluid IgG level in herpes simplex encephalitis. *JAMA*, **236**, 2092–2093.
[118] Carson, J. H., Barbarese, E., Braun, P. E. and McPherson, T. A. (1978). Components in multiple sclerosis cerebrospinal fluid that are detected by radioimmunoassay for myelin basic protein. *Proc. Natl. Acad. Sci.*, **75**, 1976–1978.
[119] Carter, M. J., Willcocks, M. M. and ter Mevlen, V. (1983). Defective translation of measles virus matrix protein in a subacute sclerosing panencephalitis cell line. *Nature*, **305**, 153–155.
[120] Casey, B. R., Wong, S. T., Mason, A. J. *et al.* (1981). The electrophoretic demonstration of unique oligoclonal immunoglobulins in cerebrospinal fluid as a diagnostic test for multiple sclerosis. *Clin. Chim. Acta*, **114**, 187–194.
[121] Caspary, E. A. (1965). Comparison of immunological specificity of gamma globulin in the cerebrospinal fluid in normal and multiple sclerosis subjects. *J. Neurol. Neurosurg. Psychiatry*, **28**, 61–64.
[122] Castaigne, P., Lhermitte, F., Schuller, E. *et al.* (1972). Valeur diagnostique de la distribution oligoclonale des gamma-globulins dans le liquide cephalo-rachidien. *Rev. Eur. Etud. Clin. Biol.*, **17**, 324–327.

[123] Castaigne, P., Lhermitte, F., Schuller, E. et al. (1972). Oligoclonal aspect of gamma globulins in CS. *Mul. Scl. Prog. Res.*, **17**, 152–158.
[124] Castaigne, P., Lhermitte, F., Schuller, E. et al. (1972). Symptomatic and prognostic value of electrophoretic analysis of cerebrospinal fluid proteins in multiple sclerosis. In: V. Leibowitz (ed.), *Progress in Multiple Sclerosis Research and Treatment*, pp. 164–185. Academic Press.
[125] Castaigne, P., Lhermitte, F., Schuller, E. and Rouques, C. (1971). Les proteines du liquide cephalo-rachidien au cours de la sclerose laterale amyotrophique. *Rev. Neurol.*, **125**, 393–400.
[126] Catterall, R. D. (1977). Neurosyphilis. *Br. J. Hosp. Med.*, **17**, 585–604.
[127] Cawley, L. P., Minard, J., Tourtellotte, W. W. et al. (1976). Immunofixation electrophoretic techniques applied to identification of proteins in serum and cerebrospinal fluid. *Clin. Chem.*, **22**, 1262–1268.
[128] Cerda, M. and Basauri, L. (1980). Isoelectric focusing of cerebrospinal fluid proteins in children with nontumoral hydrocephalus. *Child's Brain*, **7**, 170–181.
[129] Cerda, M., Vielma, J. and Basauri, L. (1978). Electroforesis en gel de poliacrilamida de proteinas del liquido cefalorraquideo. *Neurocirugia*, **36**, 330–333.
[130] Cerda, M., Vielma, J., Martinez, C. and Basauri, L. (1980). Polyacrylamide gel electrophoresis of cerebrospinal fluid proteins in children with nontumoral hydrocephalus. *Child's Brain*, **6**, 140–149.
[131] Chamoun, V. and Thompson, E. J. (1994). CSF analysis in the diagnosis of multiple sclerosis. *Postgrad. Doct.*, **17**, 116–119.
[132] Chamoun, V., Zeman, A., Blennow, K. et al. (2001). Haptoglobins as markers of blood-CSF barrier dysfunction: the findings in normal CSF. *J. Neurol. Sci.*, **182**, 117–121.
[133] Chapel, H. M., Esiri, M. M. and Wilcock, G. K. (1984). Immunoglobulin and other proteins in the cerebrospinal fluid of patients with Alzheimer's disease. *J. Clin. Pathol.*, **37**, 697–699.
[134] Chataway, J., Davies, N. W. S., Farmer, S. et al. (2004). Herpes simplex encephalitis: an audit of the use of laboratory diagnostic tests. *Q. J. Med.*, **97**, 325–330.
[135] Chazot, G., Lasne, Y., Benzerara, O. et al. (1980). Radio-immunofixation: une nouvelle technique de caracterisation des immunoglobulines dans le liquide cephalo-rachidien non concentre. *Rev. Neurol.*, **136**, 783–786.
[136] Chen, A. B., Ben Porat, T., Whitley, J. and Kaplan, A. S. (1978). Purification and characterization of proteins excreted by cells infected with herpes simplex virus and their use in diagnosis. *Virology*, **91**, 234–242.
[137] Choe, L. H., Green, A., Knight, R. S. et al. (2002). Apolipoprotein E and other cerebrospinal fluid proteins differentiate ante mortem variant Creutzfeldt-Jakob disease from ante mortem sporadic Creutzfeldt-Jakob disease. *Electrophoresis*, **23**, 2242–2246.
[138] Chou, C. H. J., Chou, F. C. H., Tourtellotte, W. W. and Kibler, R. F. (1983). Search for a multiple sclerosis-specific brain antigen. *Neurology*, **33**, 1300–1304.
[139] Chou, C. H. J., Chou, F. C. H., Tourtellotte, W. W. and Kibler, R. F. (1984). Devic's syndrome: antibody to glial fibrillary acidic protein in cerebrospinal fluid. *Neurology*, **34**, 86–88.
[140] Chou, C. H. J., Tourtellotte, W. W. and Kibler, R. F. (1983). Failure to detect antibodies to myelin basic protein or peptic fragments of myelin basic protein in CSF of patients with MS. *Neurology*, **33**, 24–28.

[141] Chrambach, A., An der L. B., Mohrmann, H. and Felgenhauer, K. (1981). Toward an improved immunoglobulin analysis by gel electrophoresis and electrofocusing. *Electrophoresis*, **2**, 279–287.

[142] Christensen, O., Clausen, J. and Fog, T. (1978). Relationships between abnormal IgG index, oligoclonal bands, acute phase reactants and some clinical data in multiple sclerosis. *J. Neurol.*, **218**, 237–244.

[143] Christensen, P., Johansson, A. and Nielsen, V. (1978). Quantitation of protein adsorbance to glass and plastics: investigation of a new tube with low adherence. *J. Immunol. Methods*, **23**, 23–28.

[144] Christenson, R. H., Behlmer, P., Howard, J. F. *et al.* (1983). Interpretation of cerebrospinal fluid protein assays in various neurologic diseases. *Clin. Chem.*, **29**, 1028–1030.

[145] Christopherson, R. I., Jones, M. E. and Finch, L. R. (1979). A simple centrifuge column for desalting protein solutions. *Anal. Biochem.*, **100**, 184–187.

[146] Chu, A. B., Sever, J. L., Madden, D. L. *et al.* (1983). Oligoclonal IgG bands in cerebrospinal fluid in various neurological diseases. *Ann. Neurol.*, **13**, 434–439.

[147] Church, A. J., Cardoso, F., Dale, R. C. *et al.* (2002). Anti-basal ganglia antibodies in acute and persistent Sydenham's chorea. *Neurology*, **59**, 227–231.

[148] Church, A. J., Dale, R. C., Cardoso, F. *et al.* (2003). CSF and serum immune parameters in Sydenham's chorea: evidence of an autoimmune syndrome? *J. Neuroimmunol.*, **136**, 149–153.

[149] Ciardi, M., Sharief, M. K., Noori, M. A. *et al.* (1993). Intrathecal synthesis of interleukin-2 and soluble IL-2 receptor in asymptomatic HIV-1 seropositive individuals. Correlation with local production of specific IgM and IgG antibodies. *J. Neurol. Sci.*, **115**, 117–122.

[150] Ciardi, M., Sharief, M. K., Thompson, E. J. *et al.* (1994). High cerebrospinal fluid and serum levels of tumor necrosis factor-alpha in asymptomatic HIV-1 seropositive individuals. Correlation with interleukin-2 and soluble IL-2 receptor. *J. Neurol. Sci.*, **125**, 175–179.

[151] Cinque, P., Cleator, G. M., Weber, T. *et al.* (1996). The role of laboratory investigation in the diagnosis and management of patients with suspected herpes simplex encephalitis: a consensus report. The EU Concerted Action on Virus Meningitis and Encephalitis. *J. Neurol. Neurosurg. Psychiatry*, **61**, 339–345.

[152] Clanet, M., Comi, G., Fernandez, O. *et al.* (1995). Guidelines for early treatment trials on patients presenting with clinical and paraclinical abnormalities which put them at high risk for conversion to multiple sclerosis. *Mult. Scler.*, **1**, 55–59.

[153] Clark, B. R. and Todd, C. W. (1982). Avidin as a precipitant for biotin-labeled antibody in a radioimmunoassay for carcinoembryonic antigen. *Anal. Biochem.*, **121**, 257–262.

[154] Clark, C. A., Downs, E. C. and Primus, F. J. (1982). An unlabeled antibody method using glucose oxidase-antiglucose oxidase complexes (GAG). *J. Histochem. Cytochem.*, **30**, 27–34.

[155] Clark, C. M., Xie, S., Chittams, J. *et al.* (2003). Cerebrospinal fluid tau and beta-amyloid. *Arch. Neurol.*, **60**, 1696–1702.

[156] Clark, C. M., Xie, S., Chittams, J. *et al.* (2003). Cerebrospinal fluid tau and beta-amyloid: how well do these biomarkers reflect autopsy-confirmed dementia diagnoses? *Arch. Neurol.*, **60**, 1696–1702.

[157] Clausen, J. (1983). Serum antibodies against cytosol antigens in multiple sclerosis. *J. Neurol. Sci.*, **60**, 205–216.

[158] Clausen, J., Fog, T. and Roboz-Einstein, E. (1969). The clinical value of assaying proteins in the cerebrospinal fluid. *Acta Neurol. Scand.*, **45**, 513–528.

[159] Cleland, P. G., MacFarlane, J. T., Baird, D. R. and Greenwood, B. M. (1979). Fibrin degradation products in the cerebrospinal fluid of patients with pneumococcal meningitis. *J. Neurol. Neurosurg. Psychiatry*, **42**, 843–846.

[160] Cloppet, H., Francina, A., Coquelin, H. *et al.* (1982). Laser nephelometry and radial immunodiffusion compared for immunoglobulin quantification in pathological sera. *Clin. Chem.*, **28**, 180–183.

[161] Cohen, S. and Bannister, R. (1967). Immunoglobulin synthesis within the central nervous system in disseminated sclerosis. *Lancet*, **1**, 366–367.

[162] Cohen, S. R., Brooks, B. R., Herndon, R. M. *et al.* (1980). A diagnostic index of active demyelination: myelin basic protein in cerebrospinal fluid. *Ann. Neurol.*, **8**, 25–31.

[163] Cohen, S. R., Brooks, B. R., Jubelt, B. *et al.* (1980). Myelin basic protein in cerebrospinal fluid. In: J. H. Wood (ed.), *Neurobiology of Cerebrospinal Fluid*, pp. 487–493. Plenum Press.

[164] Coleman, R. M., Bailey, P. D., Whitley, R. J. *et al.* (1983). ELISA for the detection of herpes simplex virus antigens in the cerebrospinal fluid of patients with encephalitis. *J. Virol. Methods*, **7**, 117–125.

[165] Colover, J., Feinberg, J. G., Temple, A. and Tooley, M. (1963). A comparsion of measurement of gamma globulin in cerebrospinal fluid by electrophoretic and immunological methods. *J. Clin. Pathol.*, **16**, 370–374.

[166] Colten, H. R., Ooi, Y. M. and Edelson, P. J. (1979). Synthesis and secretion of complement proteins by macrophages. *Ann. N Y Acad. Sci.*, **332**, 482–490.

[167] Confavreux, C., Gianazza, E., Arnaud, P. *et al.* (1982). Coloration au nitrate d'argent des proteines du liquide cephalo-rachidien non concentre. *Rev. Neurol.*, **138**, 317–325.

[168] Coren, M. E., Buchdahl, R. M., Cowan, F. M. *et al.* (1999). Imaging and laboratory investigation in herpes simplex encephalitis. *J. Neurol. Neurosurg. Psychiatry*, **67**, 243–245.

[169] Corrall, C. J., Pepple, J. M., Moxon, E. R. and Hughes, W. T. (1981). C-reactive protein in spinal fluid of children with meningitis. *J. Pediatr.*, **99**, 365–369.

[170] Coyle, P. K., Hirsch, R. L., O'Donnell, P. *et al.* (1980). Cerebrospinal-fluid lymphocyte populations and immune complexes in active multiple sclerosis. *Lancet*, **ii**, 229–233.

[171] Coyle, P. K. and Johnson, C. (1983). Optimal detection of oligoclonal bands in CSF by silver stain. *Neurology*, **33**, 1510–1512.

[172] Coyle, P. K. and Procyk-Dougherty, Z. (1984). Multiple sclerosis immune complexes: An analysis of component antigens and antibodies. *Ann. Neurol.*, **16**, 660–667.

[173] Coyle, P. K. and Wolinsky, J. S. (1981). Characterization of immune complexes in progressive rubella panencephalitis. *Ann. Neurol.*, **9**, 557–562.

[174] Cremer, N. E., Johnson, K. P., Fein, G. and Likosky, W. H. (1980). Comprehensive viral immunology of multiple sclerosis. II analysis of serum and CSF antibodies by standard serologic methods. *Arch. Neurol.*, **37**, 610–615.

[175] Crichton, R. R., Wilmet, S., Legssyer, R. and Ward, R. J. (2001). Quantitative analysis of cell death and ferritin expression in response to cortical iron: implications for hypoxia-ischemia and stroke. *Brain Res.*, **91**, 9–18.

[176] Croce, C. M., Linnenbach, A., Hall, W. *et al.* (1980). Production of human hybridomas secreting antibodies to measles virus. *Nature*, **288**, 488–489.
[177] Crow, T. J., Johnstone, E. C., Owens, D. G. C. *et al.* (1979). Characteristics of patients with schizophrenia or neurological disorder and virus-like agent in cerebrospinal fluid. *Lancet*, **i**, 842–844.
[178] Cullmann, W. and Dick, W. (1981). A chromogenic assay for evaluation of alpha 2 macroglobulin level in serum and cerebrospinal fluid. *J. Clin. Chem. Clin. Biochem.* **19**, 287–290.
[179] Cumings, J. N. (1952). The diagnostic value of the examination of the cerebrospinal fluid. *Postgrad. Med. J.*, **28**, 287–292.
[180] Cumings, J. N. (1953). The examination of the cerebrospinal fluid and cerebral cyst fluid by paper strip electrophoresis. *J. Neurol. Neurosurg. Psychiatry*, **16**, 152–157.
[181] Cumings, J. N. (1954). The cerebrospinal fluid in diagnosis. *Br. Med. J.* **1**, 1–8.
[182] Cumings, J. N., Shortman, R. C. and Tooley, M. (1969). Polyacrylamide disc electrophoresis of cerebrospinal fluid and cerebral cyst fluids. *Clin. Chim. Acta*, **27**, 29–34.
[183] Cumings, J. N., Thompson, E. J. and Goodwin, H. (1968). Sphingolipids and phospholipids in microsomes and myelin from normal and pathological brains. *J. Neurochem.*, **15**, 243–248.
[184] Cunningham, V. R. (1964). Analysis of native cerebrospinal fluid by the polyacrylamide disc electrophoresis technique. *J. Clin. Pathol.*, **17**, 143–148.
[185] Cutler, R. W. P., Deuel, R. K. and Barlow, C. F. (1967). Albumin exchange between plasma and cerebrospinal fluid. *Arch. Neurol.*, **17**, 261–270.
[186] Cutler, R. W. P., Watters, G. V. and Hammerstad, J. P. (1970). The origin and turnover rates of cerebrospinal fluid albumin and gamma-globulin in man. *J. Neurol. Sci.*, **10**, 259–268.
[187] Cutler, R. W. P., Watters, G. V., Hammerstad, J. P. and Merler, E. (1967). Origin of cerebrospinal fluid gamma globulin in subacute sclerosing leukoencephalitis. *Arch. Neurol.*, **17**, 620–628.
[188] Dalakas, M. C., Li, M., Fujii, M. and Jacobowitz, D. M. (2003). Stiff person syndrome. Quantification, specificity, and intrathecal synthesis of GAD65 antibodies. *Neurology*, **57**, 780–784.
[189] Dale, R. C., Church, A. J., Benton, S. *et al.* (2002). Post-streptococcal autoimmune dystonia with isolated bilateral striatal necrosis. *Dev. Med. Child Neurol.*, **44**, 485–489.
[190] Dale, R. C., Church, A. J., Cardoso, F. *et al.* (2001). Poststreptococcal acute disseminated encephalomyelitis with basal ganglia involvement and auto-reactive antibasal ganglia antibodies. *Ann. Neurol.*, **50**, 588–595.
[191] Dale, R. C., Church, A. J., Surtees, R. A. *et al.* (2004). Encephalitis lethargica syndrome: 20 new cases and evidence of basal ganglia autoimmunity. *Brain*, **127**, 21–33.
[192] Dale, R. C., Church, A. J., Surtees, R. A. *et al.* (2002). Post-streptococcal autoimmune neuropsychiatric disease presenting as paroxysmal dystonic choreoathetosis. *Mov. Disord.*, **17**, 817–820.
[193] Dalens, B., Bezou, M. J., Coulet, M. and Raynaud, E. J. (1981). Fibrin-fibrinogen degradation products in cerebrospinal fluid as an indicator of neonatal brain damage. *Acta Neurol. Scand.*, **64**, 81–87.

[194] Dasgupta, M. K., Warren, K. G., Johny, K. V. and Dossetor, J. B. (1982). Circulating immune complexes in multiple sclerosis: relation with disease activity. *Neurology*, **32**, 1000–1004.

[195] Dau, P. C., Petajan, J. H., Johnson, K. P. et al. (1980). Plasmapheresis in multiple sclerosis: Preliminary findings. *Neurology*, **30**, 1023–1028.

[196] Dauberschmidt, R., Marangos, P. J., Zinsmeyer, J. B. et al. (1983). Severe head trauma and the changes of concentration of neuro-specific enolase in plasma and in cerebrospinal fluid. *Clin. Chim. Acta*, **131**, 165–170.

[197] Dautry-Varsat, A., Ciechanover, A. and Lodish, H. F. (1983). pH and the recycling of transferrin during receptor-mediated endocytosis. *Proc. Natl. Acad. Sci.*, **80**, 2258–2262.

[198] Davies, G., Keir, G., Thompson, E. J. and Giovannoni, G. (2003). The clinical significance of an intrathecal monoclonal immunoglobulin band: a follow-up study. *Neurology*, **60**, 1163–1166.

[199] Davies-Jones, G. A. B. (1969). Lactate dehydrogenase and glutamic oxalacetic transaminase of the cerebrospinal fluid in tumours of the central nervous system. *J. Neurol. Neurosurg. Psychiatry*, **32**, 324–327.

[200] Davis, B. (1964). Disc Electrophoresis- II Method and application to human serum proteins. Part II clinical applications. *Ann. N Y Acad. Sci.*, 404–427.

[201] Davis, L. E. and Sperry, S. (1979). The CSF-FTA test and significance of blood contamination. *Ann. Neurol.*, **6**, 68–69.

[202] Davson, H., Welch, K., Segal, M. B. (1987). Physiology and Pathophysiology of the Cerebrospinal Fluid. pp. 1–1013. Churchill Livingstone.

[203] Day, E., Varitek, V. A. and Paterson, P. Y. (1981). Endogenous myelin basic protein-serum factors (MBP-SFs) in Lewis rats. *J. Neurol. Sci.*, **49**, 1–17.

[204] Day, E. D., Hashim, G. A., Varitek, V. A. et al. (1981). Synthetic peptides from region 65–84 of bovine myelin basic protein. *Neurochem. Res.*, **6**, 913–929.

[205] Day, E. D., Hashim, G. A., Varitek, V. A. et al. (1981). Affinity purification of an acylated and radiolabelled synthetic derivative of residues 75–83 of bovine myelin basic protein 125 I S79. *J. Neuroimmunol.*, **1**, 311–324.

[206] Day, E. D., Hashim, G. A., Varitek, V. A. and Paterson, P. Y. (1981). Equilibrium competitive inhibition analysis of synthetic peptide antigens from myelin basic protein as affected by the dual-dilution phenomenon. *J. Neuroimmunol.*, **1**, 217–226.

[207] De Vecchi, A., Montagnino, G., Massaro, L. et al. (1979). Comparison between immunofixation and crossed immunoelectrophoresis for the detection of C3 activation products. *Clin. Chim. Acta*, **97**, 27–32.

[208] Debiasi, R. L. and Tyler, K. L. (2004). Molecular methods for diagnosis of viral encephalitis. *Clin. Microbiol. Rev.*, **17**, 903–25. Table.

[209] Deisenhammer, F., Keir, G., Pfausler, B. and Thompson, E. J. (1996). Affinity of anti-GM1 antibodies in Guillain-Barre syndrome patients. *J. Neuroimmunol.*, **66**, 85–93.

[210] Deisenhammer, F., Miller, R. F., Brink, N. S. et al. (1997). Cerebrospinal fluid ferritin in HIV infected patients with acute neurological episodes. *Genitourin. Med.*, **73**, 181–183.

[211] Delacroix, D. L., Meykens, R. and Vaerman, J. P. (1982). Influence of molecular size of IgA on its immunoassay by various techniques. *Mol. Immunol.*, **19**, 297–305.

REFERENCES

[212] Delacroix, D. L. and Vaerman, J. P. (1982). Influence of molecular size of IgA on its immunoassay by various techniques III. Immunonephelometry. *J. Immunol. Methods*, **51**, 49–55.
[213] Delank, H. W. (1968). Klinische Erfahrungen mit der polyacrylamide-elektrophoretischen Analyse von Eiweiss im Liquor cerebrospinalis. *Klin. Wochenschr.*, **46**, 779–783.
[214] Delasnerie-Laupretre, N., Prevot, D., Martin-Mondiere, C. and Eizenbaum, J.-F., et al. (1981). Serum and cerebrospinal fluid C2 in multiple sclerosis. *J. Clin. Lab. Immunol.*, **6**, 23–25.
[215] Delmotte, P. (1971). Gel isoelectric focusing of cerebrospinal fluid proteins: a potential diagnostic tool. *Z. klin. Chem. klin. Biochem.*, **9**, 334–336.
[216] Delmotte, P. and Gonsette, R. (1977). Biochemical findings in multiple sclerosis. *J. Neurol.*, **215**, 27–37.
[217] Delmotte, P. and Tourtellotte, W. W. (1980). Immunonephelometric measurement of nine proteins in the CSF (and serum) of forty six normal individuals and calculation of their blood/CSF permeability coefficients. *Behr. Symp.*, **22**, 77–78.
[218] Delpech, B. and Boquet, J. (1976). The blood-CSF barrier for Ig. *Prot. Biol. Fluids*, **23**, 489–491.
[219] Delpech, B. and Lichtblau, E. (1972). Etude quantitative des immunoglobulines G et de L'albumin du liquide cephalo rachidien. *Clin. Chim. Acta*, **37**, 15–23.
[220] Dencker, S. J. and Swahn, B. (1962). Proteins of central nervous origin present in cerebrospinal fluid. *Nature*, **194**, 288–289.
[221] Derfuss, T., Gurkov, R., Then, B. F. et al. (2001). Intrathecal antibody production against Chlamydia pneumoniae in multiple sclerosis is part of a polyspecific immune response. *Brain*, **124**, 1325–1335.
[222] Dermer, G. B. and Edwards, J. J. (1983). Detection of the major non-serum derived proteins of body fluids by immunodeletion and two-dimensional gel electrophoresis. *Electrophoresis*, **4**, 212–218.
[223] Dermer, G. B., Silverman, L. M. and Chapman, J. F. (1982). Enhancement techniques for detecting trace and fluid-specific components in two-dimensional electrophoresis patterns. *Clin. Chem.*, **28**, 759–765.
[224] Detels, R., Myers, L. W., Ellison, G. W. et al. (1981). Changes in immune response during relapses in MS patients. *Neurology*, **31**, 492–495.
[225] DiChiro, G., Hammock, M. K. and Bleyer, A. (1976). Spinal descent of cerebrospinal fluid in man. *Neurology*, **26**, 1–8.
[226] Dommasch, D., Mertins, H. G. (1980). *Cerebrospinalflussigkeit-CSF*. pp. 1–284. Stuttgart: Thieme.
[227] Doran, J. F., Jackson, P., Kynoch, P. A. M. and Thompson, R. J. (1983). Isolation of PGP 9.5, a new human neurone-specific protein detected by high-resolution two-dimensional electrophoresis. *J. Neurochem.*, **40**, 1542–1547.
[228] Dorries, R. and ter Mevlen, V. (1984). Detection and identification of virus-specific, oligoclonal IgG in unconcentrated cerebrospinal fluid by immunoblot technique. *J. Neuroimmunol.*, **7**, 77–89.
[229] Dunlop, E. M., Al Egaily, S. S. and Houang, E. T. (1979). Penicillin levels in blood and CSF achieved by treatment of syphilis. *JAMA*, **241**, 2538–2540.
[230] Dunnette, S. L., Gleich, G. J., Miller, R. D., Kyle, R. A. (1977). Measurement of IgD by a double antibody radioimmunoassay: Demonstration of an apparent trimodal distribution of IgD levels in normal human sera. *J. Immunol.*, **119**, 1727–1731.

REFERENCES

[231] Dwulet, F. E. and Benson, M. D. (1984). Primary structure of an amyloid prealbumin and its plasma precursor in a heredofamilial polyneuropathy of Swedish origin. *Proc. Natl. Acad. Sci.*, **81**, 694–698.

[232] Dwyer, D. S., Bradley, R. J., Urquhart, C. K. and Kearney, J. F. (1983). Naturally occurring anti-idiotypic antibodies in myasthenia gravis patients. *Nature*, **301**, 611–614.

[233] Ebers, G. C. (1982). A study of CSF idiotypes in multiple sclerosis. *Scand. J. Immunol.*, **16**, 151–161.

[234] Ebers, G. C. and Paty, D. W. (1980). CSF electrophoresis in one thousand patients. *J. Can. Sci. Neurol.*, **7**, 275–280.

[235] Ehrenkranz, N. J., Zemel, E. S., Bernstein, C. and Slater, K. (1974). Immunoglobulin M in the cerebrospinal fluid of patients with arbovirus encephalitis and other infections of the central nervous system. *Neurology*, **24**, 976–980.

[236] Eickhoff, K. and Heipertz, R. (1977). Discrimination of elevated immunoglobulin concentrations in CSF due to inflammatory reaction of the central nervous system and blood-brain-barrier dysfunction. *Acta Neurol. Scand.*, **56**, 475–482.

[237] Eickhoff, K. and Heipertz, R. (1978). Determination of immunoglobulin content of CSF based on light chain characteristics. *Ann. Neurol.*, **3**, 509–512.

[238] Eickhoff, K., Heipertz, R. and Wikstrom, J. (1978). Determination of kappa lambda immunoglobulin light chain ratios in CSF from patients with multiple sclerosis and other neurological diseases. *Acta Neurol. Scand.*, **57**, 385–395.

[239] Eickhoff, K., Kaschka, W., Skvaril, F. et al. (1979). Determination of IgG subgroups in cerebrospinal fluid of multiple sclerosis patients and others. *Acta Neurol. Scand.*, **60**, 277–282.

[240] Eickhoff, K., Wikstrom, J., Poser, S. and Bauer, H. (1977). Protein profile of cerebrospinal fluid in multiple sclerosis with special reference to the blood brain barrier. *J. Neurol.*, **214**, 207–215.

[241] Eikelenboom, M. J., Petzold, A., Lazeron, R. H. et al. (2003). Multiple sclerosis: Neurofilament light chain antibodies are correlated to cerebral atrophy. *Neurology*, **60**, 219–223.

[242] Eilat, D., Fischel, R. and Zlotnick, A. (1981). Albumin-immunoglobulin complexes in human serum: classification and immunochemical analysis. *Scand. J. Immunol.*, **14**, 77–87.

[243] Ekstrom, B. and Berggard, I. (1977). Human alpha 1-microglobulin. *J. Biol. Chem.*, **252**, 8048–8057.

[244] Eldridge, R., McFarland, H., Sever, J. et al. (1978). Familial multiple sclerosis: clinical, histocompatibility, and viral serological studies. *Ann. Neurol.*, **3**, 72–80.

[245] Elovaara, I. (1984). Proteins in serum and cerebrospinal fluid in demented patients with Down's syndrome. *Acta Neurol. Scand.*, **69**, 302–305.

[246] Elovaara, I., Icen, A., Palo, J. and Erkinjuntti, T. (1985). CSF in Alzheimer's disease. *J. Neurol. Sci.*, **70**, 73–80.

[247] Emerson, D. L., Galbraith, R. M. and Arnaud, P. (1984). Electrophoretic demonstration of interactions between group specific component (vitamin D binding protein), actin, and 2,5-hydroxycholecalciferol. *Electrophoresis*, **5**, 22–26.

[248] Epstein, M. H., Feldman, A. M. and Brusilow, S. W. (1977). Cerebrospinal fluid production: stimulation by cholera toxin. *Science*, **196**, 1012–1013.

[249] Ericsson, J., Link, H. and Zettervall, O. (1969). Urinary excretion of a low molecular weight protein in cerebral damage. *Neurology*, **19**, 606–610.
[250] Ersmark, B. and Siden, A. (1984). Isoelectric focusing of CSF proteins and future evolution of multiple sclerosis: a clinical follow-up. *J. Neurol.*, **231**, 117–121.
[251] Esiri, M. M. (1977). Immunoglobulin-containing cells in multiple-sclerosis plaques. *Lancet*, **ii**, 478–480.
[252] Esiri, M. M. (1980). Multiple sclerosis: a quantitative and qualitative study of immunoglobulin-containing cells in the CNS. *Neuropathol. Appl. Neurobiol.*, **6**, 9–21.
[253] Esiri, M. M. (1980). Poliomyelitis: immunoglobulin-containing cells in the CNS in acute and convalescent phases of the human disease. *Clin. Exp. Immunol.*, **40**, 42–48.
[254] Esiri, M. M., Chapel, H. M., Morton, J. A. and Wilcock, G. K. (1985). Non-immunoglobin in cerebrospinal fluid of patients with Alzheimer's disease and cerebral amyloid angiopathy. *Lancet*, **ii**, 507–508.
[255] Esiri, M. M. and Gay, D. (1990). Immunological and neuropathological significance of the Virchow-Robin space. *J. Neurol. Sci.*, **100**, 3–8.
[256] Esiri, M. M., Oppenheimer, D. R., Brownell, B. and Haire, M. (1982). Distribution of measles antigen and immunoglobulin-containing cells in the CNS in subacute sclerosing panencephalitis (SSPE) and atypical measles encephalitis. *J. Neurol. Sci.*, **53**, 29–43.
[257] Esiri, M. M., Taylor, C. R. and Mason, D. Y. (1976). Application of an immunoperoxidase method to a study of the central nervous system: preliminary findings in a study of human formalin-fixed material. *Neuropathol. Appl. Neurobiol.*, **2**, 233–246.
[258] Essex Wynter, W. (1891). Four cases of tubercular meningitis in which paracentesis of the theca vertebralis was performed for the relief of fluid pressure. *Lancet*, **i**, 981–982.
[259] Evans, J. H. and Quick, D. T. (1966). Polyacrylamide gel electrophoresis of cerebrospinal fluid proteins. *Arch. Neurol.*, **14**, 64–72.
[260] Evans, J. H. and Quick, D. T. (1966). Polyacrylamide gel electrophoresis of spinal-fluid proteins. *Clin. Chem.*, **12**, 28–36.
[261] Ewan, P. W. and Lachmann, P. J. (1979). IgG synthesis within the brain in multiple sclerosis and subacute sclerosing panencephalitis. *Clin. Exp. Immunol.*, **35**, 227–235.
[262] Farell, K., Hill, A. and Chuang, S. (1981). Papilledema in Guillain-Barre syndrome. *Arch. Neurol.*, **38**, 55–57.
[263] Farrell, M. A., Kaufmann, J. C. E., Gilbert, J. J. et al. (1985). Oligoclonal bands in multiple sclerosis. *Neurology*, **35**, 212–218.
[264] Fasman, G. D. (1976). *Handbook of Biochemistry and Molecular Biology*, 3rd ed. pp. 242–253. Prot - Vol. II. Cleveland, CRC Press.
[265] Felding, P. (1984). Electroimmunoassay of prealbumin in cerebrospinal fluid. *Clin. Chem.*, **30**, 498.
[266] Felgenhauer, K. (1968). Immunological techniques following microelectrophoresis on polyacrylmide gel. *Biochim. Biophys. Acta*, **160**, 267–269.
[267] Felgenhauer, K. (1970). Quantitation and specific detection methods after disc electrophoresis of serum proteins. *Clin. Chim. Acta*, **27**, 305–312.
[268] Felgenhauer, K. (1971). *Vergleichende Disc-elektrophorese von Serum und Liquor cerebrospinalis.* Stuttgart: Thieme.

[269] Felgenhauer, K. (1972). Immunoglobulins in disc electrophoresis. *Clin. Chim. Acta*, **39**, 177–181.
[270] Felgenhauer, K. (1974). Evaluation of molecular size by gel electrophoretic techniques. *Hoppe Seylers, Z. Physiol Chem.*, **355**, 1281–1290.
[271] Felgenhauer, K. (1974). Protein size and cerebrospinal fluid composition. *Klin. Wochenschr.*, **52**, 1158–1164.
[272] Felgenhauer, K. (1979). Characterization of native multicomponent protein mixtures by one and two-dimensional gradient electrophoresis. *J. Chromatogr.*, **173**, 299–311.
[273] Felgenhauer, K. (1980). Protein filtration and secretion at human body fluid barriers. *Pflugers Arch.*, **384**, 9–17.
[274] Felgenhauer, K. (1980.) Two-dimensional protein patterns of human body fluids obtained under dissociating and non-dissociating conditions. In: B. J. Radola (ed.), *Electrophoresis '79*, pp. 647–654. Walter de Gruyter.
[275] Felgenhauer, K. (1982). Differentiation of the humoral immune response in inflammatory diseases of the central nervous system. *J. Neurol.*, **228**, 223–237.
[276] Felgenhauer, K. (1984). Local protein synthesis and passive transfer in human body fluid formation. In: H. Peeters (ed.), *Protides of the Biological Fluids*, pp. 21–24. Pergamon Press.
[277] Felgenhauer, K., Ackermann, R. and Schliep, G. (1980). The process dynamics of viral and bacterial diseases of the central nervous system. *J. Neurol. Sci.*, **47**, 21–34.
[278] Felgenhauer, K., Bach, S. and Stammler, A. (1967). Elektrophorese von serum und liquor cerebrospinalis in polyacrylamide-gel. *Klin. Wochenschr.*, **45**, 371–377.
[279] Felgenhauer, K., and Beuche, W. (1999). Labordiagnostik neurologischer Erkrankungen. pp. 1–177. Georg Thieme Verlag.
[280] Felgenhauer, K. and Hagedorn, D. (1980). Two-dimensional separation of human body fluid proteins. *Clin. Chim. Acta*, **100**, 121–132.
[281] Felgenhauer, K., Hanssen, C. H. R. and Remy, A. (1982). Laser-nephelometric evaluation of the humoral immune response in the central nervous system. *Med. Lab.*, **11**, 48–57.
[282] Felgenhauer, K., Holzgraefe, M., Prange, H. W. (1993). CNS Barriers and Modern CSF Diagnosis. pp. 1–468. Weinheim: VCH Publishers.
[283] Felgenhauer, K. and Kober, D. (1985). Apurulent bacterial meningitis (compartmental leucopenia in purulent meningitis). *J. Neurol.*, **232**, 157–161.
[284] Felgenhauer, K., Liappis, N. and Nekic, M. (1982). Low molecular solutes and the blood cerebrospinal fluid barrier. *Klin. Wochenschr.*, **60**, 1385–1392.
[285] Felgenhauer, K., Mohrmann, H. (1981). Evaluation of immunoglobulin diversity. In: R. C. Allen and P. P. Amaud (eds.), *Electrophoresis '81*, pp. 427–432. de Gruyter & Co.
[286] Felgenhauer, K., Nekic, M. and Ackermann, R. (1982). The demonstration of locally synthesized herpes simplex IgG antibodies in CSF by a sepharose 4B linked enzyme immunoassay. *J. Neuroimmunol.*, **3**, 149–158.
[287] Felgenhauer, K. and Pak, S. J. (1973). Detection of ampholine patterns. *Ann. N Y Acad. Sci.*, **209**, 147–152.
[288] Felgenhauer, K. and Renner, E. (1977). Hydrodynamic radii versus molecular weights in clearance studies of urine and cerebrospinal fluid. *Ann. Clin. Biochem.*, **14**, 100–104.

[289] Felgenhauer, K., Schadlich, H. J., Nekic, M. and Ackermann, R. (1985). Cerebrospinal fluid virus antibodies. A diagnostic indicator for multiple sclerosis? *J. Neurol. Sci.*, **71**, 291–299.
[290] Felgenhauer, K., Schliep, G. and Rapic, N. (1976). Evaluation of the blood-CSF barrier by protein gradients and the humoral immune response within the central nervous system. *J. Neurol. Sci.*, **30**, 113–128.
[291] Felgenhauer, K., Schliep, G., Rapic, N. (1976). Protein permeability of the blood-CSF barrier. In: H. Peeters (ed.), *Protides of the Biological Fluids*, pp. 481–487. Oxford Pergamon.
[292] Felgenhauer, K., Weis, A. and Glenner, G. G. (1970). The demonstration of tryptophan, tyrosine and carbohydrate-containing proteins in disc electrophoresis gels. *J. Chromatogr.*, **46**, 116–119.
[293] Fex, G., Laurell, C. B. and Thulin, E. (1977). Purification of prealbumin from human and canine serum using a two-step affinity chromatographic procedure. *Eur. J. Biochem.*, **75**, 181.
[294] Fieschi, C. and Agnoli, A. (1964). Fractional exchange rate of albumin from cerebrospinal fluid to plasma in man. *Minerva Nucl.*, **8**, 344–347.
[295] Firth, G., Rees, J. and McKeran, R. O. (1985). The value of the measurement of cerebrospinal fluid levels of lysozyme in the diagnosis of neurological disease. *J. Neurol. Neurosurg. Psychiatry*, **48**, 709–712.
[296] Fischer, C. L., Gill, C. W. (1975). Acute-phase proteins. In: S. E. Ritzmann and J. C. Daniels (eds.), *Serum Protein Abnormalities Diagnostic and Clinical Aspects*, pp. 18–43. Little, Brown and Co.
[297] Fischer-Williams, M. (1976). Recent advances in the study of cerebrospinal fluid. *Practitioner*, **217**, 108–115.
[298] Fischer-Williams, M. and Roberts, R. C. (1971). Cerebrospinal Fluid Proteins and serum immunoglobulins. *Arch. Neurol.*, **25**, 526–534.
[299] Fish, W. W., Reynolds, J. A. and Tanford, C. (1970). Gel chromatography of proteins in denaturing solvents. *J. Biol. Chem.*, **19**, 5166–5168.
[300] Fishman, R. A. (1975). Cerebrospinal fluid: A review of recent clinical advances. In: D. B. Tower (ed.), *The Nervous System*, pp. 55–60. Raven Press.
[301] Fishman, R. A. (1980). *The Cerebrospinal Fluid in Diseases of the Nervous System*. pp. 1–384. Saunders, Philadelphia.
[302] Fishman, R. A., Ransohoff, J. and Osserman, E. F. (1958). Factors influencing the concentration gradient of protein in cerebrospinal fluid. *J. Clin. Invest.*, **37**, 1419–1424.
[303] Flachenecker, P., Jung, S., Rieckmann, P. and Toyka, K. V. (2002). sICAM-1 is not a marker for disease activity in the relapse-free interval of multiple sclerosis–a cross-sectional pilot study. *J. Neurol.*, **249**, 1001–1003.
[304] Fleisher, G. A., Wakim, K. G. and Goldstein, N. P. (1957). Glutamic-oxalacetic transaminase and lactic dehydrogenase in serum and cerebrospinal fluid of patients with neurologic disorders. *Proc. Staff Mtgs. Mayo Clin.*, **32**, 188–197.
[305] Forghani, B., Cremer, N. E., Johnson, K. P. et al. (1980). Comprehensive viral immunology of multiple sclerosis. III analysis of CSF antibodies by radioimmunoassay. *Arch. Neurol.*, **37**, 616–619.
[306] Forghani, B., Cremer, N. E., Johnson, K. P. et al. (1978). Viral antibodies in cerebrospinal fluid of multiple sclerosis and control patients: comparison between radioimmunoassay and conventional techniques. *J. Clin. Microbiol.*, **7**, 63–69.

[307] Forsberg, P., Henriksson, A., Link, H. and Ohman, S. (1984). Reference values for CSF-IgM, CSF-IgM/S-IgM ratio and IgM index and its application to patients with multiple sclerosis and aseptic meningoencephalitis. *Scand. J. Clin. Lab. Invest.*, **44**, 7–12.
[308] Fortini, A. S., Sanders, E. L., Weinshenker, B. G. and Katzmann, J. A. (2004). Cerebrospinal fluid oligoclonal bands in the diagnosis of multiple sclerosis. Isoelectric focusing with IgG immunoblotting compared with high-resolution agarose gel electrophoresis and cerebrospinal fluid IgG index. *Am. J. Clin. Pathol.*, **120**, 672–675.
[309] Fossan, G. O. (1976). Reduced CSF IgG in patients treated with phenytoin (diphenylhydantoin). *Eur. Neurol.*, **14**, 426–432.
[310] Fossan, G. O. (1977). The transfer of IgG from serum to CSF, evaluated by means of a naturally occurring antibody. *Eur. Neurol.*, **15**, 231–236.
[311] Fossan, G. O. and Aarli, J. A. (1979). Immunoglobulin G in serum and cerebrospinal fluid from epileptic patients treated with phenytoin. *Eur. Neurol.*, **18**, 322–327.
[312] Fossan, G. O. and Larsen, J. L. (1978). Variation of IgG and total protein concentrations during lumbar cerebrospinal fluid collection. *Acta Neurol. Scand.*, **57**, 287–279.
[313] Fossan, G. O. and Larsen, J. L. (1979). Variation of immunoglobulin G and total protein concentrations during lumbar cerebrospinal fluid collection. *Eur. Neurol.*, **18**, 140–144.
[314] Fossard, C., Dale, G. and Latner, A. L. (1970). Separation of the proteins of cerebrospinal fluid using gel electrofocusing followed by electrophoresis. *J. Clin. Pathol.*, **23**, 586–589.
[315] Fraser, K. B. (1978). False-positive measles-specific IgM in multiple sclerosis. *Lancet*, **i**, 91.
[316] Freedman, M. S., Thompson, E. J., Deisenhammer, F. *et al.* (2005). Recommended Standard of Cerebrospinal Fluid Analysis in the Diagnosis of Multiple Sclerosis. *Arch. Neurol.*, **62**, 865–970.
[317] Frick, E. (1963). Quantitative bestimmung des transferrins im normalen und pathologischen liquor cerebrospinal. *Klin. Wochenschr.*, **41**, 75–78.
[318] Frick, E. and Scheid-Seydel, L. (1958). Untersuchungen mit J 131-markiertem albumin uber austauschvorgange zwischen plasma und liquor cerebrospinalis. *Klin. Wochenschr.*, **36**, 66–69.
[319] Frick, E. and Scheid-Seydel, L. (1958). Untersuchungen mit J 131-markiertem gamma globulin zur frage der abstammung der liquoreiweisskorper. *Klin. Wochenschr.*, **36**, 857–863.
[320] Frick, E. and Scheid-Seydel, L. (1960). Untersuchungen mit J 131-markierteum Beta-globulin zur frage der abstammung der liquoreiweisskorper. *Klin. Wochenschr.*, **38**, 1240–1243.
[321] Frick, E. and Scheid-Seydel, L. (1963). Untersuchungen mit J 131-markiertem transferrin zur frage der abstammung der liquoreiweisskorper. *Klin. Wochenschr.*, **41**, 589–593.
[322] Frick, E. and Stickl, H. (1980). Antibody-dependent lymphocyte cytotoxicity against basic protein of myelin in multiple sclerosis. *J. Neurol. Sci.*, **46**, 187–197.
[323] Frick, E. and Stickl, H. (1982). Specificity of antibody-dependent lymphocyte cytotoxicity against cerebral tissue constituents in multiple sclerosis. *Acta Neurol. Scand.*, **65**, 30–37.

[324] Fryden, A., Forsberg, P. and Link, H. (1983). Synthesis of the complement factors C3 and C4 within the central nervous system over the course of aseptic meningitis. *Acta Neurol. Scand.*, **68**, 157–163.
[325] Fryden, A. and Link, H. (1979). Predominance of oligoclonal IgG type lambda in CSF in aseptic meningitis. *Arch. Neurol.*, **36**, 478–480.
[326] Fryden, A., Link, H. and Norrby, E. (1978). Cerebrospinal fluid and serum immunoglobulins and antibody titers in mumps meningitis and aseptic meningitis of other etiology. *Infect. Immun.*, **21**, 852–861.
[327] Fukazawa, T., Kikuchi, S., Sasaki, H. et al. (1998). The significance of oligoclonal bands in multiple sclerosis in Japan: relevance of immunogenetic backgrounds. *J. Neurol. Sci.*, **158**, 209–214.
[328] Furrows, S., Hartley, J. C., Bell, J. et al. (2004). Chlamydia pneumoniae infection of the central nervous system in patients with multiple sclerosis. *J. Neurol. Neurosurg. Psychiatry*, **75**, 152–154.
[329] Gahmberg, C. G. and Andersson, L. C. (1978). Leukocyte surface origin of human alpha 1- acid glycoprotein (orosomucoid). *J. Exp. Med.*, **148**, 507–521.
[330] Gallo, P., Bracco, F., Morara, S. et al. (1985). The cerebrospinal fluid transferrin/tau proteins. *J. Neurol. Sci.*, **70**, 81–92.
[331] Gallo, P., Ollson, O. and Siden, A. (1985). Methods for chromatofocusing of cerebrospinal fluid and serum immunoglobulin G. *J. Chromatogr.*, **327**, 293–299.
[332] Gallo, P. and Siden, A. (1985). Examinations of cerebrospinal fluid immunoglobulin G by chromatofocusing. *J. Neurol.*, **232**, 231–235.
[333] Galvez, S., Farcas, A. and Monari, M. (1979). Cerebrospinal fluid and serum immunoglobulins and C3 in Creutzfeldt-Jakob disease. *Neurology*, **29**, 1610–1612.
[334] Galvez, S., Farcas, A. and Monari, M. (1979). The concentration of alpha-1-antitrypsin in cerebrospinal fluid and serum in a series of 40 intracranial tumors. *Clin. Chim. Acta*, **91**, 191–196.
[335] Ganrot, K. and Laurell, C. B. (1974). Measurement of IgG and albumin content of cerebrospinal fluid, and its interpretation. *Clin. Chem.*, **20**, 571–573.
[336] Ganrot-Norlin, K. (1978). Relative concentrations of albumin and IgG in cerebrospinal fluid in health and in acute meningitis. *Scand. J. Infect. Dis.*, **10**, 57–60.
[337] Garde, A. and Kjellin, K. G. (1971). Diagnostic significance of cerebrospinal-fluid examinations in myelopathy. *Acta Neurol. Scand.*, **47**, 555–568.
[338] Garton, M. J., Keir, G., Lakshmi, M. V. and Thompson, E. J. (1991). Age-related changes in cerebrospinal fluid protein concentrations. *J. Neurol. Sci.*, **104**, 74–80.
[339] Gauthier, F. and Chasse, M. H. (1983). A simple characterization of oligoclonal immunoglobulins in unconcentrated cerebrospinal fluid using polyacrylamide gel electrophoresis. *Clin. Chim. Acta*, **127**, 407–411.
[340] Gekle, D., Kult, J. and Roth, R. (1977). Lysozym und beta 2-mikroglobulin im liquor gesunder kinder und bei kindern mit erkrankungen des zentralnervensystems. *Klin. Wochenschr.*, **55**, 189–191.
[341] George, P. M., Lorier, M. A. and Donaldson, I. (1983). An evaluation of cerebrospinal fluid oligoclonal banding confirmed by immunofixation on agarose gel. *J. Neurol. Neurosurg. Psychiatry*, **46**, 500–504.
[342] Georgiadis, D., Berger, A., Kowatschev, E. et al. (2000). Predictive value of S-100beta and neuron-specific enolase serum levels for adverse neurologic outcome after cardiac surgery. *J. Thorac. Cardiovasc. Surg.*, **119**, 138–147.

[343] Gerbitz, K. D., Summer, J. and Thallemer, J. (1984). Brain-specific protein: solid-phase immunobioluminescence assay for neuron-specific enolase in human plasma. *Clin. Chem.*, **30**, 382–386.
[344] Gerner, R. H. and Sharp, B. (1982). CSF B-endorphin-immunoreactivity in normal, schizophrenic, depressed manic and anorexic subjects. *Brain Res.*, **237**, 244–247.
[345] Gershon, A., Steinberg, S., Greenberg, S. and Taber, L. (1980). Varicella-zoster-associated encephalitis: detection of specific antibody in cerebrospinal fluid. *J. Clin. Microbiol.*, **12**, 764–767.
[346] Gerson, B., Cohen, S. T., Gerson, I. M. and Guest, G. H. (1981). Myelin basic protein, oligoclonal bands and IgG in cerebrospinal fluid as indicators of multiple sclerosis. *Clin. Chem.*, **27**, 1974–1977.
[347] Gerson, B., Krolikowski, F. J. and Gerson, I. M. (1980). Two agarose electrophoretic systems for demonstration of oligoclonal bands in cerebrospinal fluid compared. *Clin. Chem.*, **26**, 343–345.
[348] Gerson, B. and Orr, J. M. (1980). Oligoclonal bands and quantitation of IgG in cerebrospinal fluid as indicators of multiple sclerosis. *Am. J. Clin. Pathol.* **73**, 87–91.
[349] Gerstl, B., Uyeda, C. T., Bond, P. and Smith, J. K. (1969). Soluble proteins in normal and diseased human brains. *Neurology*, **19**, 1019–1026.
[350] Gherardi, D. (1965). Le Complicanze neurologiche del morbo di Behcet. *Il Lavoro Neuropsichiatrico* **36**, 333–408.
[351] Giacomelli, S., Leone, M. G., Grima, J. *et al.* (1996). Astrocytes synthesize and secrete prostaglandin D synthetase in vitro. *Biochim. Biophys. Acta*, **1310**, 269–276.
[352] Gilden, D., Devlin, M. and Wroblewska, Z. (1978). A technique for the elution of cell-surface antibody from human brain tissue. *Ann. Neurol.*, **3**, 403–405.
[353] Gilden, D. H., Bennett, J. L., Kleinschmidt-DeMasters, B. K. *et al.* (1998). The value of cerebrospinal fluid antiviral antibody in the diagnosis of neurologic disease produced by varicella zoster virus. *J. Neurol. Sci.*, **159**, 140–144.
[354] Giles, P. D., Heath, J. P. and Wroe, S. J. (1989). Oligoclonal bands and the IgG index in multiple sclerosis: uses and limitations. *Ann. Clin. Biochem.*, **26**, 1317–1323.
[355] Gill, D. G. and Brody, M. (1979). Cerebrospinal fluid immunoglobulins in children. *Arch. Dis. Child.*, **54**, 961–967.
[356] Giovannoni, G., Bell, J. B., Feldmann, M. and Thompson, E. J. (1996). Increased urinary neopterin excretion in vaccinated adults. *Pteridines*, **7**, 82–86.
[357] Giovannoni, G., Green, A., Keir G and Thompson, E. J. (1996). Urinary myelin basic protein-like material as a correlate of the progression of multiple sclerosis. *Ann. Neurol.*, **40**, 128–129.
[358] Giovannoni, G., Green, A. J. and Thompson, E. J. (1998). Are there any body fluid markers of brain atrophy in multiple sclerosis? *Mult. Scler.*, **4**, 138–142.
[359] Giovannoni, G., Heales, S. J., Land, J. M. and Thompson, E. J. (1998). The potential role of nitric oxide in multiple sclerosis. *Mult. Scler.*, **4**, 212–216.
[360] Giovannoni, G., Heales, S. J., Silver, N. C. *et al.* (1997). Raised serum nitrate and nitrite levels in patients with multiple sclerosis. *J. Neurol. Sci.*, **145**, 77–81.
[361] Giovannoni, G., Lai, M., Kidd, D. *et al.* (1997). Daily urinary neopterin excretion as an immunological marker of disease activity in multiple sclerosis. *Brain*, **120**, 1–13.
[362] Giovannoni, G., Lai, M., Thorpe, J. *et al.* (1997). Longitudinal study of soluble adhesion molecules in multiple sclerosis: correlation with gadolinium enhanced magnetic resonance imaging. *Neurology*, **48**, 1557–1565.

[363] Giovannoni, G., Land, J. M., Keir, G. *et al.* (1997). Adaptation of the nitrate reductase and Griess reaction methods for the measurement of serum nitrate plus nitrite levels. *Ann. Clin. Biochem.*, **34**, 193–198.

[364] Giovannoni, G., Miller, D. H., Losseff, N. A. *et al.* (2001). Serum inflammatory markers and clinical/MRI markers of disease progression in multiple sclerosis. *J. Neurol.*, **248**, 487–495.

[365] Giovannoni, G., Miller, R. F., Heales, S. J.*et al.* (1998). Elevated cerebrospinal fluid and serum nitrate and nitrite levels in patients with central nervous system complications of HIV-1 infection: a correlation with blood-brain-barrier dysfunction. *J. Neurol. Sci.*, **156**, 53–58.

[366] Giovannoni, G., Silver, N. C., Good, C. D. *et al.* (2000). Immunological time-course of gadolinium-enhancing MRI lesions in patients with multiple sclerosis. *Eur. Neurol.*, **44**, 222–228.

[367] Giovannoni, G., Silver, N. C., O'Riordan, J. *et al.* (1999). Increased urinary nitric oxide metabolites in patients with multiple sclerosis correlates with early and relapsing disease. *Mult. Scler.*, **5**, 335–341.

[368] Giovannoni, G., Thompson, A. J., Miller, D. H. and Thompson, E. J. (2001). Fatigue is not associated with raised inflammatory markers in multiple sclerosis. *Neurology*, **57**, 676–681.

[369] Giovannoni, G. and Thompson, E. J. (1996). The detection and significance of cerebrospinal fluid oligoclonal IgG. In: E. J. Thompson, M. Trojano and P. Livrea (eds.), *CSF Analysis in MS*, pp. 29–39. Springer-Verlag.

[370] Giovannoni, G. and Thompson, E. J. (1998). Urinary markers of disease activity in multiple sclerosis. *Mult. Scler.*, **4**, 247–253.

[371] Giovannoni, G., Thorpe, J. W., Kidd, D. *et al.* (1996). Soluble E-selectin in multiple sclerosis: raised concentrations in patients with primary progressive disease. *J. Neurol. Neurosurg. Psychiatry*, **60**, 20–26.

[372] Glasner, H. (1975). Barrier impairment and immune reaction in the cerebrospinal fluid. *Eur. Neurol.*, **13**, 304–314.

[373] Glasner, H. (1978). Microzone electrophoresis of the unconcentrated and concentrated CSF. *J. Neurol.*, **218**, 73–76.

[374] Glasner, H., Lowenthal, A. and Karcher, D. (1979). Diagnostic value of brief microzone electrophoresis of unconcentrated CSF and agar gel electrophoresis of concentrated and unconcentrated CSF. *J. Neurol.*, **222**, 53–58.

[375] Glasner, H. and Piepgras, U. (1980). CSF circulation and blood-CSF barrier. *Eur. Neurol.*, **17**, 280–285.

[376] Glenner, G. G. (1980). Amyloid deposits and amyloidosis (Part I). *N. Engl. J. Med.*, **302**, 1283–1292.

[377] Glenner, G. G. (1980). Amyloid deposits and amyloidosis (Part II). *N. Engl. J. Med.*, **302**, 1333–1342.

[378] Glikmann, G., Svehag, S. E., Hansen, E. *et al.* (1980). Soluble immune complexes in cerebrospinal fluid of patients with multiple sclerosis and other neurological diseases. *Acta Neurol. Scand.*, **61**, 333–343.

[379] Glynn, P., Gilbert, H. M., Newcombe, J. and Cuzner, M. L. (1982). Analysis of immunoglobulin G in multiple sclerosis brain: quantitative and isoelectric focusing studies. *Clin. Exp. Immunol.*, **48**, 102–110.

[380] Goffette, S., Schluep, M., Henry, H. et al. (2004). Detection of oligoclonal free kappa chains in the absence of oligoclonal IgG in the CSF of patients with suspected multiple sclerosis. *J. Neurol. Neurosurg. Psychiatry*, **75**, 308–310.

[381] Gold, P. W., Kaye, W., Robertson, G. L. and Ebert, M. (1983). Abnormalities in plasma and cerebrospinal-fluid arginine vasopressin in patients with anorexia nervosa. *N. Engl. J. Med.*, **308**, 1117–1123.

[382] Goldman, D., Merril, C. R. and Ebert, M. H. (1980). Two-dimensional gel electrophoresis of cerebrospinal fluid proteins. *Clin. Chem.*, **26**, 1317–1322.

[383] Goldschmidt, B., Menonna, J., Fortunato, J. et al. (1980). Mycoplasma antibody in Guillain-Barre syndrome and other neurological disorders. *Ann. Neurol.*, **7**, 108–112.

[384] Gonsette, R. E., Working group for treatment trials. (1997). Guidance concerning compensation for new therapies in multiple sclerosis patients. *Eur. J. Neurol.*, **4**, 426–428.

[385] Goodland, F. C. (1982). Detection limits for protein after electrophoresis on cellulose acetate. *Ann. Clin. Biochem.*, **19**, 117–119.

[386] Goodland, F. C. and Thompson, E. J. (1983). A comparison of cellulose acetate immunofixation with polyacrylamide gel electrophoresis for the detection of oligoclonal bands in unconcentrated cerebrospinal fluid. *J. Clin. Pathol.*, **36**, 1309–1311.

[387] Goodman, M. and Vulpe, M. (1961). A quantitative immunochemical method for determining serum and cerebrospinal fluid proteins. *World Neurol.*, **2**, 589–601.

[388] Gorg, A., Postel, W., Weser, J. et al. (1985). Horizontal SDS electrophoresis in ultrathin pore-gradient gels for the analysis of urinary proteins. *Sci. Tools*, **32**, 5–9.

[389] Gosseye-Lissoir, F., Delmotte, P. and Carton, H. (1977). Biochemical findings in multiple sclerosis. *J. Neurol.*, **216**, 197–205.

[390] Green, A. J., Giovannoni, G., Hall-Craggs, M. A. et al. (2000). Cerebrospinal fluid tau concentrations in HIV infected patients with suspected neurological disease. *Sex. Transm. Infect.*, **76**, 443–446.

[391] Green, A. J., Giovannoni, G., Miller, R. F. et al. (1999). Cerebrospinal fluid S-100b concentrations in patients with HIV infection. *AIDS*, **13**, 139–140.

[392] Green, A. J., Harvey, R. J., Thompson, E. J. and Rossor, M. N. (1997). Increased S100beta in the cerebrospinal fluid of patients with frontotemporal dementia. *Neurosci. Lett.*, **235**, 5–8.

[393] Green, A. J., Harvey, R. J., Thompson, E. J. and Rossor, M. N. (1999). Increased tau in the cerebrospinal fluid of patients with frontotemporal dementia and Alzheimer's disease. *Neurosci. Lett.*, **259**, 133–135.

[394] Green, A. J., Jackman, R., Marshall, T. A. and Thompson, E. J. (1999). Increased S-100b in the cerebrospinal fluid of some cattle with bovine spongiform encephalopathy. *Vet. Rec.*, **145**, 107–109.

[395] Green, A. J., Keir, G. and Thompson, E. J. (1997). A specific and sensitive ELISA for measuring S-100b in cerebrospinal fluid. *J. Immunol. Methods*, **205**, 35–41.

[396] Green, A. J., Thompson, E. J., Stewart, G. E. et al. (2001). Use of 14-3-3 and other brain-specific proteins in CSF in the diagnosis of variant Creutzfeldt-Jakob disease. *J. Neurol. Neurosurg. Psychiatry*, **70**, 744–748.

[397] Greenwood, B. M. and Whittle, H. C. (1973). Cerebrospinal-fluid IgM in patients with sleeping-sickness. *Lancet*, **ii**, 525–527.

[398] Greer, J. (1980). Model for haptoglobin heavy chain based upon structure homology. *Proc. Natl. Acad. Sci.*, **6**, 3393–3397.

[399] Griffin, D. E. (1981). Immunoglobulins in the cerebrospinal fluid: changes during acute viral encephalitis in mice. *J. Immunol.*, **126**, 27–31.
[400] Griffin, D. E., Giffels, J. and Hughes, H. (1982). Study of protein characteristics that influence entry into the cerebrospinal fluid of normal mice and mice with encephalitis. *J. Clin. Invest.*, **70**, 289–295.
[401] Groc, W. (1970). Simultanbestimmung von proteinantigenen im liquor. *Arztl Lab*, **16**, 15–23.
[402] Grubb, A. O. (1974). Crossed immunoelectrophoresis and electroimmunoassay of IgG. *J. Immunol.*, **113**, 343–347.
[403] Grubb, A. O. (1974). Crossed immunoelectrophoresis and electroimmunoassay of IgM. *J. Immunol.*, **112**, 1420.
[404] Grubb, A. O., Jensson, O., Gudmundsson, G. *et al.* (1984). Abnormal metabolism of gamma trace alkaline microprotein. *N. Engl. J. Med.*, **24**, 1547–1549.
[405] Grubb, A. O. and Lofberg, H. (1982). Human gamma-trace, a basic microprotein: Amino acid sequence and presence in the adenohypophysis. *Proc. Natl. Acad. Sci.*, **79**, 3024–3027.
[406] Gschnait, F., Schmidt, B. L. and Luger, A. (1981). Cerebrospinal fluid immunoglobulins in neurosyphilis. *Br. J. Vener. Dis.*, **57**, 238–240.
[407] Gudmundsson, G., Kjellin, K. G., Mettinger, K. L. *et al.* (1980). Isoelectric focusing of cerebrospinal fluid proteins in ischemic cerebrovascular disease. *J. Neurol.*, **222**, 227–234.
[408] Guenther, W. (1984). Investigation of the Blood-brain barrier for IgG in inflammatory syndromes of the central nervous system. *Eur. Neurol.*, **23**, 132–136.
[409] Guiloff, R. J., McGregor, B., Thompson, E. *et al.* (1980). Motor neurone disease with elevated cerebrospinal fluid protein. *J. Neurol. Neurosurg. Psychiatry*, **43**, 390–396.
[410] Guindi, S., Mansour, M. M., Girgis, N. I. and Miner, W. F. (1980). Serum and cerebrospinal fluid proteins in tuberculous meningitis. *Eur. Neurol.*, **19**, 247–251.
[411] Gurnett, C. A., Landt, M. and Wong, M. (2003). Analysis of cerebrospinal fluid glial fibrillary acidic protein after seizures in children. *Epilepsia*, **44**, 1455–1458.
[412] Gutstein, H. S. and Cohen, S. R. (1978). Spinal fluid differences in experimental allergic encephalomyelitis and multiple sclerosis. *Science*, **199**, 301–303.
[413] Haber, E., Margolies, M. N. and Cannon, L. E. (1979). Insights gained from the study of homogeneous rabbit antibodies. In: D. K. Karcher, A. Lowenthal and A. D. Strosberg (eds.), *Humoral Immunity in Neurological Diseases*, pp. 327–359. Plenum Press.
[414] Hadler, N. M., Gerwin, R. D., Frank, M. M. *et al.* (1973). The fourth component of complement in the cerebrospinal fluid in systemic lupus erythematosus. *Arthritis Rheum.*, **16**, 507–521.
[415] Hagberg, B., Aicardi, J., Dias, K. and Ramos, O. (1983). A progressive syndrome of autism, dementia, ataxia, and loss of purposeful hand use in girls: Rett's syndrome: report of 35 cases. *Ann. Neurol.*, **14**, 471–479.
[416] Haghighi, S., Andersen, O., Oden, A. and Rosengren, L. (2004). Cerebrospinal fluid markers in MS patients and their healthy siblings. *Acta Neurol. Scand.*, **109**, 97–99.
[417] Haghighi, S., Andersen, O., Rosengren, L. *et al.* (2000). Incidence of CSF abnormalities in siblings of multiple sclerosis patients and unrelated controls. *J. Neurol.*, **247**, 616–622.

[418] Haile, R., Smith, P., Read, D. et al. (1982). A study of measles virus and canine distemper virus antibodies, and of childhood infections in multiple sclerosis patients and controls. *J. Neurol. Sci.*, **56**, 1–10.
[419] Haire, M., Fraser, K. B. and Millar, J. H. D. (1973). Measles and other virus-specific immunoglobulins in multiple sclerosis. *Br. Med. J.*, **3**, 612–615.
[420] Hall, W. W. and Choppin, P. W. (1981). Measles-virus proteins in the brain tissue of patients with subacute sclerosing panencephalitis. *N. Engl. J. Med.*, **304**, 1152–1155.
[421] Hallander, L. and Kjellin, K. G. (1980). Isotachophoretic systems for studying proteins in cerebrospinal fluid and serum. In: A. Adam and C. Scholz C (eds.), *Biochem. Biol. App. Isotachophoresis*, pp. 245–257. Elsevier.
[422] Hallgren, R., Terent, A. and Venge, P. (1983). Eosinophil cationic protein (ECP) in the cerebrospinal fluid. *J. Neurol. Sci.*, **58**, 57–71.
[423] Hamida, M. B., Mrabet, A. and El Younsi, C. (1982). Profil de type degeneratif de L'electrophorese du liquide cephalo-rachidien au cours des myopathies. *Rev. Neurol.*, **138**, 169–172.
[424] Hamida, M. B., Younsi, C. E. and Isautier, C. (1980). Electrophorese du liquide cephalo-rachidien au cours des heredodegenerescences spino-cerebelleuses. *Rev. Neurol.*, **136**, 25–32.
[425] Hampel, H., Buerger, K., Zinkowski, R. et al. (2004). Measurement of phosphorylated tau epitopes in the differential diagnosis of Alzheimer's disease. *Arch. Gen. Psychiatry*, **61**, 95–102.
[426] Hansen, K., Cruz, M. and Link, H. (1990). Oligoclonal Borrelia burgdorferi-specific IgG antibodies in cerebrospinal fluid in Lyme borreliosis. *J. Infect. Dis.*, **161**, 1194–1202.
[427] Hansen, N. E., Karle, H., Jensen, A. and Bock, E. (1977). Lysozyme activity in cerebrospinal fluid. *Acta Neurol. Scand.*, **55**, 418–424.
[428] Harbeck, R. J., Hoffman, A. A., Hoffman, S. A. and Shucard, D. W. (1979). Cerebrospinal fluid and the choroid plexus during acute immune complex disease. *Clin. Immunol. Immunopathol.*, **13**, 413–425.
[429] Harmoinen, A., Perko, M. and Gronroos, P. (1981). Rapid quantitative determination of C-reactive protein using LKB 8600 reaction rate analyser. *Clin. Chim. Acta*, **111**, 117–120.
[430] Harrington, M. G. and Merril, C. R. (1984). Two-dimensional electrophoresis and ultrasensitive silver staining of cerebrospinal fluid proteins in neurological diseases. *Clin. Chem.*, **30**, 1933–1937.
[431] Harrington, M. G., Merril, C. R., Goldman, D. et al. (1984). Two-dimensional electrophoresis of cerebrospinal fluid proteins in multiple sclerosis and various neurological disease. *Electrophoresis*, **5**, 236–245.
[432] Harrington, M. G., Merril, C. R. and Torrey, E. F. (1985). Differences in cerebrospinal fluid proteins between patients with schizophrenia and normal persons. *Clin. Chem.*, **31**, 722–726.
[433] Harris, E. N., Gharavi, A. E., Boey, M. L. et al. (1983). Anticardiolipin antibodies: detection by radioimmunoassay and association with thrombosis in systemic lupus erythematosus. *Lancet*, **ii**, 1211–1214.
[434] Hawkes, C. H., Thompson, E. J., Keir, G. et al. (1994). Iso-electric focusing of aqueous humour IgG in multiple sclerosis. *J. Neurol.*, **241**, 436–438.

[435] Hayaishi, O. and Urade, Y. (2002). Prostaglandin D2 in sleep-wake regulation: recent progress and perspectives. *Neuroscientist*, **8**, 12–15.
[436] Hayakawa, T., Morimoto, K., Ushio, Y. *et al.* (1980). Levels of astroprotein (an astrocyte-specific cerebrosprotein) in cerebrospinal fluid of patients with brain tumors. *J. Neurosurg.*, **52**, 229–233.
[437] Hayes, A., Thaker, U., Iwatsubo, T. *et al.* (2002). Pathological relationships between microglial cell activity and tau and amyloid beta protein in patients with Alzheimer's disease. *Neurosci. Lett.*, **331**, 171–174.
[438] Heimer, R., Glick, D. L. and Abruzzo, J. L. (1981). The detection of antigens in immune complexes. *Scand. J. Immunol.*, **13**, 441–446.
[439] Helenius, A., Mellman, I., Wall, D. and Hubbard, A. (1983). Endosomes. *Trends Biochem. Sci.*, **6**, 245–250.
[440] Heltberg, H. J., Zeeberg, I., Kristensen, H. *et al.* (1984). Immune complexes and the complement factors C4 and C3 in cerebrospinal fluid and serum from patients with chronic progressive multiple sclerosis. *Acta Neurol. Scand.*, **69**, 34–38.
[441] Henriksson, A., Kam-Hansen, S. and Anderson, R. (1981). Immunoglobulin-producing cells in CSF and blood from patients with multiple sclerosis and other inflammatory neurological diseases enumerated by protein-A plaque assay. *J. Neuroimmunol.*, **1**, 299–309.
[442] Herndon, R. M. and Brumback, R. A. (1989). *The Cerebrospinal Fluid.* pp. 1–306. Kluwer, Boston.
[443] Herndon, R. M. and Kasckow, J. (1978). Electron microscopic studies of cerebrospinal fluid sediment in demyelinating disease. *Ann. Neurol.*, **4**, 515–523.
[444] Herrmann, M. and Ehrenreich, H. (2003). Brain derived proteins as markers of acute stroke: their relation to pathophysiology, outcome prediction and neuroprotective drug monitoring. *Restorative Neurol. Neurosci.*, **21**, 177–190.
[445] Hershey, L. A. and Trotter, J. L. (1980). The use and abuse of the cerebrospinal fluid IgG profile in the adult: a practical evaluation. *Ann. Neurol.*, **8**, 426–434.
[446] Hesse, C., Rosengren, L., Andreasen, N. *et al.* (2001). Transient increase in total tau but not phospho-tau in human cerebrospinal fluid after acute stroke. *Neurosci. Lett.*, **297**, 187–190.
[447] Hesse, C., Rosengren, L., Vanmechelen, E. *et al.* (2000). Cerebrospinal fluid markers for Alzheimer's disease evaluated after acute ischemic stroke. *J. Alzheimers Dis.*, **2**, 199–206.
[448] Hill, N. C., Goldstein, N. P., McKenzie, B. F. *et al.* (1959). Cerebrospinal-fluid proteins, glycoproteins, and lipoproteins in obstructive lesions of the central nervous system. *Brain*, **82**, 581–593.
[449] Hill, N. C., McKenzie, B. F., McGuckin, W. F. *et al.* (1958). I. Proteins, glycoproteins and lipoproteins in the serum and cerebrospinal fluid of healthy subjects. *Proc. Staff Mtgs. Mayo Clin.*, **33**, 686–698.
[450] Hirohata, S., Inoue, T., Yamada, A. *et al.* (1984). Quantitation of IgG, IgA and IgM in the cerebrospinal fluid by a solid-phase enzyme-immunoassay. *J. Neurol. Sci.*, **63**, 101–110.
[451] Hische, E. A. H., Van Der Helm, H. J. and Out, T. (1979). Estimation of IgM in cerebrospinal fluid by fluoroimmunoassay. *Clin. Chim. Acta*, **97**, 93–95.
[452] Hische, E. A. H., Van Der Helm, H. J. and Van Walbeek, H. K. (1982). The cerebrospinal fluid immunoglobulin G index as a diagnostic aid in multiple sclerosis: A Bayesian approach. *Clin. Chem.*, **28**, 354–355.

[453] Hochberg, F. H. and Wolfson, L. (1979). Discontinuous horizontal SDS acrylamide gel electrophoresis of cerebrospinal fluid. *Sci. Tools*, **26**, 44–46.
[454] Hochwald, G. M. (1970). Influx of serum proteins and their concentration in spinal fluid along the neuraxis. *J. Neurol. Sci.*, **10**, 269–278.
[455] Hochwald, G. M. and Thorbecke, G. J. (1963). Trace proteins in cerebrospinal fluid and other biological fluids. *Clin. Chim. Acta*, **8**, 678–684.
[456] Hoffman, P. M., Robbins, D. S., Nolte, M. T. et al. (1978). Cellular immunity in Guamanians with amyotrophic lateral sclerosis and Parkinsonism-dementia. *N. Engl. J. Med.*, **299**, 680.
[457] Hoffmann, A., Nimtz, M., Getzlaff, R. and Conradt, H. S. (1995). 'Brain-type' N-glycosylation of asialo-transferrin from human cerebrospinal fluid. *FEBS Lett.*, **359**, 164–168.
[458] Hoffmann, A., Nimtz, M., Wurster, U. and Conradt, H. S. (1994). Carbohydrate structures of beta-trace protein from human cerebrospinal fluid: evidence for "brain-type" N-glycosylation. *J. Neurochem.*, **63**, 2185–2196.
[459] Hollt, V., Muller, O. A. and Fahlbusch, R. (1979). B-Endorphin in human plasma: basal and pathologically elevated levels. *Life Sci.*, **25**, 37–44.
[460] Holmberg, B., Rosengren, L., Karlsson, J. E. and Johnels, B. (1998). Increased cerebrospinal fluid levels of neurofilament protein in progressive supranuclear palsy and multiple-system atrophy compared with Parkinson's disease. *Mov. Disord.*, **13**, 70–77.
[461] Hommes, O. R., Lamers, K. J. B. and Reekers, P. (1980). Effect of intensive immunosuppression on the course of chronic progressive multiple sclerosis. *J. Neurol.*, **223**, 177–190.
[462] Hopkins, C. R. (1983). The importance of the endosome in intracellular traffic. *Nature*, **304**, 684–685.
[463] Hornig, C. R., Busse, O., Dorndorf, W. and Kaps, M. (1983). Changes in CSF blood-brain barrier parameters in ischaemic cerebral infarction. *J. Neurol.*, **299**, 11–16.
[464] Hosein, Z. Z. and Johnson, K. P. (1981). Isoelectric focusing of cerebrospinal fluid proteins in the diagnosis of multiple sclerosis. *Neurology*, **31**, 70–76.
[465] Hughes, L. E., Bonell, S., Natt, R. S. et al. (2001). Antibody responses to Acinetobacter spp. and Pseudomonas aeruginosa in multiple sclerosis: prospects for diagnosis using the myelin-acinetobacter-neurofilament antibody index. *Clin. Diagn. Lab. Immunol.*, **8**, 1181–1188.
[466] Hughes, R. A. C., Gray, I. A., Gregson, N. A. et al. (1984). Immune responses of myelin antigens in Guillain-Barre syndrome. *J. Neuroimmunol.*, **6**, 303–312.
[467] Huismans, B. D. and Felgenhauer, K. (1971). Periodic acid-Schiff staining of unsaturated serum lipoproteins following disc electrophoresis. *Biochim. Biophys. Acta*, **248**, 330–332.
[468] Hullin, D. A., Brown, K., Kynoch, A. M. et al. (1980). Purification, radioimmunoassay, and distribution of human brain 14-3-2 protein (nervous-system specific enolase) in human tissue. *Biochim. Biophys. Acta*, **628**, 98–108.
[469] Hunter, R., Thompson, T., Reynolds, C. M. and Pitcher, P. M. (1974). Fibrin/fibrinogen degradation products in cerebrospinal fluid of patients admitted to a psychiatric unit. *J. Neurol. Neurosurg. Psychiatry*, **37**, 249–251.

[470] Hutchinson, M., Martin, E. A., Maguire, P. et al. (1983). Visual evoke responses and immunoglobulin abnormalities in the diagnosis of multiple sclerosis. *Acta Neurol. Scand.*, **68**, 90–95.
[471] Iivanainen, M. V., Wallen, W., Leon, M. E. et al. (1981). Micromethod for dection of oligoclonal IgG in unconcentrated CSF by polyacrylamide gel electrophoresis. *Arch. Neurol.*, **38**, 427–430.
[472] Ingebrigtsen, T., Romner, B., Marup-Jensen, S. et al. (2000). The clinical value of serum S-100 protein measurements in minor head injury: a Scandinavian multicentre study. *Brain Inj.*, **14**, 1047–1055.
[473] Inshasi, J. S., Gledhill, R. F., Keir, G. and Thompson, E. J. (1995). Intrathecal synthesis of IgG in benign intracranial hypertension: a re-examination. *J. Neurol.* **242**, 593–595.
[474] Inuzuka, T. (2000). Autoantibodies in paraneoplastic neurological syndrome. *Am. J. Med. Sci.*, **319**, 217–226.
[475] Ishida, K., Gohara, T., Kawata, R. et al. (2003). Are serum S100beta proteins and neuron-specific enolase predictors of cerebral damage in cardiovascular surgery? *J. Cardiothorac. Vasc. Anesth.*, **17**, 4–9.
[476] Ishiguro, K., Ohno, H., Arai, H. et al. (1999). Phosphorylated tau in human cerebrospinal fluid is a diagnostic marker for Alzheimer's disease. *Neurosci. Lett.*, **270**, 91–94.
[477] Itoh, N., Arai, H., Urakami, K. et al. (2001). Large-scale, multicenter study of cerebrospinal fluid tau protein phosphorylated at serine 199 for the antemortem diagnosis of Alzheimer's disease. *Ann. Neurol.*, **50**, 150–156.
[478] Itoh, Y., Enomoto, H., Takagi, K. et al. (1983). Human alpha 1-microglobulin levels in neurological disorders. *Eur. Neurol.*, **22**, 1–6.
[479] Iwashita, H., Grunwald, F. and Bauer, H. (1974). Double ring formation in single radial immunodiffusion for kappa chains in multiple sclerosis cerebrospinal fluid. *J. Neurol.*, **207**, 45–52.
[480] Jackson, P. and Thompson, R. J. (1981). The demonstration of new human brain-specific proteins by high-resolution two-dimensional polyacrylamide gel electrophoresis. *J. Neurol. Sci.*, **49**, 429–438.
[481] Jackson, P., Thomson, V. M. and Thompson, R. J. (1982). Demonstration of basic human-brain-specific proteins by the BASO-DALT system. *Clin. Chem.*, **28**, 920–924.
[482] Jacobi, C. and Reiber, H. (1988). Clinical relevance of increased neuron-specific enolase concentration in cerebrospinal fluid. *Clin. Chim. Acta*, **177**, 49–54.
[483] Jacobi, C., Reiber, H. and Felgenhauer, K. (1986). The clinical relevance of locally produced carcinoembryonic antigen in cerebrospinal fluid. *J. Neurol.* **233**, 358–361.
[484] Jacobs, L. D., Cookfair, D. L., Rudick, R. A. et al. (1996). Intramuscular interferon beta-1a for disease progression in relapsing multiple sclerosis. The Multiple Sclerosis Collaborative Research Group (MSCRG). *Ann. Neurol.*, **39**, 285–294.
[485] Jaffe, H. W., Larsen, S. A., Peters, M. et al. (1978). Tests for treponemal antibody in CSF. *Arch. Intern. Med.*, **138**, 252–255.
[486] Jamnback, T. L., Beaty, B. J., Hildreth, S. W. et al. (1982). Capture immunoglobulin M system for rapid diagnosis of La Crosse (California encephalitis) virus infections. *J. Clin. Microbiol.*, **16**, 577–580.
[487] Janssen, J. C., Godbolt, A. K., Ioannidis, P. et al. (2004). The prevalence of oligoclonal bands in the CSF of patients with primary neurodegenerative dementia. *J. Neurol.* **251**, 184–188.

[488] Jenkins, M. A., Cheng, L. and Ratnaike, S. (2001). Multiple sclerosis: use of light-chain typing to assist diagnosis. *Ann. Clin. Biochem.*, **38**, 235–241.
[489] Jeppsson, J.-O., Laurell, C.-B. and Franzen, B. (1979). Agarose gel electrophoresis. *Clin. Chem.*, **25**, 629–638.
[490] Jeziore, Y. and Thompson, E. J. (1995). Limitation of dot-blot immunoassays in the diagnosis of Creutzfeld-Jakob disease. *Neurosci. Res. Commun.*, **16**, 163–171.
[491] Johnson, A. M. (1978). Immunofixation following electrophoresis or isoelectric focusing for identification and phenotyping of proteins. *Ann. Clin. Lab. Sci.*, **8**, 195–200.
[492] Johnson, A. M., Cejka, J., Hellsing, K. *et al.* (1982). Immunofixation electrophoresis and electrofocusing. *Clin. Chem.*, **28**, 1797–1800.
[493] Johnson, A. W., Land, J. M., Thompson, E. J. *et al.* (1995). Evidence for increased nitric oxide production in multiple sclerosis. *J. Neurol. Neurosurg. Psychiatry*, **58**, 107.
[494] Johnson, G. V., Seubert, P., Cox, T. M. *et al.* (1997). The tau protein in human cerebrospinal fluid in Alzheimer's disease consists of proteolytically derived fragments. *J. Neurochem.*, **68**, 430–433.
[495] Johnson, J. A. and Lott, J. A. (1978). Standardization of the Coomassie blue method for cerebrospinal fluid proteins. *Clin. Chem.*, **24**, 1931–1933.
[496] Johnson, K. P. (1980). Cerebrospinal fluid and blood assays of diagnostic usefulness in multiple sclerosis. *Neurology*, **30**, 106–109.
[497] Johnson, K. P., Arrigo, S. C., Nelson, B. J. and Ginsberg, A. (1977). Agarose electrophoresis of cerebrospinal fluid in multiple sclerosis. *Neurology*, **27**, 273–277.
[498] Johnson, K. P., Likosky, W. H., Nelson, B. J. and Fein, G. (1980). Comprehensive viral immunology of multiple sclerosis. *Arch. Neurol.*, **37**, 537–541.
[499] Johnson, K. P. and Nelson, B. J. (1977). Multiple sclerosis: diagnostic usefulness of cerebrospinal fluid. *Ann. Neurol.*, **2**, 425–431.
[500] Johnson, M. H. and Thompson, E. J. (1981). Freeze-dried cadaveric dural grafts can stimulate a damaging immune response in the host. *Eur. Neurol.*, **20**, 445–447.
[501] Johnson, M. H. and Thompson, E. J. (1981). Proteins of the cerebrospinal fluid. *Hospital Update*, **11**, 1155–1163.
[502] Johnson, M. H. and Thompson, E. J. (1982). Measurement of body fluid proteins by polyacrylamide gel electrophorsis. *J. Clin. Pathol.*, **35**, 1328–1333.
[503] Johnson, M. H. and Thompson, E. J. (1986). Diagnosis of neurosyphilis. *Hospital Update* **12**, 561–563.
[504] Johnson, M. H., Walker, R. W. H., Keir, G. and Thompson, E. J. (1982). A new method for identification of proteins separated in polyacrylamide gels. *Biochem. Soc. Trans.*, **10**, 32–33.
[505] Johnson, M. H., Walker, R. W. H., Keir, G. and Thompson, E. J. (1982). Identification of protein bands in polyacrylamide gel by protein printing. *Biochim. Biophys. Acta*, **718**, 121–124.
[506] Johnson, R. T. (1998). *Viral Infections of the Nervous System*. pp. 1–284. Lippincott-Raven, Philadelphia.
[507] Johnson, R. T., Griffin, D. E., Hirsch, R. L. *et al.* (1984). Measles encephalomyelitis - clinical and immunologic studies. *N. Engl. J. Med.*, **310**, 137–141.
[508] Johnson, R. T., Olson, L. C. and Buescher, E. L. (1968). Herpes simplex virus infection of the nervous system. *Arch. Neurol.*, **18**, 260–264.

[509] Jones, V. E. and Orlans, E. (1981). Isolation of immune complexes and characterisation of their constituent antigens and antibodies in some human diseases: a review. *J. Immunol. Methods*, **44**, 249–270.

[510] Jorgensen, O. S. and Bock, E. (1975). Synaptic plasma membrane antigen D2 measured in human cerebrospinal fluid by rocket-line immunoelectrophoresis, determination in psychiatric and neurological patients. *Scand. J. Immunol.*, **4**, 25–30.

[511] Jornvall, H., Carlstrom, A., Pettersson, T. et al. (1981). Structural homologies between prealbumin, gastrointestinal prohormones and other proteins. *Nature*, **291**, 261–263.

[512] Julkunen, I., Koskiniemi, M., Lehtokoski-Lehtiniemi, E. et al. (1985). Chronic mumps virus encephalitis mumps antibody levels in cerebrospinal fluid. *J. Neuroimmunol.*, **8**, 167–175.

[513] Kabat, E. A., Freedman, D. A., Murray, J. P. and Knaub, V. (1950). A study of the crystalline albumin, gammaglobulin and total protein in the cerebrospinal fluid of one hundred cases of multiple sclerosis and in other diseases. *Am. J. Med. Sci.*, **219**, 55–64.

[514] Kabat, E. A., Glusman, M. and Knaub, V. (1948). Quantitative estimation of the albumin and gamma globulin in normal and pathologic cerebrospinal fluid by immunochemical methods. *Am. J. Med.*, **4**, 653–662.

[515] Kabat, E. A., Moore, D. H. and Landow, H. (1942). An electrophoretic study of the protein components in cerebrospinal fluid and their relationship to the serum proteins. *J. Clin. Invest.*, **21**, 571–577.

[516] Kahn, S. N., Riches, P. G. and Kohn, J. (1980). Paraproteinaemia in neurological disease: incidence, associations, and classification of monoclonal immunoglobulins. *J. Clin. Pathol.*, **33**, 617–621.

[517] Kahn, S. N., Shortman, R. C., Khan, R. A. and Thompson, E. J. (1980). Effect of sample preparation on cerebrospinal fluid protein patterns in polyacrylamide gels. *J. Clin. Chem. Clin. Biochem.* **18**, 23–26.

[518] Kahn, S. N. and Thompson, E. J. (1976). New ultramicromethod for concentration of cerebrospinal fluid. *Lancet*, **i**, 1275–1276.

[519] Kahn, S. N. and Thompson, E. J. (1978). Effect of sample preparation on protein patterns in polyacrylamide gels. *Clin. Chem.*, **24**, 1014.

[520] Kahn, S. N. and Thompson, E. J. (1978). Rapid quantitative surface immunofixation of proteins in polyacrylamide gels. *Clin. Chim. Acta*, **89**, 253–265.

[521] Kaiser, R. (1991). Affinity immunoblotting: rapid and sensitive detection of oligoclonal IgG, IgA and IgM in unconcentrated CSF by agarose isoelectric focusing. *J. Neurol. Sci.*, **103**, 216–225.

[522] Kaiser, R., Czygau, M., Kaufmann, R. and Lucking, C. H. (1995). [Intrathecal IgG synthesis: when is determination of oligoclonal bands necessary?]. *Der Nervenarzt*, **66**, 618–623.

[523] Kaiser, R. and Lucking, C. H. (1993). Intrathecal synthesis of IgM and IgA in neurological diseases: comparison of two formulae with isoelectric focusing. *Clin. Chim. Acta*, **216**, 39–51.

[524] Kalimo, K. O., Marttila, R. J., Granfors, K. and Viljanen, M. K. (1977). Solid-phase radioimmunoassay of human immunoglobulin M and immunoglobulin G antibodies against herpes simplex virus type 1 capsid, envelope , and excreted antigens. *Infect Immun.*, **15**, 883–889.

[525] Kalimo, K. O. K., Marttila, R. J., Ziola, B. R. et al. (1977). Radiommunoassay of herpes-simplex and measles virus antibodies in serum and cerebrospinal fluid of patients without infectious or demyelinating diseases of the central nervous system. *J. Med. Microbiol.*, **10**, 431–438.

[526] Kalimo, K. O. K., Ziola, B. R., Viljanen, M. K. et al. (1977). Solid-phase radioimmunoassay of herpes simplex virus IgG and IgM antibodies. *J. Immunol. Methods*, **14**, 183–195.

[527] Kalish, S. B., Radin, R. C., Levitz, D. et al. (1983). The enzyme-linked immunosoarbent assay method for IgG antibody to purified protein derivative in cerebrospinal fluid of patients with tuberculous meningitis. *Ann. Intern. Med.*, **99**, 630–633.

[528] Kamp, H. H., Luderer, T. K. J., Muller, H. J. and Sopjes-Kruk, A. (1981). Rapid immunoturbidimetric assay of albumin and immunoglobulin G in serum and cerebrospinal fluid with an automatic discrete analyser. *Clin. Chim. Acta*, **114**, 195–205.

[529] Kanai, M., Matsubara, E., Isoe, K. et al. (1998). Longitudinal study of cerebrospinal fluid levels of tau, A beta1-40. and A beta1-42(43) in Alzheimer's disease: a study in Japan. *Ann. Neurol.*, **44**, 17–26.

[530] Kaplan, A. (1980). RIA for HSV in cerebrospinal fluid (CSF). *J. Infect. Dis.*, **142**, 797.

[531] Kaplan, E., Bigelow, D., Vatassery, G. and Ansari, K. (1982). Glutathione peroxidase in human cerebrospinal fluid. *Brain Res.*, **252**, 391–393.

[532] Kaplan, S. L. (1983). Antigen detection in cerebrospinal fluid – pros and cons. *Am. J. Med.*, **75**, 109–118.

[533] Karcher, D., Lowenthal, A. and Van Soom, G. (1979). Cerebrospinal fluid proteins electrophoresis without prior concentration. *Acta Neurol. Belg.*, **79**, 335–337.

[534] Karcher, D., Matthyssens, G. and Lowenthal, A. (1972). Isolation and characterization of IgG globulins in subacute sclerosing panencephalitis. *Immunology*, **23**, 93–99.

[535] Karcher, D., Noppe, M. and Lowenthal, A. (1977). A heat stable serum inhibitor of an antigen antibody reaction of subacute sclerosing panencephalitis. *J. Neurol.*, **216**, 51–56.

[536] Karcher, D., Van Sande, M. and Lowenthal, A. (1959). Micro-electrophoresis in agar gel of proteins of the cerebrospinal fluid and central nervous system. *J. Neurochem.*, **4**, 135–140.

[537] Karcher, D., Zeman, W., Lowenthal, A. and Chamoles, N. (1970). Studies on alpha albumin in nervous tissue. 1. Biochemical investigations. *Brain Res.*, **17**, 307–314.

[538] Karlsson, B. and Alling, C. (1980). A comparative study of three approaches to the routine quantitative determination of spinal fluid total proteins. *Clin. Chim. Acta*, **105**, 65–73.

[539] Karlsson, B. and Alling, C. (1982). Radioimmunoassay of myelin basic protein. A methodological evaluation. *J. Immunol. Methods*, **55**, 51–61.

[540] Karlsson, B. and Alling, C. (1984). Molecular size of myelin basic protein immunoactivity in spinal fluid. *J. Neuroimmunol.*, **6**, 141–150.

[541] Kaufmann, P. and Thompson, E. J. (1980). Oligoclonal gamma globulin patterns in acrylamide gel electrophoresis in patients with multiple sclerosis. In: D. Dommesch and H.G. Mertens (eds.), *Cerebrospinal-flussigkeit-CSF*, pp. 97–99. Georg Thieme Verlag.

[542] Kay, A., Petzold, A., Kerr, M. *et al.* (2003). Decreased cerebrospinal fluid apolipoprotein E after subarachnoid haemorrhage: correlation with injury severity and clinical outcome. *Stroke*, **34**, 637–642.
[543] Kay, A., Petzold, A., Kerr, M. *et al.* (2003). Temporal alterations in cerebrospinal fluid amyloid (beta)-protein and apolipoprotein E after subarachnoid haemorrhage. *Stroke*, **34**, 240–243.
[544] Kay, A. D., Petzold, A., Kerr, M. *et al.* (2003). Alterations in cerebrospinal fluid apolipoprotein E and amyloid beta-protein after traumatic brain injury. *J. Neurotrauma*, **20**, 943–952.
[545] Kay, A. D., Petzold, A., Kerr, M. *et al.* (2003). Cerebrospinal fluid apolipoprotein E concentration decreases after traumatic brain injury. *J. Neurotrauma*, **20**, 243–250.
[546] Keck, K., Grossberg, A. L. and Pressman, D. (1973). Specific characterization of isoelectrofocused immunoglobulins in polyacrylamide gel by reaction with 125 I-labeled protein antigens or antibodies. *Eur. J. Immunol.*, **3**, 99–102.
[547] Keeffe, E. B., Bardana, E. J., Harbeck, R. J. *et al.* (1974). Antibody to deoxyribonucleic acid (DNA) and DN. *Ann. Intern. Med.*, **80**, 58–60.
[548] Keir, G., Chowhan, M. R. and Thompson, E. J. (1985). High electroendosmotic agarose electrophoresis and nitrocellulose immobilisation for the detection of oligoclonal bands in unconcentrated cerebrospinal fluid. *Ann. Clin. Biochem.*, **22**, 381–386.
[549] Keir, G., Luxton, R. W. and Thompson, E. J. (1990). Isoelectric focusing of cerebrospinal fluid immunoglobulin G: an annotated update. *Ann. Clin. Biochem.*, **27**, 436–443.
[550] Keir, G., Tasdemir, N. and Thompson, E. J. (1993). Cerebrospinal fluid ferritin in brain necrosis: evidence for local synthesis. *Clin. Chim. Acta*, **216**, 153–166.
[551] Keir, G. and Thompson, E. J. (1986). Immunoglobulins in CSF. In: M. A. H. French (ed.), *Immunoglobulins in Health and Disease*, pp. 173–187. MTP Press Ltd.
[552] Keir, G. and Thompson, E. J. (1986). Proteins as parameters in the discrimination between different blood-CSF barriers. *J. Neurol. Sci.*, **75**, 245–253.
[553] Keir, G. and Thompson, E. J. (1988). Ultrasensitive staining for proteins after agarose electrophoresis of unconcentrated cerebrospinal fluid. *Ann. Clin. Biochem.*, **25**, 116–117.
[554] Keir, G., Walker, R. W. H., Johnson, M. H. and Thompson, E. J. (1982). Nitrocellulose immunofixation following agarose electrophoresis in the study of immunoglobulin G subgroups in unconcentrated cerebrospinal fluid. *Clin. Chim. Acta*, **121**, 231–236.
[555] Keir, G., Walker, R. W. H. and Thompson, E. J. (1982). Oligoclonal immunoglublin M in cerebrospinal fluid from multiple sclerosis patients. *J. Neurol. Sci.*, **57**, 281–185.
[556] Keir, G., Zeman, A., Brookes, G. *et al.* (1992). Immunoblotting of transferrin in the identification of cerebrospinal fluid otorrhoea and rhinorrhoea. *Ann. Clin. Biochem.*, **29**, 210–213.
[557] Kempe, C. H., Takabayashi, K., Miyamoto, H. *et al.* (1973). Elevated cerebrospinal fluid vaccinia antibodies in multiple sclerosis. *Arch. Neurol.*, **28**, 278–279.
[558] Kennedy, P. G. E. (1984). Herpes simplex virus and the nervous system. *J. Postgrad. Med.*, **60**, 13–19.

[559] Kerenyi, L. and Gallyas, F. (1972). A highly sensitive method for demonstrating proteins in electrophoretic immunophoretic and immunodiffusion preparations. *Clin. Chim. Acta*, **38**, 465–467.
[560] Kerenyi, L., Hegedus, K. and Palffy, G. (1975). Characteristic gamma globulin subfractions of native CSF in multiple sclerosis. *Brain Res.*, **87**, 123–125.
[561] Kidd, D., Duncan, J. S. and Thompson, E. J. (1998). Pontine inflammatory lesion due to shingles. *J. Neurol. Neurosurg. Psychiatry*, **65**, 208.
[562] Kiessling, W. R., Yung, L. L., Hall, W. W. and Ter, M. V. (1977). Measles-virus-specific immunoglobulin-M response in subacute sclerosing panencephalitis. *Lancet*, **i**, 324–327.
[563] Killingsworth, L. M. (1982). Clinical applications of protein determinations in biological fluids other than blood. *Clin. Chem.*, **28**, 1093–1102.
[564] Kimball, J. W., Pappenheimer, A. M. and Jaton, J.-C. (1971). The response in rabbits to prolonged immunization with type III pneumococci. *J. Immunol.*, **106**, 1177–1184.
[565] Kinnman, J., Fryden, A., Eriksson, S. *et al.* (1981). Tuberculous meningitis: immune reactions within the central nervous system. *Scand. J. Immunol.*, **13**, 289–296.
[566] Kinnman, J. and Link, H. (1984). Intrathecal production of oligoclonal IgM and IgG in CNS sarcoidosis. *Acta Neurol. Scand.*, **69**, 97–106.
[567] Kinnman, J., Link, H. and Fryden. (1981). Characterization of antibody activity in oligoclonal immunoglobulin G synthesized within the central nervous system in a patient with tuberculous meningitis. *J. Clin. Microbiol.*, **13**, 30–35.
[568] Kinnman, J., Link, H., Moller, E. and Norrby, E. (1978). Influence of measles virus antigen on leukocyte migration in multiple sclerosis and controls. *Acta Neurol. Scand.*, **58**, 261–271.
[569] Kjellin, K. G. and Hallander, L. (1979). Isotachophoresis in capillary tubes of CSF proteins - especially gammaglobulins. *J. Neurol.*, **221**, 235–244.
[570] Kjellin, K. G. and Hallander, L. (1979). Isotachophoresis of CSF proteins in gel tubes especially gammaglobulins. *J. Neurol.*, **221**, 225–233.
[571] Kjellin, K. G. and Hallander, L. B. (1982). High-voltage isoelectric focusing in ultrathin gels and enzyme-amplified immunoassay: a new method for analysis of cerebrospinal fluid protein. *J. Neurol.*, **228**, 49–57.
[572] Kjellin, K. G. and Siden, A. (1977). Aberrant CSF protein fractions found by electrofocusing in multiple sclerosis. *Eur. Neurol.*, **15**, 40–50.
[573] Kjellin, K. G. and Siden, A. (1977). Electrofocusing and electrophoresis of cerebrospinal fluid proteins in CNS disorders of known or probable infectious etiology. *Eur. Neurol.*, **16**, 79–89.
[574] Kjellin, K. G. and Stibler, H. (1974). CSF-protein patterns in extrapyramidal diseases. *Eur. Neurol.*, **12**, 186–194.
[575] Kjellin, K. G. and Stibler, H. (1974). Protein pattern of cerebrospinal fluid in spasmodic torticollis. *J. Neurol. Neurosurg. Psychiatry*, **37**, 1128–1132.
[576] Kjellin, K. G. and Stibler, H. (1975). Protein patterns of cerebrospinal fluid in hereditary ataxias and hereditary spastic paraplegia. *J. Neurol. Sci.*, **25**, 65–74.
[577] Kjellin, K. G. and Stibler, H. (1976). Isoelectric focusing and electrophoresis of cerebrospinal fluid proteins in muscular dystrophies and spinal muscular atrophies. *J. Neurol. Sci.*, **27**, 45–57.
[578] Kjellin, K. G. and Vesterberg, O. (1974). Isoelectric focusing of CSF proteins in neurological diseases. *J. Neurol. Sci.*, **23**, 199–213.

[579] Klapper, P. E., Laing, I. and Longson, M. (1981). Rapid non-invasive diagnosis of herpes encephalitis. *Lancet*, **ii**, 607–609.
[580] Klapper, P. E. and Longson, M. (1981). Acute Viral Encephalitis. *Br. Med. J.*, **283**, 1544–1545.
[581] Klatzo, I., Miquel, J., Ferris, P. J. and Prokop, J. D. (1964). Observations on the passage of the fluorescein labeled serum proteins (FLSP) from the cerebrospinal fluid. *J. Neuropathol. Exp. Neurol.*, **23**, 18–35.
[582] Kleine, T. O. (1979). New Developments in the Diagnosis of cerebrospinal fluid. *J. Clin. Chem. Clin. Biochem.* **17**, 505–511.
[583] Kleine, T. O. and Merten, B. (1980). Rapid manual immunoturbidimetric and immunonephelometric assays of prealbumin, albumin, IgG, IgA, and IgM in cerebrospinal fluid. *J. Clin. Chem. Clin. Biochem.* **18**, 245–254.
[584] Kleine, T. O. and Stroh, J. (1974). Neue Mikroelektrophorese fur nativen lumballiquor. Unterschiede im pherogramm von nativen und konzentrierten proteinen. *Z. Klin. Chem. Klin. Biochem.* **12**, 73–80.
[585] Kleine, T. O., Stroh, M. and Stroh, J. (1974). Vergleichende untersuchungen zur anreicherung von proteinen im gepoolten lumballiquor. *Z. Klin. Chem. Klin. Biochem.* **12**, 66–72.
[586] Knopf, P. M., Harling-Berg, C. J., Cserr, H. F. *et al.* (1998). Antigen-dependent intrathecal antibody synthesis in the normal rat brain: tissue entry and local retention of antigen-specific B cells. *J. Immunol.*, **161**, 692–701.
[587] Kobatake, K., Shinohara, Y. and Yoshimura, S. (1980). Immunoglobulins in cerebrospinal fluid. *J. Neurol. Sci.*, **47**, 273–283.
[588] Kolar, O. J. (1975). Differential diagnostic aspects in malignant lymphomas involving the central nervous system. *Acta Neuropathol.*, **Suppl VI**, 181–186.
[589] Kolar, O. J. (1977). Light chains in cerebrospinal fluid in multiple sclerosis. *Lancet*, **ii**, 1030.
[590] Kolar, O. J., Rice, P. H., Jones, F. H. *et al.* (1980). Cerebrospinal fluid immunoelectrophoresis in multiple sclerosis. *J. Neurol. Sci.*, **47**, 221–230.
[591] Koppel, H., Riederer, P. (1977). Electrophoresis of unconcentrated natural cerebrospinal fluid. *J. Neural. Transm.*, **41**, 313–318.
[592] Koskiniemi, M., Rantalaiho, T., Piiparinen, H. *et al.* (2001). Infections of the central nervous system of suspected viral origin: a collaborative study from Finland. *J. Neurovirol.*, **7**, 400–408.
[593] Koskiniemi, M. L. and Vaheri, A. (1982). Diagnostic value of cerebrospinal fluid antibodies in herpes simplex virus encephalitis. *J. Neurol. Neurosurg. Psychiatry*, **45**, 239–242.
[594] Koskiniemi, M. L., Vaheri, A., Manninen, V. and Nikki, P. (1982). Ascending myelitis with high antibody titer to herpes simplex virus in the cerebrospinal fluid. *J. Neurol.*, **227**, 187–191.
[595] Koskiniemi, M. L., Vaheri, A., Manninen, V. *et al.* (1980). Herpes Simplex virus encephalitis. *Arch. Neurol.*, **37**, 763–767.
[596] Kostulas, V. K. and Link, H. (1982). Agarose isoelectric focusing of unconcentrated CSF and radioimmunofixation for detection of oligoclonal bands in patients with multiple sclerosis and other neurological diseases. *J. Neurol. Sci.*, **54**, 117–127.

REFERENCES

[597] Kostulas, V. K., Link, H. and Lefvert, A.-K. (1987). Oligoclonal IgG bands in cerebrospinal fluid: Principles for demonstration and interpretation based on findings in 1114 neurological patients. *Arch. Neurol.*, **44**, 1041–1044.
[598] Krakauer, M., Schaldemose, H., Nielson, J. *et al.* (1998). Intrathecal synthesis of free immunoglobulin light chains in multiple sclerosis. *Acta Neurol. Scand.*, **98**, 161–165.
[599] Krambovitis, E., McIllmurray, M. B., Lock, P. E. *et al.* (1984). Rapid diagnosis of tuberculous meningitis by latex particle agglutination. *Lancet*, **ii**, 1229–1231.
[600] Krause, V. H. D., Wisser, H. and Pirke, K. M. (1975). Methodische Untersuchungen zur liquorelektrophorese. *Z. Klin. Chem. Klin. Biochem.* **13**, 79–84.
[601] Kruger, H., Englert, D. and Pflughaupt, K. W. (1981). Demonstration of oligoclonal immunoglobulin G in Guillain-Barre syndrome and lymphocytic meningoradiculitis by isoelectric focusing. *J. Neurol.*, **226**, 15–24.
[602] Krzalic, L. (1982). Prealbumin content of cerebrospinal fluid and sera in persons with multiple sclerosis. *Clin. Chim. Acta*, **124**, 339–341.
[603] Kun-yu W. K., Jacobsen, C. D. and Hoak, J. C. (1973). Plasminogen in normal and abnormal human cerebrospinal fluid. *Arch. Neurol.*, **28**, 64–66.
[604] Kurosky, A., Barnett, D. R., Lee, T.-H. *et al.* (1980). Covalent structure of human haptoglobulin: A serine protease homolog. *Proc. Natl. Acad. Sci.*, **77**, 3388–3392.
[605] Kurtz, J. B. (1974). Specific IgG and IgM antibody responses in herpes-simplex-virus infections. *J. Med. Microbiol.* **7**, 333–341.
[606] Kuruvilla, T., Bharucha, N. E., Thompson, E. J. *et al.* (1999). Stiff-man syndrome–diagnostic criteria and pitfalls. *J. Assoc. Physicians India*, **47**, 836.
[607] Kutt, H., Hurwitz, L. J., Ginsburg, S. M. and McDowell, F. (1961). Cerebrospinal fluid protein in diabetes mellitus. *Arch. Neurol.*, **4**, 31–36.
[608] Kutt, H., McDowell, F., Chapman, L. *et al.* (1960). Abnormal protein fractions of cerebrospinal fluid demonstrated by starch gel electrophoresis. *Neurology*, **10**, 1064–1067.
[609] Laemmli, U. K. and Favre, M. (1973). Maturation of the head of bacteriophage T-4. *J. Mol. Biol.*, **80**, 575–599.
[610] Lafer, E. M., Rauch, J., Andrzejewski, C. *et al.* (1981). Polyspecific monolonal lupus autoantibodies reactive with both polynucleotides and phospholipids. *J. Exp. Med.*, **153**, 897–909.
[611] Laman, J. D., Thompson, E. J. and Kappos, L. (1998). Balancing the Th1/Th2 concept in multiple sclerosis. *Immunol. Today*, **19**, 489–490.
[612] Laman, J. D., Thompson, E. J. and Kappos, L. (1998). Body fluid markers to monitor multiple sclerosis: the assays and the challenges. *Mult. Scler.*, **4**, 266–269.
[613] Lamoureux, G., Jolicoeur, R., Giard, N. *et al.* (1975). Cerebrospinal fluid proteins in multiple sclerosis. *Neurology*, **25**, 537–546.
[614] Lascelles, P. T., Thompson, E. J. and Warner, D. S. (1983). Chemistry of the cerebrospinal fluid in health and disease. In: V. Marks (ed.), *Scientific Foundations of Clinical Biochemistry*, pp. 445–455. Heineman Medical.
[615] Lasne, Y., Benzerara, O., Chazot, G. and Creyssel, R. (1981). A sensitive method for characterization of oligoclonal immunoglobulins in unconcentrated cerebrospinal fluid. *J. Neurochem.*, **36**, 1872–1874.
[616] Laterre, E. C. (1965). Les proteins du cephalorachidien a l'etat normal et pathologique. pp. 1–285. Arscia.

[617] Laterre, E. C., Callewaert, A., Heremans, J. F. and Sfaello, Z. (1970). Electrophoretic morphology of gamma globulins in cerebrospinal fluid of multiple sclerosis and other diseases of the nervous system. *Neurology*, **20**, 982–990.

[618] Laterre, E. C., Heremans, J. F. and Carbonara, A. (1964). Immunological comparison of some proteins found in cerebrospinal fluid, urine and extracts from brain and kidney. *Clin. Chim. Acta*, **10**, 197–209.

[619] Laterre, E. C. and Heulle, H. (1972). Resultats compares de l'electrophorese en agar et du dosage immunochimique des gammaglobulines du liquide cephalo-rachidien dans la sclerose en plaques. *Acta Neurol. Belg.*, **72**, 240–253.

[620] Latner, A. L., Marshall, T. and Gambie, M. (1980). A simplified technique of high resolution two-dimensional electrophoresis: serum immunoglobulins. *Clin. Chim. Acta*, **103**, 51–59.

[621] Laurell, C.-B. (1972). Composition and variation of the gel electrophoretic fractions of plasma, cerebrospinal fluid and urine. *Scand. J. Clin. Lab. Invest.*, **29**, 71–82.

[622] Laurell, C.-B. (1972). Electroimmuno assay. *Scand. J. Clin. Lab. Invest.*, **29**, 21–37.

[623] Laurell C-B and McKay, E. J. M. (1981) Electroimmunoassay. In: J. J. Lagone and H. Van Vunakis (eds.), *Methods in Enzymology: Immunological Methods*, pp. 340–350. Academic Press.

[624] Laurent, T. C. and Killander, J. (1964). A theory of gel filtration and its experimental verification. *J. Chromatogr.*, **14**, 317–330.

[625] Laurenzi, M. A. and Link, H. (1978). Comparison between agarose gel electrophoresis and isoelectric focusing of CSF for demonstration of oligoclonal immunoglobulin bands in neurological disorders. *Acta Neurol. Scand.*, **58**, 148–156.

[626] Laurenzi, M. A. and Link, H. (1978). Localization of the immunoglobulins G, A and M, beta-trace protein and gamma-trace protein on isoelectric focusing of serum and cerebrospinal fluid by immunofixation. *Acta Neurol. Scand.*, **58**, 141–147.

[627] Laurenzi, M. A. and Link, H. (1979). Characterisation of the mobility on isoelectric focusing of individual proteins in CSF and serum by immunofixation. *J. Neurol. Neurosurg. Psychiatry*, **42**, 368–372.

[628] Laurenzi, M. A., Mavra, M., Kam-Hansen, S. and Link, H. (1980). Oligoclonal IgG and free light chains in multiple sclerosis demonstrated by thin-layer polyacrylamide gel isoelectric focusing and immunofixation. *Ann. Neurol.*, **8**, 241–247.

[629] Leary, S. M., McLean, B. N. and Thompson, E. J. (2000). Local synthesis of IgA in the cerebrospinal fluid of patients with neurological diseases. *J. Neurol.* **247**, 609–615.

[630] Leclerc, G., Giroux, M., Birry, A. and Kasatiya, S. (1978). Study of fluorescent treponemal antibody test on cerebrospinal fluid using monospecific anti-immunoglobulin conjugates IgG, IgM, and IgA. *Br. J. Vener. Dis.*, **54**, 303–308.

[631] Lems-Van Kan, P., Verspaget, H. W. and Pena, A. S. (1983). ELISA Assay for quantitative measurement of human immunoglobulins IgA IgG and IgM in nanograms. *J. Immunol. Methods*, **57**, 51-57.

[632] Lennon, V. and Mackay, I. R. (1972). Binding of 125 I-myelin basic protein by serum and cerebrospinal fluid. *Clin. Exp. Immunol.*, **11**, 595–603.

[633] Leonardi, A., Abbruzzese, G., Arata, L. et al. (1984). Cerebrospinal fluid (CSF) findings in amyotrophic lateral sclerosis. *J. Neurol.*, **231**, 75–78.

[634] Leonardi, A., Gandolfo, C., Caponnetto, C. et al. (1985). The integrity of the blood-brain barrier in Alzheimer's type and multi-infarct dementia evaluated by the study of albumin and IgG in serum and cerebrospinal fluid. *J. Neurol. Sci.*, **67**, 253–261.

[635] Levine, D. P., Lauter, C. B. and Lerner, A. M. (1978). Simultaneous serum and CSF antibodies in herpes simplex virus encephalitis. *JAMA*, **240**, 356–360.

[636] Levine, J. E., Povlishock, J. T. and Becker, D. P. (1982). The morphological correlates of primate cerebrospinal fluid absorption. *Brain Res.*, **214**, 31–41.

[637] Levine, M. C. (1973). Changes in the CSF during pneumoencephalography. *Neuroradiology*, **5**, 1–6.

[638] Lietze, A., Sinclair, C. and Rowe, A. (1970). An intrinsic inaccuracy of radial immunodiffusion measurements of incomplete antigens. *Clin. Biochem.*, **3**, 335–338.

[639] Lim, E. T., Grant, D., Pashenkov, M. *et al.* (2004). Cerebrospinal fluid levels of brain specific proteins in optic neuritis. *Mult. Scler.*, **10**, 261–265.

[640] Lin, F. H. and Thormar, H. (1980). Absence of M protein in a cell-associated subacute sclerosing panencephalitis virus. *Nature*, **285**, 490–492.

[641] Link, H. (1967). Immunoglobulin G and low molecular weight proteins in human cerebrospinal fluid – chemical and immunological characterisation with special reference to multiple sclerosis. *Acta Neurol. Scand.*, **43**, 1–136.

[642] Link, H. (1967). The mobility of beta-trace protein gamma-trace protein and immunoglobulin G from cerebrospinal fluid on agar gel electrophoresis. *Acta Neurol. Scand.*, **43**, 39–45.

[643] Link, H. (1973). Comparison of electrophoresis on agar gel and agarose gel in the evaluation of gamma-globulin abnormalities in cerebrospinal fluid and serum in multiple sclerosis. *Clin. Chim. Acta*, **46**, 383–389.

[644] Link, H. (1975). Demonstration of oligoclonal immunoglobulin G in Guillain-Barre syndrome. *Acta Neurol. Scand.*, **52**, 111–120.

[645] Link, H. (1978). Characteristics of the immune response within the CNS in neurological disorders. *Acta Neurol. Scand.*, **57**, Sp67.

[646] Link, H. and Kostulas, V. (1983). Utility of isoelectric focusing of cerebrospinal fluid and serum on agarose evaluated for neurological patients. *Clin. Chem.*, **29**, 810–815.

[647] Link, H. and Laurenzi, M. A. (1979). Immunoglobulin class and light chain type of oligoclonal bands in CSF in multiple sclerosis determined by agarose gel electrophoresis and immunofixation. *Ann. Neurol.*, **6**, 107–110.

[648] Link, H., Laurenzi, M. A. and Fryden, A. (1981). Viral antibodies in oligoclonal and polyclonal IgG synthesized within the central nervous system over the course of mumps meningitis. *J. Neuroimmunol.*, **1**, 287–298.

[649] Link, H., Moller, E., Muller, R. *et al.* (1977). Immunoglobulin abnormalities in spinal fluid in multiple sclerosis. *Acta Neurol. Scand.*, **55**, 173–206.

[650] Link, H. and Muller, R. (1971). Immunoglobulins in multiple sclerosis and infections of the nervous system. *Arch. Neurol.*, **25**, 326–344.

[651] Link, H., Norrby, E. and Olsson, J.-E. (1976). Immunoglobulin abnormalities and measles antibody response in chronic myelopathy. *Arch. Neurol.*, **33**, 26–32.

[652] Link, H. and Olsson, J.-E. (1972). Beta-trace protein concentration in CSF in neurological disorders. *Acta Neurol. Scand.*, **48**, 57–68.

[653] Link, H. and Tibbling, G. (1977). Principles of albumin and IgG analyses in neurological disorder II. Relation of the concentration of the proteins in serum and cerebrospinal fluid. *Scand. J. Clin. Lab. Invest.*, **37**, 391–396.

[654] Link, H. and Tibbling, G. (1977). Principles of albumin and IgG analyses in neurological disorders. III Evaluation of IgG synthesis within the central nervous system in multiple sclerosis. *Scand. J. Clin. Lab. Invest.*, **37**, 397–401.

[655] Link, H. and Zettervall, O. (1970). Multiple sclerosis: disturbed kappa:lambda chain ratio of immunoglobulin G in cerebrospinal fluid. *Clin. Exp. Immunol.*, **6**, 435–438.
[656] Link, H., Zettervall, O. and Blennow, G. (1972). Individual cerebrospinal fluid (CSF) proteins in the evaluation of increased CSF total protein. *J. Neurol.*, **203**, 119–132.
[657] Lippincott, S. W., Korman, S., Lax, L. C. and Corcoran, C. (1965). Transfer rates of gamma globulin between cerebrospinal fluid and blood plasma (Results obtained on a series of multiple sclerosis patients). *J. Nucl. Med.*, **6**, 632–644.
[658] Lisak, R. P., Behan, P. O., Zweiman, B. and Shetty, T. (1974). Cell-mediated immunity to myelin basic protein in acute disseminated encephalomyelitis. *Neurology*, **24**, 560–564.
[659] Lisak, R. P. and Zweiman, B. (1977). In vitro cell-mediated immunity of cerebrospinal-fluid lymphocytes to myelin basic protein in primary demyelinating diseases. *N. Engl. J. Med.*, **297**, 850–853.
[660] Lisak, R. P., Zweiman, B. and Whitaker, J. N. (1981). Spinal fluid basic protein immunoreactive material and spinal fluid lymphocyte reactivity to basic protein. *Neurology*, **31**, 180–182.
[661] Liskiewicz, W. (1982). The nature of 'oligoclonal' bands in cerebrospinal fluid of patients with multiple sclerosis. *Clin. Chim. Acta*, **123**, 145–152.
[662] Livingston, D. M. (1974). Immunoaffinity chromatography of proteins. In: W. B. Jakoby and M. Wilchek (eds.), *Methods in Enzymology*, pp. 723–731. Academic Press.
[663] Livrea, P., Simone, I. L., Trojano, M. et al. (1987). Cerebrospinal fluid (CSF) parameters and clinical course of multiple sclerosis. *Riv. neurol.*, **57**, 189–196.
[664] Livrea, P., Trojano, M., Simone, I. L. et al. (1981). Intrathecal IgG synthesis in multiple sclerosis: Comparison between isoelectric focusing and quantitative estimation of cereborspinal fluid IgG. *J. Neurol.*, **224**, 159–169.
[665] Livrea, P., Trojano, M., Simone, I. L. et al. (1984). Heterogeneous models for blood-cerebrospinal fluid barrier permeability to serum proteins in normal and abnormal cerebrospinal fluid/serum protein concentration gradients. *J. Neurol. Sci.*, **64**, 245–258.
[666] Livrea, P., Trojano, M., Zimatore, G. B. et al. (1978). Isoelectric focusing and crossed immunoelectrofocusing of cerebrospinal fluid proteins in neurological disorders. *Acta. Neurol. (Napoli)*, **33**, 501–517.
[667] Locoge, M. and Cumings, J. N. (1958). Cerebrospinal fluid in various diseases. *Br. Med. J.* **i**, 618–620.
[668] Lofberg, H. and Grubb, A. O. (1979). Quantitation of gamma trace in human biological fluids: indications for production in the central nervous system. *Scand. J. Clin. Lab. Invest.*, **39**, 619–626.
[669] Lofberg, H., Grubb, A. O., Sveger, T. and Olsson, J.-E. (1980). The cerebrospinal fluid and plasma concentrations of gamma-trace and beta 2-microglobulin at various ages and in neurological disorders. *J. Neurol.*, **223**, 159–170.
[670] Lord, R. M., Goldblum, R. M., Forman, P. M. et al. (1973). Cerebrospinal-fluid IgM in the absence of serum-IgM in combined immunodeficiency. *Lancet*, **ii**, 528–529.
[671] Lowenthal, A. (1964). *Agar Gel Electrophoresis in Neurology*. Elsevier.
[672] Lowenthal, A. (1979). Restricted heterogeneity of the IgG in neurology. In: D. Karcher, A. Lowenthal and A. D. Strosberg (eds.), *Humoral Immunity in Neurological Diseases (NATO Advanced Study Inst)*, pp. 1281–1288. Plenum Press.

[673] Lowenthal, A., Adriaenssens, K., Colfs, B. et al. (1972). Oligoclonal gammopathy in ataxia-telangiectasia. *Z. Neurol.*, **202**, 58–63.

[674] Lowenthal, A., Crols, R., De Schutter, E. et al. (1984). Cerebrospinal fluid proteins in neurology. *Int. Rev. Neurobiol.*, **25**, 95–138.

[675] Lowenthal, A. and Karcher, D. (1980). *Cerebrospinal Fluid in Clinical Neurology*. 39 ed. Alan R Liss.

[676] Lowenthal, A., Noppe, M., Gheuens, J. and Karcher, D. (1978). Alpha-albumin (glial fibrillary acidic protein) in normal and pathological human brain and cerebrospinal fluid. *J. Neurol.*, **219**, 87–91.

[677] Lowenthal, A., Van Sande, M. and Karcher, D. (1960). The differential diagnosis of neurological diseases by fractionating electrophoretically the CSF gamma globulins. *J. Neurochem.*, **6**, 51-56.

[678] Lubahn, D. B. and Silverman, L. M. (1984). A rapid silver-stain procedure for use with routine electrophoresis of cerebrospinal fluid on agarose gels. *Clin. Chem.*, **30**, 1689–1691.

[679] Luc, C. L., Pianeta, C. M. and Depieds, R. C. (1979). Presence of free light chains in urine of patients suffering from subacute sclerosing panencephalitis. *J. Neuropathol. Exp. Neurol.*, **38**, 392–400.

[680] Luger, A. (1981). Diagnosis of syphilis. *Bull. World Health Organ.*, **59**, 647–654.

[681] Luger, A., Schmidt, B. L., Steyrer, K. and Schonwald, E. (1981). Diagnosis of neurosyphilis by examination of the cerebrospinal fluid. *Br. J. Vener. Dis.*, **57**, 232–237.

[682] Lund, G. A., Arnadottir, T., Hukkanen, V. et al. (1983). Characterization of immune complexes in multiple sclerosis by an antigen-specific immune complex radioimmunoassay. *J. Neuroimmunol.*, **4**, 253–264.

[683] Luxton, R. W., McLean, B. N. and Thompson, E. J. (1990). Isoelectric focusing versus quantitative measurements in the detection of intrathecal local synthesis of IgG. *Clin. Chim. Acta*, **187**, 297–308.

[684] Luxton, R. W., Patel, P., Keir, G. and Thompson, E. J. (1989). A micro-method for measuring total protein in cerebrospinal fluid by using benzethonium chloride in microtiter plate wells. *Clin. Chem.*, **35**, 1731–1734.

[685] Luxton, R. W. and Thompson, E. J. (1989). Differential oligoclonal band patterns on polyvinyldifluoride membranes. *J. Immunol. Methods*, **121**, 269–274.

[686] Luxton, R. W. and Thompson, E. J. (1990). Affinity distributions of antigen-specific IgG in patients with multiple sclerosis and in patients with viral encephalitis. *J. Immunol. Methods*, **131**, 277–282.

[687] Luxton, R. W., Zeman, A., Holzel, H. et al. (1995). Affinity of antigen-specific IgG distinguishes multiple sclerosis from encephalitis. *J. Neurol. Sci.*, **132**, 11–19.

[688] Lycke, J. N., Karlsson, J. E., Andersen, O. and Rosengren, L. E. (1998). Neurofilament protein in cerebrospinal fluid: a potential marker of activity in multiple sclerosis. *J. Neurol. Neurosurg. Psychiatry* **64**, 402–404.

[689] Lyman, D. J. (1973). Polymers in medicine. In: H. G. Elias (ed.), *Trends in Macro Molecular Science*, pp. 55–69. Gordon & Breech.

[690] Ma, B. I., Joseph, B. S., Walsh, M. J. et al. (1981). Multiple sclerosis serum and cerebrospinal fluid immunoglobulin binding to Fc receptors of oligodendrocytes. *Ann. Neurol.*, **9**, 371–377.

[691] MacCallum, F. O., Chinn, I. J. and Gostling, J. V. T. (1974). Antibodies to herpessimplex virus in the cerebrospinal fluid of patients with herpetic encephalitis. *J. Med. Microbiol.*, **7**, 325–331.
[692] MacGillivray, R. T. A., Mendez, E., Sinha, S. K. *et al.* (1982). The complete amino acid sequence of human serum transferrin. *Proc. Natl. Acad. Sci.*, **79**, 2504–2508.
[693] MacNamara, E. M. and Whicher, J. T. (1990). Electrophoresis and densitometry of serum and urine in investigation and significance of monoclonal immunoglobulins. *Electrophoresis*, **11**, 376–381.
[694] MacPherson, C. F. C. and Cosgrove, J. B. R. (1961). Immunochemical evidence for a gamma globulin peculiar to cerebrospinal fluid. *Can. J. Biochem. Physiol.*, **39**, 1567–1574.
[695] Maida, E. and Kristoferitsch, W. (1982). Cerebrospinal fluid findings in mycoplasma pneumoniae. *Acta Neurol. Scand.*, **65**, 524–538.
[696] Maidment, B. W., Papsidero, L. D. and Chu, T. M. (1980). Isoelectric focusing - A new approach to the study of immune complexes. *J. Immunol. Methods*, **35**, 297–306.
[697] Maidment, B. W., Papsidero, L. D., Gamarra, M. *et al.* (1981). Isoelectric focusing analysis of soluble immune complexes bound to protein A-sepharose. *Anal. Biochem.*, **111**, 336–342.
[698] Malmestrom, C., Haghighi, S., Rosengren, L. *et al.* (2003). Neurofilament light protein and glial fibrillary acidic protein as biological markers in MS. *Neurology*, **61**, 1720–1725.
[699] Manabe, T., Takahashi, Y., Okuyama, T. *et al.* (1983). Two-dimensional electrophoresis of cerebrospinal fluid proteins in the absence of denaturing agent and immunochemical identification after parallel nitrocellulose blotting. In: H. Hirai (ed.), *Electrophoresis' 83*, pp. 180–187. Walter de Gruyter.
[700] Mancini, G., Carbonara, A. O. and Heremans, J. F. (1965). Immunochemical quantitation of antigens by single radial immunodiffusion. *Immunochemistry*, **2**, 235–254.
[701] Mandler, R., Goren, H. and Valenzuela, R. (1982). Value of central nervous system IgG daily synthesis determination in the diagnosis of multiple sclerosis. *Neurology*, **32**, 296–298.
[702] Mansour, M. M., Guindi, S. and Girgis, N. I. (1981). Levels of individual serum and cerebrospinal fluid proteins in purulent and tuberculous meningitis. *Eur. Neurol.*, **20**, 40–45.
[703] Mar, P., Gradl, T. and Dorner, C. (1979). A longitudinal study of immunological parameters in multiple sclerosis. *J. Neurol. Sci.*, **41**, 369–377.
[704] Marangos, P. J., Schmechel, D., Parma, A. M. *et al.* (1979). Measurement of neuronspecific (NSE) and non-neuronal (NNE) isoenzymes of enolase in rat, monkey and human nervous tissue. *J. Neurochem.*, **33**, 319–329.
[705] Marshall, T. (1984). Sodium dodecyl sulfate polyacrylamide gel electrophoresis of serum after protein denaturation in the presence or absence of 2-mercaptoethanol. *Clin. Chem.*, **30**, 475–479.
[706] Martin, J. R., Goudswaard, J., Palsson, P. A. *et al.* (1982). Cerebrospinal fluid immunoglobulins in sheep with visna, a slow virus infection of the central nervous system. *J. Neuroimmunol.*, **3**, 139–148.
[707] Martinez-Yelamos, A., Saiz, A., Sanchez-Valle, R. *et al.* (2001). 14-3-3 protein in the CSF as prognostic marker in early multiple sclerosis. *Neurology*, **57**, 722–724.

[708] Martino, G., Grimaldi, L. M., Moiola, L. et al. (1990). Discontinuous distribution of IgG oligoclonal bands in cerebrospinal fluid from multiple sclerosis patients. *J. Neuroimmunol.*, **30**, 129–134.

[709] Marttila, R. J., Rinne, U. K. and Tiilikainen, A. (1982). Virus antibodies in Parkinson's disease. *J. Neurol. Sci.*, **54**, 227–238.

[710] Mase, M., Yamada, K., Shimazu, N. et al. (2003). Lipocalin-type prostaglandin D synthase (beta-trace) in cerebrospinal fluid: a useful marker for the diagnosis of normal pressure hydrocephalus. *Neurosci. Res.*, **47**, 455–459.

[711] Massaro, A. R. (1978). Modifications of the cerebrospinal fluid IgG concentrations in patients with mulitple sclerosis treated with intrathecal steroids. *J. Neurol.*, **219**, 221–226.

[712] Masters, C. L., Harris, J. O., Gajdusek, D. C. et al. (1979). Creutzfeldt-Jakob disease: patterns of worldwide occurrence and the significance of familial and sporadic clustering. *Ann. Neurol.*, **5**, 177–188.

[713] Matikainen, M. T. (1981). Solid-phase immunoassay methods for quantitation of IgG and viral antibodies in cerebrospinal fluid and its electrophoretic fractions. *J. Neurosci. Methods*, **4**, 277–286.

[714] Matthews, R. C., Burnie, J. P. and Tabaqchali, S. (1984). Immunoblot analysis of the serological response in systemic candidosis. *Lancet*, **ii**, 1415–1418.

[715] Mattson, D. H., Roos, R. P. and Arnason, B. G.W. (1980). Immunoperoxidase staining of cerebrospinal fluid IgG in isoelectric focusing gels: a sensitive new technique. *J. Neurosci. Methods*, **3**, 67–75.

[716] Mattson, D. H., Roos, R. P. and Arnason, B. G. W. (1980). Isoelectric focusing of IgG eluted from multiple sclerosis and subacute sclerosing panencephalitis brains. *Nature*, **287**, 335–337.

[717] Mattson, D. H., Roos, R. P. and Arnason, B. G. W. (1981). Comparison of agar gel electrophoresis and isoelectric focusing in multiple sclerosis and subacute sclerosing panencephalitis. *Ann. Neurol.*, **9**, 34–41.

[718] Mattson, D. H., Roos, R. P. and Arnason, B. G.W. (1982). Oligoclonal IgG in multiple sclerosis and subacute sclerosing panencephalitis brains. *J. Neuroimmunol.*, **2**, 261–276.

[719] Mattson, D. H., Roos, R. P., Hooper, J. E. and Arnason, B. G. W. (1981). Light chain composition of oligoclonal IgG bands in multiple sclerosis (MS) and subacute sclerosing panencephalitis (SSPE). *Neurology*, **31**, 145.

[720] Mattson, D. H., Roos, R. P., Hopper, J. E. and Arnason, B. G. W. (1982). Light chain composition of CSF oligoclonal IgG bands in multiple sclerosis and subacute sclerosing panencephalitis. *J. Neuroimmunol.*, **3**, 63–76.

[721] Maurer, H. R. and Allen, R. C. (1972). Polyacrylamide gel electrophoresis in clinic chemistry: problems of standardization and performance. *Clin. Chim. Acta*, **40**, 359–370.

[722] Mavra, M., Chadha, G., Umamahsewaran, I. and Thompson, E. J. (1998). Detection of herpes simplex virus and herpes simplex virus antibodies in encephalitis and similar infectious diseases. *Kuwait Med. J.*, **30**, 190–195.

[723] Mavra, M., Drulovic, J., Levic, Z. et al. (1999). CNS tumours: oligoclonal immunoglobulin D in cerebrospinal fluid and serum. *Acta Neurol. Scand.*, **100**, 117–118.

[724] Mavra, M., Luxton, R., Keir, G. et al. (1991). A new qualitative method for detecting IgD in unconcentrated cerebrospinal fluid. *J. Immunol. Methods*, **144**, 63–67.
[725] Mavra, M., Luxton, R., Keir, G. and Thompson, E. J. (1992). Oligoclonal immunoglobulin D in the cerebrospinal fluid of neurologic patients. *Neurology*, **42**, 1244–1245.
[726] Mavra, M., Luxton, R. and Thompson, E. J. (1992). IgG paraproteins in neurological diseases: lack of association with neurotropic viral/bacterial antigens. *Acta Neurol. Scand.*, **86**, 596–598.
[727] Mavra, M., Newcombe, J., Keir, G. and Thompson, E. J. (1990). Spleen versus CNS immunoglobulin G in multiple sclerosis: an isoelectric focusing study. *Acta Neurol. Scand.*, **81**, 125–127.
[728] Mavra, M., Thompson, E. J., Nikolic, J. et al. (1990). The occurrence of oligoclonal IgG in tears from patients with MS and systemic immune disorders. *Neurology*, **40**, 1259–1262.
[729] May, C., Kaye, J. A., Atack, J. R. et al. (1990). Cerebrospinal fluid production is reduced in healthy aging. *Neurology*, **40**, 500–503.
[730] McDonald, W. I., Compston, A., Edan, G. et al. (2001). Recommended diagnostic criteria for multiple sclerosis: guidelines from the International Panel on the diagnosis of multiple sclerosis. *Ann. Neurol.*, **50**, 121–127.
[731] McDonald, W. I. and Halliday, A. M. (1977). Diagnosis and classification of multiple sclerosis. *Br. Med. Bull.*, **33**, 4–8.
[732] McGarry, R. C., Helfand, S. L., Quarles, R. H. and Roder, J. C. (1983). Recognition of myelin-associated glycoprotein by the monoclonal antibody HNK-1. *Nature*, **306**, 376–378.
[733] McLean, B. N., Luxton, R. W. and Thompson, E. J. (1990). A study of immunoglobulin G in the cerebrospinal fluid of 1007 patients with suspected neurological disease using isoelectric focusing and the Log IgG-Index. A comparison and diagnostic applications. *Brain*, **113**, 1269–1289.
[734] McLean, B. N., Miller, D. and Thompson, E. J. (1995). Oligoclonal banding of IgG in CSF, blood-brain barrier function, and MRI findings in patients with sarcoidosis, systemic lupus erythematosus, and Behcet's disease involving the nervous system. *J. Neurol. Neurosurg. Psychiatry*, **58**, 548–554.
[735] McLean, B. N., Mitchell, D. N. and Thompson, E. J. (1990). Local synthesis of specific IgG in the cerebrospinal fluid of patients with neurosarcoidosis detected by antigen immunoblotting using Kveim material. *J. Neurol. Sci.*, **99**, 165–175.
[736] McLean, B. N., Rudge, P. and Thompson, E. J. (1989). Cyclosporin A curtails the progression of free light chain synthesis in the CSF of patients with multiple sclerosis. *J. Neurol. Neurosurg. Psychiatry*, **52**, 529–531.
[737] McLean, B. N., Rudge, P. and Thompson, E. J. (1989). Viral specific IgG and IgM antibodies in the CSF of patients with tropical spastic paraparesis. *J. Neurol.* **236**, 351–352.
[738] McLean, B. N. and Thompson, E. J. (1989). Antibodies against the paramyxovirus SV5 are not specific for cerebrospinal fluid from multiple sclerosis patients. *J. Neurol. Sci.*, **92**, 261–266.
[739] McLean, B. N., Zeman, A. Z., Barnes, D. and Thompson, E. J. (1993). Patterns of blood-brain barrier impairment and clinical features in multiple sclerosis. *J. Neurol. Neurosurg. Psychiatry*, **56**, 356–360.

[740] McLester, J. and Leung, F. Y. (1983). Assessment of a computerized diagnostic program for serum protein electrophoresis. *Clin. Chem.*, **29**, 2000–2001.
[741] McMillan, S. A. and Haire, M. (1979). The specificity of IgG-and IgM-Class smooth muscle antibody in the sera of patients with multiple sclerosis and active chronic hepatitis. *Clin. Immunol. Immunopathol.*, **14**, 256–263.
[742] McPherson, T. A., Gilpin, A. and Seland, T. P. (1972). Radioimmunoassay of CSF for encephalitogenic basic protein: A diagnostic test for MS ? *Can. Med. Assoc. J.*, **107**, 856–859.
[743] McPherson, T. A., Liburd, E. M. and Seland, T. P. (1975). Binding of 125 I-labelled encephalitogenic basic protein to normal lymphocytes. *Clin. exp. Immunol.*, **19**, 451–458.
[744] Meco, C., Oberascher, G., Arrer, E. et al. (2003). Beta-trace protein test: new guidelines for the reliable diagnosis of cerebrospinal fluid fistula. *Otolaryngol. Head Neck Surg.*, **129**, 508–517.
[745] Mehta, P. D., Frisch, S., Thormar, H. et al. (1981). Bound antibody in multiple sclerosis brains. *J. Neurol. Sci.*, **49**, 91–98.
[746] Mehta, P. D., Kane, A. and Thormar, H. (1978). Futher characterization of bound measles-specific IgG eluted from SSPE brains. *Ann. Neurol.*, **3**, 552–555.
[747] Mehta, P. D., Mehta, S. P. and Patrick, B. A. (1984). Silver staining of unconcentrated cerebrospinal fluid in agarose gel (Panagel) electrophoresis. *Clin. Chem.*, **30**, 735–736.
[748] Mehta, P. D., Miller, J. A. and Tourtellotte, W. W. (1982). Oligoclonal IgG bands in plaques from multiple sclerosis brains. *Neurology*, **32**, 372–376.
[749] Mehta, P. D., Patrick, B. A., isoelectric focusing and silver staining. (1983). Detection of oligoclonal bands in unconcentrated CS. *Neurology*, **33**, 1365–1367.
[750] Mehta, P. D., Patrick, B. A. and Miller, J. A. (1984). Absence of oligoclonal IgA in CSF and serum of multiple sclerosis patients. *J. Neuroimmunol.*, **6**, 67–69.
[751] Mehta, P. D., Patrick, B. A. and Wisniewski, H. M. (1981). Isoelectric focusing and immunofixation of cerebrospinal fluid and serum in multiple sclerosis. *J. Clin. Immunol.*, **6**, 17–22.
[752] Mehta, P. D., Thormar, H. and Wisniewski, H. M. (1980). Quantitation of measles-specific IgG. Its presence in CSF and brain extracts of patients with multiple sclerosis. *Arch. Neurol.*, **37**, 607–609.
[753] Menonna, J., Galantowicz, D., Dowling, P. and Cook, S. (1977). Rapid fluorometric assay for cerebrospinal fluid immunoglobulin G. *Neurology*, **27**, 481–483.
[754] Merril, C. R., Goldman, D., Sedman, S. A. and Ebert, M. H. (1981). Ultrasensitive stain for proteins in polyacrylamide gels shows regional variation in cerebrospinal fluid proteins. *Science*, **211**, 1437–1438.
[755] Merril, C. R., Switzer, R. C. and Van Keuren, M. L. (1979). Trace polypeptides in cellular extracts and human body fluids detected by two-dimensional electrophoresis and a highly sensitive silver stain. *Proc. Natl. Acad. Sci.*, **76**, 4335–4339.
[756] Mertin, J., Knight, S. C., Rudge, P. et al. (1980). Double-blind, controlled trial of immunosuppression in treatment of multiple sclerosis. *Lancet*, **2**, 949–951.
[757] Mertin, J., Kremer, M., Knight, S. C. et al. (1982). Double-blind controlled trial of immunosuppression in the treatment of multiple sclerosis: final report. *Lancet*, **ii**, 351–354.

REFERENCES

[758] Mertin, J. and Thompson, E. J. (1981). Immunity and the nervous system. In: A. N. Davison and E. J. Thompson (eds.), *The Molecular Basis of Neuropathology*, pp. 14–43. Edward Arnold Ltd.

[759] Michetti, F., Massaro, A. and Murazio, M. (1979). The nervous system-specific S-100 antigen in cerebrospinal fluid of multiple sclerosis patients. *Neurosci. Lett.*, **11**, 171–175.

[760] Michetti, F., Massaro, A., Russo, G. and Rigon, G. (1980). The S-100 antigen in cerebrospinal fluid as a possible index of cell injury in the nervous system. *J. Neurol. Sci.*, **44**, 259–263.

[761] Miles, L. E. M., Simmons, J. E. and Chrambach, A. (1972). Instability of pH gradients in isoelectric focusing on polyacrylamide gel. *Anal. Biochem.*, **49**, 109–117.

[762] Miller, B., Goldberg, M. A., Heiner, D. *et al.* (1984). A new immunologic test for CNS cysticercosis. *Neurology*, **34**, 695–697.

[763] Miller, J. R., Burke, A. M. and Bever, C. T. (1983). Occurrence of oligoclonal bands in multiple sclerosis and other CNS diseases. *Ann. Neurol.*, **13**, 53–58.

[764] Miller, R. F., Green, A. J., Giovannoni, G. and Thompson, E. J. (2000). Detection of 14-3-3 brain protein in cerebrospinal fluid of HIV infected patients. *Sex. Transm. Infect.*, **76**, 408.

[765] Milstein, C. (1981). Monoclonal antibodies from hybrid myelomas. *Proc. R. Soc. Lond. B*, **211**, 393–412.

[766] Mingioli, E. S., Strober, W., Tourtellotte, W. W. *et al.* (1978). Quantitation of IgG, IgA and IgM in the CSF by radioimmunoassay. *Neurology*, **28**, 991–995.

[767] Missler, U., Wiesmann, M., Friedrich, C. and Kaps, M. (1997). S-100 protein and neuron-specific enolase concentrations in blood as indicators of infarction volume and prognosis in acute ischemic stroke. *Stroke*, **28**, 1956–1960.

[768] Mohr, J. A., Griffiths, W., Jackson, R. *et al.* (1976). Neurosyphilis and penicillin levels in cerebrospinal fluid. *JAMA*, **236**, 2208–2209.

[769] Mokuno, K., Kato, K., Kawai, K. *et al.* (1983). Neuron-specific enolase and S-100 protein levels in cerebrospinal fluid of patients with various neurological diseases. *J. Neurol. Sci.*, **60**, 443–451.

[770] Monseu, G. and Cumings, J. N. (1965). Polyacrylamide disc electrophoresis of the proteins of cerebrospinal fluid and brain. *J. Neurol. Neurosurg. Psychiatry*, **28**, 56–60.

[771] Monteyne, P., Albert, F., Weissbrich, B. *et al.* (1997). The detection of intrathecal synthesis of anti-herpes simplex IgG antibodies: comparison between an antigen-mediated immunoblotting technique and antibody index calculations. *J. Med. Virol.*, **53**, 324–331.

[772] Moore, B. W. and McGregor, D. (1965). Chromatographic and electrophoretic fractionation of soluble proteins of brain and liver. *J. Biol. Chem.*, **240**, 1647–1653.

[773] Moore, B. W. and Perez, V. J. (1968). Specific acidic proteins of the nervous system. In: F. O. Carlson (ed.), *Physiological and Biochemical Assays of Nervous Integration*, pp. 343–360. Prentice-Hall.

[774] Morgan, B. P., Campbell, A. K. and Compston, D. A. S. (1984). Terminal component of complement (C9) in cerebrospinal fluid of patients with multiple sclerosis. *Lancet*, **ii**, 251–255.

[775] Morishima, T., Miyazu, M., Ozaki, T. *et al.* (1980). Local immunity in mumps meningitis. *Am. J. Dis. Child*, **134**, 1060–1063.

[776] Morrissey, J. H. (1981). Silver stain for proteins in polyacrylamide gels: a modified procedure with enhanced uniform sensitivity. *Anal. Biochem.*, **117**, 307–310.
[777] Moticka, E. J. (1974). The non-specific stimulation of immunoglobulin secretion following specific stimulation of the immune system. *Immunology*, **27**, 401–412.
[778] Motomura, S., Tabira, T. and Kuroiwa, Y. (1980). Clinical comparative study of multiple sclerosis and neuro-Behcet's syndrome. *J. Neurol. Neurosurg. Psychiatry*, **43**, 210–213.
[779] Moulin, D., Paty, D. W. and Ebers, G. C. (1983). The predictive value of cerebrospinal fluid electrophoresis in 'possible' multiple sclerosis. *Brain*, **106**, 809–816.
[780] Moyle, S., Keir, G. and Thompson, E. J. (1984). Viral immunoblotting: A sensitive method for detecting viral-specific oliogoclonal bands in unconcentrated cerebrospinal fluid. *Biosci. Rep.*, **4**, 505–510.
[781] Moyle, S. P. and Thompson, E. J. (1985). Intrathecal herpes-specific oligoclonal immunoglobulin IgG in herpes simplex encephalitis. *Biochem. Soc. Trans.*, **13**, 964–965.
[782] Moyle, S. P. and Thompson, E. J. (1985). Viral immunoblotting of measles-specific oligoclonal immunoglobulin IgG kappa and lambda light chains in subacute sclerosing panencephalitis. *Biochem. Soc. Trans.*, **13**, 902–903.
[783] Muir, A. and Hensley, W. J. (1979). Rapid measurement of total protein, albumin and IgG in cerebrospinal fluid. *Clin. Chim. Acta*, **98**, 277–279.
[784] Muir, P., Nicholson, F., Sharief, M. K. et al. (1995). Evidence for persistent enterovirus infection of the central nervous system in patients with previous paralytic poliomyelitis. *Ann. N Y Acad. Sci.*, **753**, 219–232.
[785] Mulder, J., Sloots, L. C. E. and Verhaar, M. A. T. (1972). Molecular heterogeneity effects of IgM immuno-globulins in radial immunodiffusion. *J. Immunol. Methods*, **2**, 89–98.
[786] Mulder, J., Sloots, L. C. E. and Verhaar, M. A. T. (1972). Molecular heterogeneity effects of immunoglobulin classes in radial immunodiffusion. *J. Immunol. Methods*, **1**, 211–213.
[787] Mulder, J. and Verhaar, M. A. T. (1973). Multiple precipitation of sera containing monoclonal IgG proteins in radial immunodiffusion. *Clin. Chim. Acta*, **45**, 325–333.
[788] Muller, F., Kruska, M. and Hippius, H. (1970). Uber beziehungen zwischen immobilisierenden antikorpern in serum and liquor cerebrospinalis. *Z. Neurol.*, **198**, 237–255.
[789] Muller, F. and Lindenschmidt, E.-G. (1982). Demonstration of specific 19S(IgM) antibodies in untreated and treated syphilis. *Br. J. Vener. Dis.*, **58**, 12–17.
[790] Muller, F. and Moskophidis, M. (1983). Estimation of the local production of antibodies to treponema pallidum in the central nervous system of patients with neurosyphilis. *Br. J. Vener. Dis.*, **59**, 80–84.
[791] Muller, F., Moskophidis, M. and Prange, H. W. (1984). Demonstration of locally synthesized immunoglobulin M antibodies to treponema pallidum in the central nervous system of patients with untreated neurosyphilis. *J. Neuroimmunol.*, **7**, 43–54.
[792] Muller, F. and Oelerich, S. (1979). Identification of low molecular weight IgM antibody with treponema pallidum specificity in sera of patients with chronic syphilis. *Klin. Wochenschr.*, **57**, 667–671.

[793]   Muller, F. and Ritter, G. (1978). Bedeutung treponemenspezifischer antikorper im liquor cerebrospinalis fur die diagnose und therapie der neurosyphilis. *Nervenarzt*, **49**, 185–188.
[794]   Nagelkerken, L. M., Aalberse, R. C., Van Walbeek, H. K. and Out, T. A. (1980). Preparation of antisera directed against the idiotype(s) of immunoglobulin G from the cerebrospinal fluid of patients with multiple sclerosis. *J. Immunol.*, **125**, 384.
[795]   Nahmias, A. J., Whitley, R. J., Visintine, A. N. *et al.* (1982). Herpes simplex virus encephalitis: laboratory evaluations and their diagnostic significance. *J. Infect. Dis.*, **145**, 829–836.
[796]   Nakano, T. and Nagata, A. (2003). ELISAs for free light chains of human immunoglobulins using monoclonal antibodies: comparison of their specificity with available polyclonal antibodies. *J. Immunol. Methods*, **275**, 9–17.
[797]   Nakao, K., Oki, S., Tanaka, I. *et al.* (1980). Immunoreactive B-Endorphin and adrenocorticotropin in human cerebrospinal fluid. *J. Clin. Invest.*, **66**, 1383–1390.
[798]   Nakashima, I. and Kato, N. (1974). Non-specific stimulation of immunoglobulin synthesis in mice by capsular polysaccharide of Klebsiella pneumoniae. *Immunology*, **27**, 179–193.
[799]   Nargessi, R. D., Shine, B. and Landon, J. (1984). Immunoassays for serum C-reactive protein employing fluorophore-labelled reactants. *J. Immunol. Methods*, **71**, 17–24.
[800]   Narita, M., Yamada, S., Matsuzono, Y. *et al.* (1996). Immunoglobulin G avidity testing in serum and cerebrospinal fluid for analysis of measles virus infection. *Clin. Diagn. Lab. Immunol.*, **3**, 211–215.
[801]   Nerenberg, S. T. and Prasad, R. (1975). Radioimmunoassays for Ig classes G, A, M, D, and E in spinal fluids: normal values of different age groups. *J. Lab. Clin. Med.*, **86**, 887–898.
[802]   Nerenberg, S. T., Prasad, R. and Rothman, M. E. (1978). Cerebrospinal fluid IgG, IgA, IgM, IgD, IgE levels in central nervous system disorders. *Neurology*, **28**, 988–990.
[803]   Neumann, B., Ritter, K. and Felgenhauer, K. (1990). Fine specificities of antibodies in sera and cerebrospinal fluid in herpes virus infections of the central nervous system as detected by the antigen variable immunoblot technique. *J. Neuroimmunol.*, **28**, 111–118.
[804]   Newcombe, J., Glynn, P. and Cuzner, M. L. (1982). Analysis by transfer electrophoresis of reactivity of IgG with brain proteins in multiple sclerosis. *J. Neurochem.*, **39**, 1192–1194.
[805]   Nikoskelainen, E., Frey, H. and Salmi, A. (1981). Prognosis of optic neuritis with special reference to cerebrospinal fluid immunoglobulins and measles virus antibodies. *Ann. Neurol.*, **9**, 545–550.
[806]   Nilsson, K. and Olsson, J.-E. (1978). Analysis for cerebrospinal fluid proteins by isoelectric focusing on polyacrylamide gel: methodological aspects and normal values, with special reference to the alkaline region. *Clin. Chem.*, **24**, 1134–1139.
[807]   Noori, M. A., Ciardi, M., Sharief, M. K. *et al.* (1992). Free circulating ICAM-1 in serum and cerebrospinal fluid of HIV-1 infected patients correlate with TNF-alpha and blood-brain barrier damage. *Mediat. Inflamm.*, **1**, 235–240.
[808]   Nordal, H. J., Vandvik, B. and Norrby, E. (1978). Demonstration of electrophoretically restricted virus-specific antibodies in serum and cerebrospinal fluid by imprint electroimmunofixation. *Scand. J. Immunol.*, **7**, 381–388.

[809] Nordal, H. J., Vandvik, B. and Norrby, E. (1978). Multiple sclerosis: local synthesis of electrophoretically restricted measles, rubella mumps and herpes simplex virus antibodies in the central nervous system. *Scand. J. Immunol.*, **7**, 473–479.

[810] Norgren, N., Karlsson, J. E., Rosengren, L. and Stigbrand, T. (2002). Monoclonal antibodies selective for low molecular weight neurofilaments. *Hybrid Hybridomics*, **21**, 53–59.

[811] Norgren, N., Rosengren, L. and Stigbrand, T. (2003). Elevated neurofilament levels in neurological diseases. *Brain Res.*, **987**, 25–31.

[812] Norgren, N., Sundstrom, P., Svenningsson, A. et al. (2004). Neurofilament and glial fibrillary acidic protein in multiple sclerosis. *Neurology*, **63**, 1586–1590.

[813] Noronha, A. B. C., Mattson, D. H., Sundaresan, N. et al. (1981). Cerebrospinal fluid (CSF) oligoclonal IgG in central nervous system (CNS) neoplasia. *Neurology*, **31**, 43.

[814] Norrby, E., Link, H., Olsson, J.-E. et al. (1974). Comparsion of antibodies against different viruses in cerebrospinal fluid and serum sample from patients with multiple sclerosis. *Infect. Immun.*, **10**, 688–694.

[815] Norrby, E. and Vandvik, B. (1975). Relationship between measles virus-specific antibody activities and oligoclonal IgG in the central nervous system of patients with subacute sclerosing panencephalitis and multiple sclerosis. *Med. Microbiol. Immunol.*, **162**, 63–72.

[816] Nunez, E. J. and Rivero, I. (1982). C-Reactive Protein in cerebrospinal fluid. *Eur. Neurol.*, **21**, 380–383.

[817] Nye, L. and Roitt, I. M. (1980). Isoelectric focusing of human antibodies directed against a high molecular weight antigen. *J. Immunol. Methods*, **35**, 97–103.

[818] Nylen, K., Karlsson, J. E., Blomstrand, C. et al. (2002). Cerebrospinal fluid neurofilament and glial fibrillary acidic protein in patients with cerebral vasculitis. *J. Neurosci. Res.*, **67**, 844–851.

[819] O'Connor, A. F., Luxon, L. M., Shortman, R. C. et al. (1982). Electrophoretic separation and identification of perilymph proteins in cases of acoustic neuroma. *Acta Otolaryngol.*, **93**, 195–200.

[820] O'Connor, K. C., Chitnis, T., Griffin, D. E. et al. (2003). Myelin basic protein-reactive autoantibodies in the serum and cerebrospinal fluid of multiple sclerosis patients are characterized by low affinity interactions. *J. Neuroimmunol.*, **136**, 140–148.

[821] O'Duffy, J. D. and Goldstein, N. P. (1976). Neurologic involvement in seven patients with Behcet's disease. *Am. J. Med.*, **61**, 170–178.

[822] O'Farrell, P. H. (1975). High resolution two-dimensional electrophoresis of protein. *J. Biol. Chem.*, **250**, 4007–4021.

[823] O'Riordan, J. I., Gallagher, H. L., Thompson, A. J. et al. (1996). Clinical, CSF, and MRI findings in Devic's neuromyelitis optica. *J. Neurol. Neurosurg. Psychiatry*, **60**, 382–387.

[824] Oakley, B. R., Kirsch, D. R. and Morris, N. R. (1980). A simplified ultrasensitive silver stain for detecting proteins in polyacrylamide gel. *Anal. Biochem.*, **105**, 361–363.

[825] Ochi, Y., Fujiyama, Y., Hosoda, S. et al. (1982). Immunological similarity of CEA with alpha 1-acid glycoprotein (orosomucoid). *Clin. Chim. Acta*, **122**, 145–160.

[826] Oehmichen, M., Domasch, D. and Wietholter, H. (1982). Origin, proliferation, and fate of cerebrospinal fluid cells. *J. Neurol.*, **227**, 145–150.

[827] Offner, H., Clausen, J. and Fog, T. (1974). Precipitation of myelin basic protein by beta-lipoprotein of human serum. *Acta Neurol. Scand.*, **50**, 221–226.

[828] Ohman, S. (1991). Is isoelectric focusing a "gold standard" for evaluation of formulas for calculation of intrathecal immunoglobulin synthesis? *J. Neurol. Sci.*, **106**, 230–232.

[829] Ohman, S., Ernerudh, J., Forsberg, P. *et al.* (1992). Comparison of seven formulae and isoelectrofocusing for determination of intrathecally produced IgG in neurological diseases. *Ann. Clin. Biochem.*, **29**, 417–418.

[830] Olsson, J.-E. (1980). Brain and CSF proteins in Creutzfeldt-Jakob disease. *Eur. Neurol.*, **19**, 85–90.

[831] Olsson, J.-E., Blomstrand, C. and Haglid, K. G. (1974). Cellular distribuation of beta-trace protein in CNS and brain tumours. *J. Neurol. Neurosurg. Psychiatry*, **37**, 302–311.

[832] Olsson, J.-E. and Link, H. (1973). Immunoglobulin abnormalities in multiple sclerosis. Relation to clinical parameters: exacerbations and remissions. *Arch. Neurol.*, **28**, 392–399.

[833] Olsson, J.-E., Link, H. and Muller, R. (1976). Immunoglobulin abnormalities in multiple sclerosis. Relation to clinical parameters: disability, duration and age of onset. *J. Neurol. Sci.*, **27**, 233–245.

[834] Olsson, J.-E. and Nilsson, K. (1978). Isoelectric focusing of alkaline CSF proteins in normal and pathological conditions. *Acta Neurol. Scand.*, **57**, 231–232.

[835] Olsson, J.-E. and Nilsson, K. (1979). Gamma globulins of CSF and serum in multiple sclerosis: isoelectric focusing on polyacrylamide gel and agar gel electrophoresis. *Neurology*, **29**, 1383–1391.

[836] Olsson, J.-E. and Pettersson, B. (1976). A comparison between agar gel electrophoresis and CSF serum quotients of IgG and albumin in neurological diseases. *Acta Neurol. Scand.*, **53**, 308–322.

[837] Olsson, T., Kostulas, V. and Link, H. (1984). Improved detection of oligoclonal IgG in cerebrospinal fluid by isoelectric focusing in agarose, double-antibody peroxidase labeling, and avidin-biotin amplification. *Clin. Chem.*, **30**, 1246–1249.

[838] Olsson, Y. (1971). Studies on vascular permeability in peripheral nerves. *Acta Neuropathol.*, **17**, 114–126.

[839] Oxelius, V.-A., Rorsman, H. and Laurell, A.-B. (1969). Immunoglobulins of cerebrospinal fluid in syphilis. *Br. J. Vener. Dis.*, **45**, 121–124.

[840] Page, N., Perruisseau, G. and Steck, A. J. (1983). Binding properties of cerebrospinal fluid IgG in multiple sclerosis and other neurological diseases. *J. Neurol. Sci.*, **60**, 23–30.

[841] Palfreyman, J. W., Johnston, R. V., Ratchiffe, J. G. *et al.* (1979). Radioimmunoassay of serum myelin basic protein and its application to patients with cerebrovascular accident. *Clin. Chim. Acta*, **92**, 403–409.

[842] Palfreyman, J. W. and Ratcliffe, J. G. (1977). Cerebrospinal-fluid changes in demyelinating disorders. *N. Engl. J. Med.*, **296**, 883–884.

[843] Palfreyman, J. W., Thomas, D. G. T. and Ratcliffe, J. G. (1977). Radioimmunoassay for human myelin basic protein in biological fluids and tissue extract. *Biochem. Soc. Trans.*, **5**, 1422–1425.

[844] Palfreyman, J. W., Thomas, D. G. T. and Ratcliffe, J. G. (1978). Radioimmunoassay of human myelin basic protein in tissue extract, cerebrospinal fluid and serum and its clinical application to patients with head injury. *Clin. Chim. Acta*, **82**, 259–270.

REFERENCES

[845] Paluch, U. H., Keir, G., Moyle, S. and Thompson, E. J. (1984). Quantification of bands produced by isoelectric focusing using immunoperoxidase. *J. Clin. Pathol.*, **37**, 1172–1176.

[846] Panitch, H. S., Hooper, C. J. and Johnson, K. P. (1980). CSF antibody to myelin basic protein. *Arch. Neurol.*, **37**, 206–209.

[847] Paolino, E., Fainardi, E., Ruppi, P. *et al.* (1996). A prospective study on the predictive value of CSF oligoclonal bands and MRI in acute isolated neurological syndromes for subsequent progression to multiple sclerosis. *J. Neurol. Neurosurg. Psychiatry*, **60**, 572–575.

[848] Papadopoulos, N. M., LeWitt, P. A., Newman, R. P. *et al.* (1983). A unique protein in normal human cerebrospinal fluid. *Clin. Chem.*, **29**, 1842–1844.

[849] Parma, A. M., Marangos, P. J. and Goodwin, F. K. (1981). A more sensitive radioimmunoassay for neuron-specific enolase suitable for cerebrospinal fluid determinations. *J. Neurochem.*, **36**, 1093–1096.

[850] Patel, Y. C., Rao, K. and Reichlin, S. (1977). Somatostatin in human cerebrospinal fluid. *N. Engl. J. Med.* **296**, 529–533.

[851] Paterson, P. Y., Day, E. D., Whitacre, C. C. *et al.* (1981). Endogenous myelin basic protein-serum factors (MBP-SFs) and anti-MBP antibodies in humans. *J. Neurol. Sci.*, **52**, 37–51.

[852] Paterson, P. Y. and Whitacre, C. C. (1981). The enigma of oligoclonal immunoglobulin G in cerebrospinal fluid from multiple sclerosis patients. *Immunol. Today*, **2**, 111–117.

[853] Paty, D. W., Blume, W. T., Brown, W. F. *et al.* (1979). Chronic progressive myelopathy: investigation with CSF electrophoresis, evoked potentials, and CT scan. *Ann. Neurol.*, **6**, 419–424.

[854] Paty, D. W., Donnelly, M. and Bernardo, M. E. (1978). CSF electrophoresis: An adaptation using cellulose acetate for the identification of oligoclonal banding. *J. Can. Sci. Neurol.*, **5**, 297–299.

[855] Patzold, U., Haller, P., Baruth, B. *et al.* (1980). Immune complexes in multiple sclerosis. *J. Neurol.*, **222**, 249–260.

[856] Pearl, G. S., Check, I. J. and Hunter, R. L. (1984). Agarose electrophoresis and immunonephelometric quantitation of cerebrospinal fluid immunoglobulins: Criteria for application in the diagnosis of neurologic disease. *Am. J. Pathol.*, **81**, 575–580.

[857] Pedersen, N. S., Kam-Hansen, S., Link, H. and Mavra, M. (1982). Specificity of immunoglobulins synthesized within the central nervous system in neurosyphilis. *Acta Pathol. Microbiol, Immunol. Scand.*, **90**, 97–104.

[858] Peltola, H. O. (1982). C-reactive protein for rapid monitoring of infections of the central nervous system. *Lancet*, **i**, 980–983.

[859] Pepe, A. J. and Hochwald, G. M. (1967). Trace proteins in biological fluids III. Quantitation of gamma-trace and major spinal fluid proteins including beta-trace. *Proc. Soc. Exp. Biol. Med.*, **126**, 630–633.

[860] Percy, M. E., Wong, S., Bauer, S. *et al.* (1998). Iron metabolism and human ferritin heavy chain cDNA from adult brain with an elongated untranslated region: new findings and insights. *The Analyst*, **123**, 41–50.

[861] Perini, J. M., Lebas, J., Roussel, P. and Biserte, G. (1979). Evidence for heterogeneous or incomplete immunoglobulins in oligoclonal CSF Studied by electroimmunofixation. *Clin. Chim. Acta*, **96**, 205–214.

[862] Perkin, G. D., Sethi, K. and Muller, B. R. (1983). IgG ratios and oligoclonal IgG in multiple sclerosis and other neurological disorders. *J. Neurol. Sci.*, **60**, 325–336.
[863] Peterslund, N. A., Ipsen, J., Schonheyder, H. *et al.* (1981). Acyclovir in herpes zoster. *Lancet*, **ii**, 827.
[864] Peterslund, N. A. and Pedersen, B. (1982). Liquor : serum quotients of IgG and albumin in patients with meningism, meningitis and multiple sclerosis. *Acta Neurol. Scand.*, **66**, 25–33.
[865] Petzold, A., Baker, D., Pryce, G. *et al.* (2003). Quantification of neurodegeneration by measurement of brain-specific proteins. *J. Neuroimmunol.*, **138**, 45–48.
[866] Petzold, A., Brassat, D., Mas, P. *et al.* (2004). Treatment response in relation to inflammatory and axonal surrogate marker in multiple sclerosis. *Mult. Scler.*, **10**, 281–283.
[867] Petzold, A., Eikelenboom, M. J., Gveric, D. *et al.* (2002). Markers for different glial cell responses in multiple sclerosis: clinical and pathological correlations. *Brain*, **125**, 1462–1473.
[868] Petzold, A., Green, A. J., Keir, G. *et al.* (2002). Role of serum S100B as an early predictor of high intracranial pressure and mortality in brain injury: a pilot study. *Crit. Care Med.*, **30**, 2705–2710.
[869] Petzold, A., Jenkins, R., Watt, H. C. *et al.* (2003). Cerebrospinal fluid S100B correlates with brain atrophy in Alzheimer's disease. *Neurosci. Lett.*, **336**, 167–170.
[870] Petzold, A., Keir, G., Green, A. J. *et al.* (2003). A specific ELISA for measuring neurofilament heavy chain phosphoforms. *J. Immunol. Methods*, **278**, 179–190.
[871] Petzold, A., Keir, G., Green, A. J. *et al.* (2004). An ELISA for glial fibrillary acidic protein. *J. Immunol. Methods*, **287**, 169–177.
[872] Petzold, A., Keir, G., Lim, D. *et al.* (2003). Cerebrospinal fluid (CSF) and serum S100B: release and wash-out pattern. *Brain Res. Bull.*, **61**, 281–285.
[873] Pfeiffer, F. E., Homburger, H. A. and Yanagihara, T. (1983). Creatine kinase BB isoenzyme in CSF in neurologic diseases. *Arch. Neurol.*, **40**, 169–172.
[874] Pitt-Rivers, R. and Impiombato, F. S. A. (1968). The binding of sodium dodecyl sulphate to various proteins. *Biochem. J.*, **109**, 825–830.
[875] Pizzolato, M. A. and Goni, F. (1984). Thin layer gel filtration immunofixation: identification of abnormal molecular weight immunoglobulins or related fragments. *J. Immunol. Methods*, **72**, 91–95.
[876] Poloni, M., Rocchelli, B., Scelsi, R. and Pinelli, P. (1979). Intrathecal IgG synthesis in multiple sclerosis and other neurological diseases: a comparative evaluation by IgG-index and isoelectric focusing. *J. Neurol.*, **221**, 245–255.
[877] Porter, K. G., Sinnamon, D. G. and Gillies, R. R. (1977). Cryptococcus neoformans-specific oligoclonal immunoglobulins in cerebrospinal fluid in cryptococcal meningitis. *Lancet*, **i**, 1262.
[878] Poser, C. M., Paty, D. W., Scheinberg, L. *et al.* (1983). New diagnostic criteria for multiple sclerosis: guidelines for research protocols. *Ann. Neurol.*, **13**, 227–231.
[879] Poser, S., Raun, N. E. and Poser, W. (1982). Age at onset, initial symptomatology and the course of multiple sclerosis. *Acta Neurol. Scand.*, **66**, 355–362.
[880] Poskitt, D. C., Frost, H., Cahill, R. N. P. and Trnka, Z. (1977). The appearance of non-specific antibody-forming cells in the efferent lymph draining antigen-stimulated single lymph nodes. *Immunology*, **22**, 81–89.

[881] Poulik, M. D. and Sekine, T. (1985). Low molecular weight (LMW) urinary proteins. In: H. Peeters (ed.), *Protides of the Biological Fluids*, pp. 3–13. Pergamon.

[882] Prange, H. W., Moskophidis, M., Schipper, H. I. and Muller, F. (1983). Relationship between neurological features and intrathecal synthesis of IgG antibodies to treponema pallidum in untreated and treated human neurosyphilis. *J. Neurol.*, **230**, 241–252.

[883] Pras, M., Prelli, F., Franklin, E. C. and Frangione, B. (1983). Primary structure of an amyloid prealbumin variant in familial polyneuropathy of Jewish origin. *Proc. Natl. Acad. Sci.*, **80**, 539–542.

[884] Prasad, R. (1983). Cerebrospinal fluid immunoglobulin D in myasthenia gravis. *Clin. Chem.*, **29**, 1561–1562.

[885] Price, P. and Cuzner, M. L. (1980). Cerebrospinal fluid complement proteins in neurological diseases. *J. Neurol. Sci.*, **46**, 49–54.

[886] Prineas, J. W. (1979). Multiple sclerosis: presence of lymphatic capillaries and lymphoid tissue in the brain and spinal cord. *Science*, **203**, 1123–1125.

[887] Prineas, J. W. and Wright, R. G. (1978). Macrophages, lymphocytes and plasma cells in the perivascular compartment in chronic multiple sclerosis. *Lab. Invest.*, **38**, 409–421.

[888] Pronovost, A. D., Baumgarten, A. and Andiman, W. A. (1982). Chemiluminescent immunoenzymatic assay for rapid diagnosis of viral infections. *J. Clin. Microbiol.*, **16**, 345–349.

[889] Prosiegel, M., Neu, I. S., Pelka, R. B. and Fateh-Moghadam, A. (1983). Multivariate analysis of the serum-cerebrospinal fluid-protein-relation for the diagnosis of neurological diseases of the central nervous system. *Acta Neurol. Scand.*, **68**, 405–412.

[890] Pullicino, P., Thompson, E. J., Moseley, I. F. et al. (1979). Cystic intracranial tumours. Cyst fluid, biochemical changes and computerised tomographic findings. *J. Neurol. Sci.*, **44**, 77–85.

[891] Putnam, F. W. (1975). The Plasma Proteins: structure, function and genetic control. pp. 1–248. Academic Press.

[892] Putnam, F. W. Progress in plasma proteins. Orlando: Acad. Press, 1984.

[893] Qin, Y., Duquette, P., Zhang, Y. et al. (2003). Intrathecal B-cell clonal expansion, an early sign of humoral immunity, in the cerebrospinal fluid of patients with clinically isolated syndrome suggestive of multiple sclerosis. *Lab. Invest.*, **83**, 1081–1088.

[894] Quincke, H. (1891). Die Lumbalpunction des Hydrocephalus. *Klin. Wochenschr.*, **38**, 929–933.

[895] Rand, K. H., Houck, H., Denslow, N. D. and Heilman, K. M. (2000). Epstein-Barr virus nuclear antigen-1 (EBNA-1) associated oligoclonal bands in patients with multiple sclerosis. *J. Neurol. Sci.*, **173**, 32–39.

[896] Rapoport, S. I. and Pettigrew, K. D. (1979). A heterogenous, pore-vesicle membrane model for protein transfer from blood to cerebrospinal fluid at the choroid plexus. *Microvasc. Res.*, **18**, 105–119.

[897] Rastogi, S. C., Clausen, J. and Fog, T. (1978). Multiple sclerosis specific antigens in MS brains. *Acta Neurol. Scand.*, **57**, 438–442.

[898] Rastogi, S. C., Clausen, J., Offner, H. et al. (1979). Partial purification of MS specific brain antigens. *Acta Neurol. Scand.*, **59**, 281–296.

[899] Reese, T. S. and Karnovsky, M. J. (1967). Fine structural localization of a blood-brain barrier to exogenous peroxidase. *J. Cell Biol.*, **34**, 207–217.

[900] Rehfeld, J. F. and Kruse-Larsen, C. (1978). Gastrin and cholecystokinin in human cerebrospinal fluid. Immunochemical determination of concentrations and molecular heterogeneity. *Brain Res.*, **155**, 19–26.
[901] Reiber, H. (1979). Quantitative bestimmung der lokal im zentralnervensystem synthetisertend immunoglobulin G-Fraktion des liquors. *J. Clin. Chem. Clin. Biochem.* **17**, 587–591.
[902] Reiber, H. (1980). The discrimination between different blood-CSF barrier dysfunctions and inflammatory reactions of the CNS by a recent evaluation graph for the protein profile of cerebrospinal fluid. *J. Neurol.*, **224**, 89–99.
[903] Reiber, H. (1991). Liquorprotein-diagnostik. In: L. Thomas, A. Fateh-Moghadam, W. G. Guder *et al.* (eds.), *Proteindiagnostik*, pp. 140–167. Behringwerke.
[904] Reiber, H. (1994). Flow rate of cerebrospinal fluid (CSF)–a concept common to normal blood-CSF barrier function and to dysfunction in neurological diseases. *J. Neurol. Sci.*, **122**, 189–203.
[905] Reiber, H. (2001). Dynamics of brain-derived proteins in cerebrospinal fluid. *Clin. Chim. Acta*, **31**, 173–186.
[906] Reiber, H. and Felgenhauer, K. (1987). Protein transfer at the blood cerebrospinal fluid barrier and the quantitation of the humoral immune response within the central nervous system. *Clin. Chim. Acta*, **163**, 319–328.
[907] Reiber, H., Jacobi, C. and Felgenhauer, K. (1986). Sensitive quantitation of carcinoembryonic antigen in cerebrospinal fluid and its barrier-dependent differentiation. *Clin. Chim. Acta*, **156**, 259–269.
[908] Reiber, H. and Peter, J. B. (2001). Cerebrospinal fluid analysis: disease-related data patterns and evaluation programs. *J. Neurol. Sci.*, **184**, 101–122.
[909] Reiber, H. and Thiele, P. (1983). Species-dependent variables in blood cerebrospinal fluid barrier function for proteins. *J. Clin. Chem. Clin. Biochem.* **21**, 199–202.
[910] Reiber, H., Thompson, E. J., Grimsley, G. *et al.* (2003). Quality assurance for cerebrospinal fluid protein analysis: international consensus by an Internet-based group discussion. *Clin. Chem. Lab. Med.*, **41**, 331–337.
[911] Reiber, H., Ungefehr, S. and Jacobi, C. (1998). The intrathecal, polyspecific and oligoclonal immune response in multiple sclerosis. *Mult. Scler.*, **4**, 111–117.
[912] Reid, A. C. and Morton, J. J. (1982). Arginine vasopressin levels in cerebrospinal fluid in neurological disease. *J. Neurol. Sci.*, **54**, 295–301.
[913] Reimer, C. B. and Maddison, S. E. (1976). Standardization of human immunoglobulin quantitation: A review of current status and problems. *Clin. Chem.*, **5**, 577–582.
[914] Reinertsen, J. L., Klippel, J. H., Johnson, A. H. *et al.* (1978). B-lymphocyte alloantigens associated with systemic lupus erythematosus. *N. Engl. J. Med.*, **299**, 515.
[915] Rejdak, K., Eikelenboom, M. J., Petzold, A. *et al.* (2004). CSF nitric oxide metabolites are associated with activity and progression of multiple sclerosis. *Neurology*, **63**, 1439–1445.
[916] Rejdak, K., Petzold, A., Sharpe, M. A. *et al.* (2004). Cerebrospinal fluid nitrite/nitrate correlated with oxyhemoglobin and outcome in patients with subarachnoid hemorrhage. *J. Neurol. Sci.*, **219**, 71–76.
[917] Rejdak, K., Petzold, A., Sharpe, M. A. *et al.* (2003). Serum and urine nitrate and nitrite are not reliable indicators of intrathecal nitric oxide production in acute brain injury. *J. Neurol. Sci.*, **208**, 1–7.

[918] Reunanen, M., Salonen, R. and Salmi, A. (1982). Intrathecal immune responses in mumps meningitis patients. *Scand. J. Immunol.*, **15**, 419–426.
[919] Reynolds, J. A. and Tanford, C. (1970). Binding of dodecyl sulfate to proteins at high binding ratios. Possible implications for the state of proteins in biological membranes. *Proc. Natl. Acad. Sci.*, **66**, 1002–1003.
[920] Riberi, M., Bernard, D. and Depieds, R. (1975). Evidence for the presence of lambda chain dimers in cerebrospinal fluid of patients suffering from subacute sclerosing panencephalitis. *Clin. Exp. Immunol.*, **19**, 45–53.
[921] Rice, G. P. A., Armstrong, H. and Ebers, G. C. (1982). Variation in immunoglobulin G and albumin concentrations during lumbar CSF removal: A reapprasisal. *Neurology*, **32**, 893–894.
[922] Rice, J. D. and Bleakney, M. (1965). Electrophoresis of unconcentrated cerebrospinal fluid using cellulose acetate strips and the dye nigrosine. *Clin. Chim. Acta*, **12**, 343–348.
[923] Riches, P. G., Sheldon, J., Smith, A. M. and Hobbs, J. R. (1991). Overestimation of monoclonal immunoglobulin by immunochemical methods. *Ann. Clin. Biochem.*, **28**, 253–259.
[924] Riddoch, D. and Thompson, R. A. (1970). Immunoglobulins levels in the cerebrospinal fluid. *Br. Med. J.*, **i**, 396–399.
[925] Rieckmann, P., Altenhofen, B., Riegel, A. *et al.* (1998). Correlation of soluble adhesion molecules in blood and cerebrospinal fluid with magnetic resonance imaging activity in patients with multiple sclerosis. *Mult. Scler.*, **4**, 178–182.
[926] Rieder, H. P. and Jegge, S. (1979). Isoelektrische Fokussierung und Agarelektrophorese des Liquor cerebrospinalis bei neurologischen Patienten. *Schweiz. Med. Wochenschr.*, **109**, 1411–1419.
[927] Riemenschneider, M., Wagenpfeil, S., Vanderstichele, H. *et al.* (2003). Phospho-tau/total tau ratio in cerebrospinal fluid discriminates Creutzfeldt-Jakob disease from other dementias. *Mol. Psychiatry*, **8**, 343–347.
[928] Rifai, N., Christenson, R. H., Gelman, B. B. and Silverman, L. M. (1987). Changes in cerebrospinal fluid IgG and apolipoprotein E indices in patients with multiple sclerosis during demyelination and remyelination. *Clin. Chem.*, **33**, 1155–1157.
[929] Rijcken, C. A., Thompson, E. J. and Teelken, A. W. (1997). An improved, ultrasensitive method for the detection of IgM oligoclonal bands in cerebrospinal fluid. *J. Immunol. Methods*, **203**, 167–169.
[930] Ritchie, R. F. (1979). Serum protein profile analysis and interpretation: some basic information. In: W. R. Diko and E. S. Tucker III (eds.), *Immunoassays in the Clinical Laboratory*, pp. 227–242. Alan R Liss Inc.
[931] Ritchie, R. F. (1983). Basics in practical applications and standardization of quantitative immunochemistry. In: R. M. Aloisi and J. Hyun (eds.), *Laboratory and Research Methods in Biology and Medicine*, pp. 21–48. Alan R Liss.
[932] Ritchie, R. F., Smith, D. E., Turgeon, P. and Palomaki, G. (1982). Interpretation of serum protein values. In: S. E. Ritzman and J. C. Daniels (eds.), *Physiology of Immunoglobulins: Diagnostic Clinical Aspects*, pp. 159–190. Allan R Liss.
[933] Ritchie, R. F. and Smith, R. (1976). Immunofixation. 1. General principles and application to agarose gel electrophoresis. *Clin. Chem.*, **22**, 497–499.

[934] Rivier, E., Crousaz, G., Steck, A. and Regli, F. (1978). Utilite diagnostique du rapport immunoglobuline sur albumine comparee a celle de l'electrophorese dans la sclerose en plaques. *Schweiz. Med. Wochenschr.*, **108**, 2066–2068.
[935] Roberts-Thompson, P. J., Esiri, M. M., Young, A. C. and MacLennan, I. C.M. (1976). Cerebrospinal fluid immunoglobulin quotients, kappa/lambda ratios, and viral antibody titres in neurological disease. *J. Clin. Pathol.*, **29**, 1105–1115.
[936] Roboz-Einstein, E. (1982). *Proteins of the Brain and CSF in Health and Disease.* Thomas, Springfield. 1–308.
[937] Roboz-Einstein, E. and Macrae, D. (1965). Spinal fluid analysis in amyotrophic lateral sclerosis. In: J. Norris and L. T. Kurland (eds.), *Motor Neuron Dis: Research Amyotrophic Lat Scl & Related Disorders*, pp. 175–178. Grune & Stratton.
[938] Rocchelli, B., Poloni, M., Mazzarello, P. and Delodovici, M. (1981). Identification of the kappa and lambda light chains within the CSF immunoglobulin region in multiple sclerosis and subacute sclerosing panencephalitis by immunofixation after isoelectric focusing. *J. Neurol.*, **226**, 169–179.
[939] Roelandse, F. W., Amons, R., ter Braak, E. P. *et al.* (1998). The high alkaline fraction on isoelectric focusing of cerebrospinal fluid is cystatin C. *J. Neurol. Sci.*, **157**, 105–108.
[940] Roheim, P. S., Carey, M., Forte, T. and Vega, G. L. (1979). Apolipoproteins in human cerebrospinal fluid. *Proc. Natl. Acad. Sci.*, **76**, 4646–4649.
[941] Romette, J., Levy, G., Dicostanzo, J. *et al.* (1979). Dosage des fractions proteiques du liquide cephalorachidien de l'adulte par deux methodes immunochimiques. *Clin. Chim. Acta*, **94**, 121–124.
[942] Ronquist, G. and Terent, A. (1982). Cerebrospinal fluid markers of disturbed brain cell metabolism. *Prog. Neurobiol.*, **18**, 167–180.
[943] Rook, G. A., Ristori, G., Salvetti, M. *et al.* (2000). Bacterial vaccines for the treatment of multiple sclerosis and other autoimmune disorders. *Immunol. Today*, **21**, 503–508.
[944] Rosen, A., Ek, K. and Aman, P. (1979). Agarose isoelectric focusing of native human immunoglobulin M and a2-Macroglobulin. *J. Immunol. Methods*, **28**, 1–11.
[945] Rosen, H., Karlsson, J. E. and Rosengren, L. (2004). CSF levels of neurofilament is a valuable predictor of long-term outcome after cardiac arrest. *J. Neurol. Sci.*, **221**, 19–24.
[946] Rosengren, L. E., Ahlsen, G., Belfrage, M. *et al.* (1992). A sensitive ELISA for glial fibrillary acidic protein: application in CSF of children. *J. Neurosci. Methods*, **44**, 113–119.
[947] Rosengren, L. E., Karlsson, J. E., Karlsson, J. O. *et al.* (1996). Patients with amyotrophic lateral sclerosis and other neurodegenerative diseases have increased levels of neurofilament protein in CSF. *J. Neurochem.*, **67**, 2013–2018.
[948] Rosengren, L. E., Karlsson, J. E., Sjogren, M. *et al.* (1999). Neurofilament protein levels in CSF are increased in dementia. *Neurology*, **52**, 1090–1093.
[949] Rosengren, L. E., Lycke, J. and Andersen, O. (1995). Glial fibrillary acidic protein in CSF of MS patients: relation to neurological deficit. *J. Neurol. Sci.*, **133**, 61–65.
[950] Rosengren, L. E., Wikkelso, C. and Hagberg, L. (1994). A sensitive ELISA for glial fibrillary acidic protein: application in CSF of adults. *J. Neurosci. Methods*, **51**, 197–204.
[951] Rosenthal, F. D. and Soothill, J. F. (1962). An immunochemical study of the proteins in cerebrospinal fluid. *J. Neurol. Neurosurg. Psychiatry*, **25**, 177–181.

[952] Rostrom, B. (1982). Antibodies against viruses and structural brain components in oligoclonal IgG obtained from multiple sclerosis brain. *J. Neurol.*, **226**, 255–263.
[953] Rostrom, B. and Link, H. (1981). Oligoclonal immunoglobulins in cerebrospinal fluid in acute cerebrovascular disease. *Neurology*, **31**, 590–596.
[954] Rostrom, B., Link, H., Laurenzi, M. A. *et al.* (1981). Viral antibody activity of oligoclonal and polyclonal immunoglobulins synthesized within the central nervous system in multiple sclerosis. *Ann. Neurol.*, **9**, 569–574.
[955] Rostrom, B., Link, H. and Norrby, E. (1981). Antibodies in oligoclonal immunoglobulins in CSF from patients with acute cerebrovascular disease. *Acta Neurol. Scand.*, **64**, 225–240.
[956] Rougemont, D., Bousser, M. G., Wechsler, B. *et al.* (1982). Manifestations neurologiques de la maladie de Behcet. *Rev. Neurol.*, **138**, 493–505.
[957] Royds, J. A., Davies-Jones, A. B., Lewtas, N. A. *et al.* (1983). Enolase isoenzymes in the cerebrospinal fluid of patients with diseases of the nervous system. *J. Neurol. Neurosurg. Psychiatry*, **46**, 1031–1036.
[958] Royds, J. A., Taylor, C. B. and Timperley, W. R. (1985). Enolase isoenzymes as diagnostic markers. *Neuropathol. App. Neurobiol.*, **11**, 1–16.
[959] Royds, J. A., Timperley, W. R. and Taylor, C. B. (1981). Levels of enolase and other enzymes in the cerebrospinal fluid as indices of pathological change. *J. Neurol. Neurosurg. Psychiatry*, **44**, 1129–1135.
[960] Ruchel, R. and Brager, M. D. (1975). Scanning electron microscopic observations of polyacrylamide gel. *Anal. Biochem.*, **68**, 415–428.
[961] Rudick, R. A., Medendorp, S. V., Namey, M. *et al.* (1995). Multiple sclerosis progression in a natural history study: predictive value of cerebrospinal fluid free kappa light chains. *Mult. Scler.*, **1**, 150–155.
[962] Ryberg, B. (1976). Complement-fixing antibrain antibodies in multiple sclerosis. *Acta Neurol.*, **54**, 1–12.
[963] Ryberg, B. (1978). Multiple specificities of antibrain antibodies in multiple sclerosis and chronic myelopathy. *J. Neurol. Sci.*, **38**, 357–382.
[964] Ryberg, B. (1980). Intrathecal and extrathecal production of antibrain antibodies in multiple sclerosis. *J. Neurol. Sci.*, **48**, 1–8.
[965] Ryberg, B. (1982). A longitudinal study of antibrain antibodies in multiple sclerosis. *J. Neurol. Sci.*, **54**, 263–270.
[966] Ryberg, B. (1982). Antibrain antibodies in multiple sclerosis. *J. Neurol. Sci.*, **54**, 239–261.
[967] Ryberg, B. and Baumann, N. A. (1983). Different types of antibrain antibodies in multiple sclerosis. *J. Neurol. Sci.*, **58**, 351–355.
[968] Ryberg, B. and Kronvall, G. (1981). Immunoglobulin characterization by bacterial absorption of antibrain antibodies in multiple sclerosis. *Eur. Neurol.*, **20**, 374–379.
[969] Sada, E., Ruiz-Palacios, G. M., Lopez-Vidal, Y. and Ponce, D. L. (1983). Detection of mycobacterial antigens in cerebrospinal fluid of patients with tuberculous meningitis by enzyme-linked immunosorbent assay. *Lancet*, **ii**, 651–652.
[970] Sadaba, M. C., Gonzalez Porque, P., Masjuan, J. *et al.* (2004). An ultrasensitive method for the detection of oligoclonal IgG bands. *J. Immunol. Methods*, **284**, 141–145.
[971] Salier, J. P., Goust, J. M., Pandey, J. P. and Fudenberg, H. H. (1981). Preferential synthesis of the G1m(1) allotype of IgG1 in the central nervous system of multiple sclerosis patients. *Science*, **213**, 1400–1402.

[972] Salmi, A., Arnadottir, T., Reunanen, M. et al. (1980). Longitudinal studies of viral antibody synthesis in MS patients. In: H. J. Bauer, S. Poser and B. Ritter (eds.), *Prog. in MS Res.*, pp. 278–288. Springer-Verlag.
[973] Salmi, A., Viljanen, M. and Reunanen, M. (1981). Intrathecal synthesis of antibodies to diphtheria and tetanus toxoids in multiple sclerosis patients. *J. Neuroimmunol.*, **1**, 333–341.
[974] Salmi, A., Ziola, B., Hovi, T. and Reunanen, M. (1982). Antibodies to coronaviruses OC 43 and 229E in multiple sclerosis patients. *Neurology*, **32**, 292–295.
[975] Salmi, A., Ziola, B., Reunanen, M. et al. (1982). Immune complexes in serum and cerebrospinal fluid of multiple sclerosis patients and patients with other neurological diseases. *Acta Neurol. Scand.*, **66**, 1–15.
[976] Sandberg-Wollheim, M. (1974). Immunoglobulin systhesis in vitro by cerebrospinal fluid cells in patients with multiple sclerosis. *Scand. J. Immunol.*, **3**, 717–730.
[977] Sandberg-Wollheim, M. (1975). Optic neuritis: studies on the cerebrospinal fluid in relation to clinical course in 61 patients. *Acta Neurol. Scand.*, **52**, 167–178.
[978] Sandberg-Wollheim, M. (1976). Immunoglobulin synthesis in vitro by cerebrospinal fluid cells in patients with meningo-encephalitis of presumed viral origin. *Scand. J. Immunol.*, **4**, 617–622.
[979] Saul, A. and Don, M. (1984). A rapid method of concentrating proteins in small volumes with high recovery using Sephadex G-25. *Anal. Biochem.*, **138**, 451–453.
[980] Scarna, H., Delafosse, B., Steinberg, R. et al. (1982). Neuron-specific enolase as a marker of neuronal lesions during various comas in man. *Neurochem. Int.*, **4**, 411–419.
[981] Schadlich, H.-J., Nekic, M. and Felgenhauer, K. (1980). The detection of activated cerebrospinal fluid B lymphocytes by peroxidase conjugated antibodies. *J. Neurol.*, **224**, 77–87.
[982] Scherer, R. and Ruhenstroth-Bauer, G. (1976). Plasma protein profiling: the diagnostic evaluation of disorders in plasma protein composition by a new immunoelectrophoretic method. *Clin. Chim. Acta*, **66**, 417–433.
[983] Schieven, G. L., Blank, A. and Dekker, C. A. (1982). Ribonucleases of human cerebrospinal fluid: Detection of altered glycosylation relative to their serum counterparts. *Biochemistry*, **21**, 5148–5155.
[984] Schinko, H. and Tschabitscher, H. (1959). Der gamma-Quotient als differentialdiagnostisches Kriterium zwischen Multiple Sklerose und degenerativen Erkrankungen des Nervensystems unter besonderer Beruckischtingung der Krankheitsdauer. *Wiener Klin. Wochenschr.*, **71**, 417–422.
[985] Schipper, H. I., Neumayer, H. and Poser, S. (1984). Prognostischer Wert der lokalen IgG-Produktion bei monosymptomatischer optikusneuritis. *Akt Neurol.*, **11**, 73–76.
[986] Schipper, H. I., Poser, S. and Wuzel Behrens-Baumann, W. (1984). Prognostic value of CSF IgG in monosymptomatic optic reuritis. In: R. E. Gonsette and P. Delmotte (eds.), *Immunological and Clinical Aspects of Multiple Sclerosis*, pp. 278–279. MTP Press Ltd.
[987] Schliep, G. and Felgenhauer, K. (1974). The alpha 2 macroglobulin level in cerebrospinal fluid: a parameter for the condition of the blood-CSF barrier. *J. Neurol.*, **207**, 171–181.

[988] Schliep, G. and Felgenhauer, K. (1978). Rapid determination of proteins in serum and cerebrospinal fluid by laser-nephelometry. *J. Clin. Chem. Clin. Biochem.* **16**, 631–635.
[989] Schliep, G. and Felgenhauer, K. (1978). Serum-CSF protein gradients, the blood-CSF barrier and the local immune response. *J. Neurol.*, **218**, 77–96.
[990] Schliep, G. and Felgenhauer, K. (1979). Die lokale humorale immunantwort bei der neurolues. In: H. Reisner and G. Schnaberth (eds.), *Fortschritte der Technischen Medizin in der Neurol Diag und Therapie*, pp. 613–617. Wien.
[991] Schliep, G., Rapic, N. and Felgenhauer, K. (1974). Quantitation of high-molecular proteins in cerebrospinal fluid. *Z. Klin. Chem. Klin. Biochem.* **12**, 367–369.
[992] Schmidt, B. L. (1980). Crossed immunoelectrophoresis of native cerebrospinal fluid on cellulose acetate membranes. *Clin. Chim. Acta*, **102**, 253–256.
[993] Schmidt, R., Rieder, H. P. and Wuthrich, R. (1977). The course of multiple sclerosis cases with extremely high gamma globulin values in the cerebrospinal fluid. *Eur. Neurol.*, **15**, 241–248.
[994] Schoneshofer, M., Fenner, A. and Molnar, I. (1981). Heterogeneity of corticotropin-immunoreactive compounds in human body fluids. *Clin. Chem.*, **27**, 1875–1877.
[995] Schreiber, G. (2002). The evolution of transthyretin synthesis in the choroid plexus. *Clin. Chem. Lab. Med.*, **40**, 1200–1210.
[996] Schuller, E. and Allinquant, B. (1973). Determination of C-reactive protein by electroimmunodiffusion in blood and CSF of neurological patients. *Eur. Neurol.*, **9**, 216–223.
[997] Schuller, E., Allinquant, B., Garcia, M. *et al.* (1971). Electro-immunodiffusion des proteines du liquide cephalo-rachidien. *Clin. Chim. Acta*, **33**, 5–11.
[998] Schuller, E., Delasnerie, N., Deloghe, G. and Loridan, M. (1973). Multiple sclerosis: a two-phase disease. *Acta Neurol. Scand.*, **49**, 453–460.
[999] Schuller, E., Delasnerie, N., Helary, M. and Lefevre, M. (1978). Serum and cerebrospinal fluid IgM in 203 neurological patients. *Eur. Neurol.*, **17**, 77–82.
[1000] Schuller, E. and Helary, M. (1983). Determination in the nanogram range of Clq in serum and unconcentrated CSF by electro-immunodiffusion. *J. Immunol. Methods*, **56**, 159–165.
[1001] Schuller, E., Lefevre, M. and Tompe, L. (1972). Electroimmunodiffusion of a2M, IgA and IgM in nanogram quantities with a hydroxyethylcellulose-agarose gel: application to unconcentrated CSF. *Clin. Chim. Acta*, **42**, 5–13.
[1002] Schuller, E. and Sagar, H. J. (1981). Local synthesis of CSF immunoglobulins. *J. Neurol. Sci.*, **51**, 361–370.
[1003] Schuller, E. and Sagar, H. J. (1983). Central nervous system IgG synthesis in multiple sclersos. *Acta Neurol. Scand.*, **67**, 365–371.
[1004] Schuller, E. and Tompe, L. (1973). Determination of IgG heavy chains by electroimmunodiffusion in the nanogram range: simultaneous application to serum and CSF of neurological patients. *Clin. Chim. Acta*, **44**, 287–294.
[1005] Schuller, E. and Tompe, L. (1974). Electroimmunodiffusion of IgG heavy chains in nanogram quantities with a carboxymethyl-cellulose-agarose gel. *Clin. Chim. Acta*, **54**, 131–133.
[1006] Schuller, E., Tompe, L., Lefevre, M. and Moreno, P. (1970). Electroimmunodiffusion des proteines du liquide cephalo-rachidien. *Clin. Chim. Acta*, **30**, 73–82.

[1007] Schwartz, S., Rieder, H. P. and Wuthrich, R. (1970). The protein fractions in cerebrospinal fluid in the various states of multiple sclerosis. *Eur. Neurol.*, **4**, 267–282.
[1008] Scott, B. J. and Burnett, D. (1978). The effect of the protein content of diluents on peak height in rocket immunoelectrophoresis. *Clin. Chim. Acta*, **89**, 475–478.
[1009] Scully, C., Boyle, P. and Yap, P. L. (1982). Immunoglobulins G,M,A,D and E in Behcet's syndrome. *Clin. Chim. Acta*, **120**, 237–242.
[1010] Sekine, T., Poulik, M. D. (1982). Post-gamma globulin: tissue distribution and physicochemical characteristics of dog post-gamma globulin. *Clin. Chim. Acta*, **120**, 225–235.
[1011] Sellebjerg, F., Christiansen, M., Rasmussen, L. S. et al. (1996). The cerebrospinal fluid in multiple sclerosis: quantitative assessment of intrathecal immunoglobulin synthesis by empirical formulae. *Eur. J. Neurol.*, **3**, 548–559.
[1012] Seneterre, J.-B., Grimaud, D., Quincy, C. I. and Pegon, Y. (1974). Electrophorese sur gel de polyacrylamide des proteines du liquide cephalo-rachidien. *Ann. Biol. Clin.*, **32**, 521–527.
[1013] Shapiro, H. D., Miller, K. D. and Harris, A. H. (1967). Low-pH disc electrophoresis of spinal fluid: changes in multiple sclerosis. *Exp. Mol. Pathol.*, **7**, 362–365.
[1014] Sharief, M. K., Ciardi, M. and Thompson, E. J. (1992). Blood-brain barrier damage in patients with bacterial meningitis: association with tumor necrosis factor-alpha but not interleukin-1 beta. *J. Infect. Dis.*, **166**, 350–358.
[1015] Sharief, M. K., Ciardi, M., Thompson, E. J. et al. (1992). Tumor necrosis factor alpha mediates blood-brain barrier damage in patients with HIV-1 infections of the central nervous system. *Mediat. Inflamm.*, **1**, 163–168.
[1016] Sharief, M. K., Green, A., Dick, J. P. et al. (1999). Heightened intrathecal release of proinflammatory cytokines in Creutzfeldt-Jakob disease. *Neurology*, **52**, 1289–1291.
[1017] Sharief, M. K., Hentges, R., Ciardi, M. and Thompson, E. J. (1993). In vivo relationship of interleukin-2 and soluble IL-2 receptor to blood-brain barrier impairment in patients with active multiple sclerosis. *J. Neurol.* **240**, 46–50.
[1018] Sharief, M. K., Hentges, R. and Thompson, E. J. (1991). The relationship of interleukin-2 and soluble interleukin-2 receptors to intrathecal immunoglobulin synthesis in patients with multiple sclerosis. *J. Neuroimmunol.*, **32**, 43–51.
[1019] Sharief, M. K., Hentges, R. and Thompson, E. J. (1992). Determination of interleukin-2 in cerebrospinal fluid by a sensitive enzyme-linked immunosorbent assay. *J. Immunol. Methods*, **147**, 51–56.
[1020] Sharief, M. K., Hentges, R. and Thompson, E. J. (1992). Significance of CSF immunoglobulins in monitoring neurological activity in Behcet's disease. *Neurology*, **41**, 1398–1401.
[1021] Sharief, M. K., Ingram, D. A., Swash, M. and Thompson, E. J. (1999). I.v. immunoglobulin reduces circulating proinflammatory cytokines in Guillain-Barre syndrome. *Neurology*, **52**, 1833–1838.
[1022] Sharief, M. K., Keir, G. and Thompson, E. J. (1989). Glutaraldehyde-enhanced immunofixation: a sensitive new method for detecting oligoclonal immunoglobulin M. *J. Neuroimmunol.*, **23**, 149–156.
[1023] Sharief, M. K., Keir, G. and Thompson, E. J. (1990). Intrathecal synthesis of IgM in neurological diseases: a comparison between detection of oligoclonal bands and quantitative estimation. *J. Neurol. Sci.*, **96**, 131–142.

[1024] Sharief, M. K., McLean, B. and Thompson, E. J. (1993). Elevated serum levels of tumor necrosis factor-alpha in Guillain-Barre syndrome. *Ann. Neurol.*, **33**, 591–596.
[1025] Sharief, M. K., Noori, M. A., Ciardi, M. *et al.* (1993). Increased levels of circulating ICAM-1 in serum and cerebrospinal fluid of patients with active multiple sclerosis. Correlation with TNF-alpha and blood-brain barrier damage. *J. Neuroimmunol.*, **43**, 15–21.
[1026] Sharief, M. K. and Thompson, E. J. (1989). Immunoglobulin M in the cerebrospinal fluid: an indicator of recent immunological stimulation. *J. Neurol. Neurosurg. Psychiatry*, **52**, 949–953.
[1027] Sharief, M. K. and Thompson, E. J. (1990). A sensitive ELISA system for the rapid detection of virus specific IgM antibodies in the cerebrospinal fluid. *J. Immunol. Methods*, **130**, 19–24.
[1028] Sharief, M. K. and Thompson, E. J. (1991). Intrathecal immunoglobulin M synthesis in multiple sclerosis. Relationship with clinical and cerebrospinal fluid parameters. *Brain*, **114**, 181–195.
[1029] Sharief, M. K. and Thompson, E. J. (1991). The predictive value of intrathecal immunoglobulin synthesis and magnetic resonance imaging in acute isolated syndromes for subsequent development of multiple sclerosis. *Ann. Neurol.*, **29**, 147–151.
[1030] Sharief, M. K. and Thompson, E. J. (1992). Distribution of cerebrospinal fluid oligoclonal IgM bands in neurological diseases: a comparison between agarose electrophoresis and isoelectric focusing. *J. Neurol. Sci.*, **109**, 83–87.
[1031] Sharief, M. K. and Thompson, E. J. (1992). In vivo relationship of tumor necrosis factor-alpha to blood-brain barrier damage in patients with active multiple sclerosis. *J. Neuroimmunol.*, **38**, 27–33.
[1032] Sharief, M. K. and Thompson, E. J. (1993). Correlation of interleukin-2 and soluble interleukin-2 receptor with clinical activity of multiple sclerosis. *J. Neurol. Neurosurg. Psychiatry*, **56**, 169–174.
[1033] Shashoua, V. E., Hesse, G. W. and Moore, B. W. (1984). Proteins of the brain extracellular fluid: evidence for release of S-100 protein. *J. Neurochem.*, **42**, 1536–1541.
[1034] Sherwin, R. M. and Moore, G. H. (1971). Microzone electrophoresis of unconcentrated cerebrospinal fluid using cellulose acetate strips and nigrosin dye. *Am. Clin. Pathol.*, **55**, 705–712.
[1035] Shoenfeld, Y., Rauch, J., Massicotte, H. *et al.* (1983). Polyspecificity of monoclonal lupus autoantibodies produced by human-human hybridomas. *N. Engl. J. Med.*, **308**, 414–420.
[1036] Shoji, M., Matsubara, E., Kanai, M. *et al.* (1998). Combination assay of CSF tau, A-beta 1-40 and A-beta 1-42(43) as a biochemical marker of Alzheimer's disease. *J. Neurol. Sci.*, **158**, 134–140.
[1037] Shorr, J., Rostrom, B. and Link, H. (1981). Antibodies to viral and non-viral antigens in subacute sclerosing panencephalitis and multiple sclerosis demonstrated by thin-layer polyacrylamide gel isoelectric focusing, antigen immunofixation and autoradiograph. *J. Neurol.*, **49**, 99–108.
[1038] Shulman, G. (1979). Crossed Immuno-electrodiffusion in the diagnosis of immunoglobulin fragment abnormalities. *Clin. Biochem.*, **12**, 93–97.
[1039] Shulman, G. (1980). Quality of commercially available controls in laser immunonephelometry. *Clin. Biochem.*, **17**, 178–182.

[1040] Siden, A. (1977). Crossed immunoelectrofocusing of cerebrospinal fluid immunoglobulins. *J. Neurol.*, **217**, 103–109.
[1041] Siden, A. (1979). Isoelectric focusing and crossed immunoelectrofocusing of CSF immunoglobulins in MS. *J. Neurol.*, **221**, 39–51.
[1042] Siden, A. (1980). Abnormal CSF immunoglobulin components detected by isoelectric focusing. *J. Neurol.*, **224**, 133–144.
[1043] Siden, A. and Kjellin, K. G. (1977). Isoelectric focusing of CSF and serum proteins in neurological disorders combined with benign and malignant proliferations of reticulocytes, lymphocytes and plasmocytes. *J. Neurol.*, **216**, 251–264.
[1044] Siden, A. and Kjellin, K. G. (1978). CSF protein examinations with thin-layer isoelectric focusing in multiple sclerosis. *J. Neurol. Sci.*, **39**, 131–146.
[1045] Siden, A. and Kjellin, K. G. (1979). Isoelectric focusing of CSF proteins in known or probable infectious neurological diseases and the Guillain-Barre syndrome. *J. Neurol. Sci.*, **42**, 139–153.
[1046] Siegert, M. and Siemes, H. (1977). Agarose gel electrophoresis of cerebrospinal fluid proteins and analysis of the pherogram profiles by analog computer. *J. Clin. Chem. Clin. Biochem.* **15**, 635–644.
[1047] Siemes, H., Siegert, M., Hanefeld, F. et al. (1977). Oligoclonal gamma-globulin banding of cerebrospinal fluid in patients with subacute sclerosing panencephalitis. *J. Neurol. Sci.*, **32**, 395–409.
[1048] Silva, C. A., Rio, M. E. and Cruz, C. (1981). Protein patterns of the cerebrospinal fluid of 30 patients with subacute sclerosing panencephalitis (SSPE). *Acta Neurol. Scand.*, **63**, 255–266.
[1049] Silva, C. A., Rio, M. E., Maia-Goncalves, A. et al. (1980). Oligoclonal gamma-globulin of cerebrospinal fluid in neurobrucellosis. *Acta Neurol. Scand.*, **61**, 42–48.
[1050] Silva, C. A. and Sa, M. J. (1978). Electrophoretic pattern of cerebrospinal fluid proteins in non-neoplastic infantile hydrocephalus. *Acta Neurol. Scand.*, **57**, 317–328.
[1051] Simionescu, N., Simionescu, M. and Palade, G. E. (1981). Differentiated microdomains on the luminal surface of the capillary endothelium 1. Preferential distribution of anionic sites. *J. Cell. Biol.*, **90**, 605–613.
[1052] Sindic, C. J. and Laterre, E. C. (1991). Oligoclonal free kappa and lambda bands in the cerebrospinal fluid of patients with multiple sclerosis and other neurological diseases. An immunoaffinity-mediated capillary blot study. *J. Immunol. Methods*, **33**, 63–72.
[1053] Sindic, C. J. M. (1985). *Cerebrospinal Fluid Proteins in Diseases of the Nervous System*. University Cath. 1–240.
[1054] Sindic, C. J. M., Cambiaso, C. L., Depre, A. et al. (1982). The concentration of IgM in the cerebrospinal fluid of neurological patients. *J. Neurol. Sci.*, **55**, 339–350.
[1055] Sindic, C. J. M., Cambiaso, C. L., Depre, A. et al. (1984). Immune complexes in cerebrospinal fluid and serum of neurological patients. Possible intrathecal formation in bacterial meningitis and herpetic encephalitis. *J. Neuroimmunol.*, **6**, 9–18.
[1056] Sindic, C. J. M., Cambiaso, C. L., Masson, P. L. and Laterre, E. C. (1980). The binding of myelin basic protein to the Fc region of aggregated IgG and to immune complexes. *Clin. Exp. Immunol.*, **41**, 1–7.
[1057] Sindic, C. J. M., Cambiaso, C. L., Masson, P. L. and Laterrre, E. C. (1980). The elution of IgG from subacute sclerosing panencephalitis and multiple sclerosis brains. *Clin. Exp. Immunol.*, **41**, 8–12.

[1058] Sindic, C. J. M., Chalon, M. P., Cambiaso, C. L. et al. (1982). Assessment of damage to the central nervous system by determination of S-100 protein in the cerebrospinal fluid. *J. Neurol. Neurosurg. Psychiatry*, **45**, 1130–1135.

[1059] Sindic, C. J. M., Collet-Cassart, D., Cambiaso, C. L. et al. (1981). The clinical relevance of ferritin concentration in the cerebrospinal fluid. *J. Neurol. Neurosurg. Psychiatry*, **44**, 329–333.

[1060] Sindic, C. J. M., Collet-Cassart, D., Depare, A. et al. (1984). C-Reactive protein in serum and cerebrospinal fluid in various neurological disorders. *J. Neurol. Sci.*, **63**, 339–344.

[1061] Sindic, C. J. M., Delacroix, D. L., Vaerman, J. P. et al. (1984). Study of IgA in the cerebrospinal fluid of neurological patients with special reference to size subclass and local production. *J. Neuroimmunol.*, **7**, 65–75.

[1062] Sindic, C. J. M., Kevers, L., Chalon, M. P. et al. (1985). Monitoring and tentative diagnosis of herpetic encephalitis by protein analysis of cerebrospinal fluid. Particular relevance of the assays of ferritin and S-100. *J. Neurol. Sci.*, **67**, 359–369.

[1063] Sindic, C. J. M., Magnusson, C. G. M., Laterre, E. C. and Masson, P. L. (1984). IgE in the cerebrospinal fluid. *J. Neruoimmunol.*, **6**, 319–324.

[1064] Sindic, C. J. M., van Antwerpen, M. P. and Goffette, S. (2001). The intrathecal humoral immune response: laboratory analysis and clinical relevance. *Clin. Chem. Lab. Med.*, **39**, 333–340.

[1065] Sjogren, M., Blomberg, M., Jonsson, M., et al. (2001). Neurofilament protein in cerebrospinal fluid: a marker of white matter changes. *J. Neurosci. Res.*, **66**, 510–516.

[1066] Sjogren, M., Rosengren, L., Minthon, L. et al. (2000). Cytoskeleton proteins in CSF distinguish frontotemporal dementia from AD. *Neurology*, **54**, 1960–1964.

[1067] Skinner, M. and Cohen, A. S. (1981). The prealbumin nature of the amyloid protein in familial amyloid polyneuropathy (FAP)- Swedish variety. *Biochem. Biophys. Res. Commun.*, **99**, 1326–1332.

[1068] Skoldenberg, B., Alestig, K., Burman, L. et al. (1984). Acyclovir versus vidarabine in herpes simplex encephalitis. *Lancet*, **ii**, 707–711.

[1069] Skoldenberg, B., Carlstrom, A., Forsgren, M. and Norrby, E. (1976). Transient appearance of oligoclonal immunoglobulins and measles virus antibodies in the cerebrospinal fluid in a case of acute measles encephalitis. *Clin. Exp. Immunol.*, **23**, 451–455.

[1070] Skoldenberg, B., Kalimo, K., Carlstrom, A. et al. (1981). Herpes simplex encephalitis: a serological follow-up study. *Acta Neurol. Scand.*, **63**, 273–285.

[1071] Slater, L. (1975). IgG, IgA and IgM by formylated rocket immunoelectrophoresis. *Ann. Clin. Biochem.*, **12**, 19–24.

[1072] Smith, C. C., Bradford, H. F., Thompson, E. J. and MacDermot, J. (1980). Actions of beta-bungarotoxin on amino acid transmitter release. *J. Neurochem.*, **34**, 487–494.

[1073] Smith, J. E. and Goodman, D. S. (1979). Retinol-binding protein and the regulation of vitamin A transport. *Fed. Proc.*, **38**, 2504–2509.

[1074] Sobczyk, W., Polna, I., Kulczycki, J. and Horbowska, H. (1982). Measles antibodies in the saliva of children with subacute sclerosing panencephalitis. *J. Neurol.*, **228**, 219–222.

[1075] Soininen, H., Pitkanen, A., Halonen, T. and Riekkinen, P. (1984). Dopamine-B-hydroxylase and acetylcholinesterase activities of cerebrospinal fluid in Alzheimer's disease. *Acta Neurol. Scand.*, **69**, 29–34.

[1076] Sorensen, P. S., Gjerris, F., Ibsen, S. and Bock, E. (1983). Low cerebrospinal fluid concentration of brain-specific protein D2 in patients with normal pressure hydrocephalus. *J. Neurol. Sci.*, **62**, 59–65.
[1077] Sotelo, J., Gibbs, C. J. and Gajdusek, D. C. (1980). Autoantibodies against axonal neurofilaments in patients with Kuru and Creutzfeldt-Jakob disease. *Science*, **210**, 190–193.
[1078] Sotrel, A., Rosen, S., Ronthal, M. and Ross, D. B. (1983). Subacute sclerosing panencephalitis: an immune complex disease? *Neurology*, **33**, 885–890.
[1079] Sottrup-Jenson, L., Stepanik, T. M., Kristensen, T. *et al.* (1985). Common evolutionary origin of alpha-2-macroglobulin and complement components C3 and C4. *Proc. Natl. Acad. Sci.*, **82**, 9–13.
[1080] Southern, E. M. (1975). Detection of specific sequences among DNA fragments separated by gel electrophoresis. *J. Mol. Biol.*, **98**, 503–517.
[1081] Souverijn, J. H., Serree, H. M., Peet, R. *et al.* (1991). Intrathecal immunoglobulin synthesis. Comparison of various formulae with the 'gold standard' of isoelectric focusing. (Erratum in: J. Neurol. Sci. 1991 Jun;103(2):241). *J. Neurol. Sci.*, **102**, 11–16.
[1082] Souverijn, J. H. M. Reply to the letter by Ohman (J. Neurol. Sci., 106 (1991) 230–231). *J. Neurol. Sci.*, **106**, 232. 1991.
[1083] Statz, A. and Felgenhauer, K. (1983). Development of the blood-CSF barrier. *Dev. Med. Child Neurol.*, **25**, 152–161.
[1084] Statz, A., Wenzel, D. and Felgenhauer, K. (1979). Blut-liquor-schranke und lokale immunantwort im verlauf von meningitiden im kindesalter. *Neuropediatrics*, **10**, 281–289.
[1085] Statz, T., Statz, A. and Felgenhauer, K. (1985). Physiologische proteinurie bei neugeborenen. *Dtsch. Med. Wochenschr.*, **110**, 55–58.
[1086] Steck, A. J. and Link, H. (1984). Antibodies against oligodendrocytes in serum and CSF in multiple sclerosis and other neurological diseases: 125 I-protein A studies. *Acta Neurol. Scand.*, **69**, 81–89.
[1087] Steere, A. C., Green, J., Schoen, R. T. *et al.* (1985). Successful parenteral penicillin therapy of established Lyme arthritis. *N. Engl. J. Med.*, **312**, 869–874.
[1088] Steinberg, R., Gueniau, C., Scarna, H. *et al.* (1984). Experimental brain ischemica: Neuron-specific enolase level in cerebrospinal fluid as an index of neuronal damage. *J. Neurochem.*, **43**, 19–24.
[1089] Steinberg, R., Scarna, H., Keller, A. and Pujol, J. F. (1983). Release of neuron specific enolase (NSE). *Neurochem. Int.*, **5**, 145–151.
[1090] Stelzmann, J. (1980). Two-dimensional separation of human body fluid proteins. *Clin. Chim. Acta*, **100**, 121–132.
[1091] Stendahl, L., Link, H., Moller, E. and Norrby, E. (1976). Relation between genetic markers and oligoclonal IgG in CSF in optic neuritis. *J. Neurol. Sci.*, **27**, 93–98.
[1092] Stendahl-Brodin, L. and Link, H. (1980). Relation between benign course of multiple sclerosis and low-grade humoral immune reponse in cerebrospinal fluid. *J. Neurol. Neurosurg. Psychiatry*, **43**, 102–105.
[1093] Stendahl-Brodin, L. and Link, H. (1983). Optic neuritis: oligoclonal bands increase the risk of multiple sclerosis. *Acta Neurol. Scand.*, **67**, 301–304.
[1094] Stendahl-Brodin, L., Link, H., Moller, E. and Norrby, E. (1978). Optic neuritis and distribution of genetic markers of the HLA system. *Acta Neurol. Scand.*, **57**, 418–431.

[1095] Stendahl-Brodin, L., Link, H., Moller, E. and Norrby, E. (1979). Genetic basis of multiple sclerosis: HLA antigens, disease progression, and oligoclonal IgG in CSF. *Acta Neurol. Scand.*, **59**, 297–308.
[1096] Stensland, E., Sandberg, S., Berge, R. *et al.* (1981). Cerebrospinal fluid enzymes in neurological diseases. *Acta Neurol. Scand.*, **63**, 51–56.
[1097] Stephenson, J. R., ter Mevlen, V., Kiessling, W. (1980). Search for canine-distemper-virus antibodies in multiple sclerosis. *Lancet*, **ii**, 772–776.
[1098] Stibler, H. (1977). Crossed immunoelectrofocusing for identification of normal and abnormal cerebrospinal fluid proteins. *J. Neurol. Sci.*, **32**, 331–336.
[1099] Stibler, H. (1978). The normal cerebrospinal fluid proteins identified by means of thin-layer isoelectric focusing and crossed immunoelectrofocusing. *J. Neurol. Sci.*, **36**, 273–288.
[1100] Stibler, H. (1979). Direct immunofixation after isoelectric focusing. *J. Neurol. Sci.*, **42**, 275–281.
[1101] Stiernstedt, G. T., Granstrom, M., Hederstedt, B. and Skoldenberg, B. (1985). Diagnosis of spirochetal meningitis by enzyme-linked immunosorbent assay and indirect immunofluorescence assay in serum and cerebrospinal fluid. *J. Clin. Microbiol.*, **21**, 819–825.
[1102] Stockley, R. A., Afford, S. C. and Burnett, D. (1980). A method for the study of local IgA production using radial immunodiffusion and thin layer chromatography. *J. Immunol. Methods*, **38**, 151–159.
[1103] Strony, L. P., Wagner, K. and Keshgegian, A. A. (1982). Demonstration of cerebrospinal fluid oligoclonal banding in neurologic diseases by agarose gel electrophoresis and immunofixation. *Clin. Chim. Acta*, **122**, 203–212.
[1104] Strosberg, A. D., Karcher, D. and Lowenthal, A. (1975). Structural homogeneity of human subacute sclerosing panencephalitis antibodies. *J. Immunol.*, **115**, 157–160.
[1105] Sun, T. (1982). High-resolution agarose electrophoresis. In: S. E. Ritzmann and J. C. Daniels (eds.), *Physiology of Immunoglobulins*, pp. 29–63. Alan R. Liss.
[1106] Sun, T., Fleming, J. O., Beresford, H. R. and Lien, Y. (1981). Synthesis of immunoglobulin within the central nervous system in multiple sclerosis and other neurological diseases. Detection by analysis of CSF/Serum IgG ratio. *Am. J. Clin. Pathol.*, **76**, 458–461.
[1107] Sun, T., Lien, Y. Y. and Gross, S. (1978). Clinical application of a high-resolution electrophoresis system. *Ann. Clin. Lab. Sci.*, **8**, 219–227.
[1108] Sundquist, J., Forsling, M. L., Olsson, J. and Akerlund, M. (1983). Cerebrospinal fluid arginine vasopressin in degenerative disorders and other neurological diseases. *J. Neurol. Neurosurg. Psychiatry*, **46**, 14–17.
[1109] Sussmuth, S. D., Reiber, H. and Tumani, H. (2001). Tau protein in cerebrospinal fluid (CSF): a blood-CSF barrier related evaluation in patients with various neurological diseases. *Neurosci. Lett.*, **300**, 95–98.
[1110] Sweet, W. H., Brownell, G. L., Scholl, J. A. *et al.* (1954). The formation flow and absorption of cerebrospinal fluid: newer concepts based on studies with isotopes. *Res. Publ. Assoc. Res. Nerv. Ment. Dis.*, **34**, 101–159.
[1111] Szilagyi, A. K. (1983). Comparative study of the proteinase inhibitor pattern in serum and cerebrospinal fluid. *J. Neurol. Sci.*, **58**, 305–313.
[1112] Takeoka, T., Gotoh, F., Furumi, K. and Mori, K. (1976). Polyacrylamide-gel disc electrophoresis of native cerebrospinal fluid proteins with special

reference to immunoglobulins and some clinical applications. *J. Neurol. Sci.,* **29**, 213–239.
[1113] Takeoka, T., Shinohara, Y., Furumi, K. and Mori, K. (1980). Characteristic protein fractions of cerebrospinal fluid disc electrophoretic analysis. *Brain Res.,* **198**, 147–156.
[1114] Takeoka, T., Shinohara, Y., Furumi, K. and Mori, K. (1983). Impairment of blood-cerebrospinal fluid barrier in multiple sclerosis. *J. Neurochem.,* **214**, 1102–1108.
[1115] Taskinen, E., Koskiniemi, M. L. and Vaheri, A. (1984). Herpes simplex virus encephalitis. *J. Neurol. Sci.,* **63**, 331–338.
[1116] Tate, J., Gill, D., Cobcroft, R. and Hickman, P. E. (2003). Practical considerations for the measurement of free light chains in serum. *Clin. Chem.,* **49**, 1252–1257.
[1117] Tejler, L. and Grubb, A. O. (1976). A complex-forming glycoprotein heterogeneous in charge and present in human plasma, urine and cerebrospinal fluid. *Biochim. Biophys. Acta,* **439**, 82–94.
[1118] Tenhunen, R., Iivanainen, M. and Kovanen, J. (1978). Cerebrospinal fluid beta 2-microglobulin in neurological disorders. *Acta Neurol. Scand.,* **57**, 366–373.
[1119] Terent, A., Hallgren, R., Venge, P. and Bergstrom, K. (1981). Lactoferrin, lysozyme, and beta 2-microglobulin in cerebrospinal fluid. *Stroke,* **12**, 40–46.
[1120] Thomas, D. G. T., Palfreyman, J. W. and Ratcliffe, J. G. (1978). Serum-myelin-basic-protein assay in diagnosis and prognosis of patients with head injury. *Lancet,* **i**, 113–115.
[1121] Thomas, D. G. T. and Rowan, T. D. (1967). Lactic dehydrogenase isoenzymes following head injury. *Injury,* **7**, 258–262.
[1122] Thompson, E. J. (1966). Phospholipids and sphingolipids in mammalian brain membrane fractions and body fluids: normal and abnormal. University of London. 1–221.
[1123] Thompson, E. J. (1976). Brain-specific antigens: biochemical role in selective pathogenesis. In: A. N. Davison (ed.), *Biochemistry and Neurological Disease,* pp. 278–316. Blackwell Scientific.
[1124] Thompson, E. J. (1977). Laboratory diagnosis of multiple sclerosis: immunological and biochemical aspects. *Br. Med. Bull.,* **33**, 28–33.
[1125] Thompson, E. J. (1979). Immunochemistry of CSF proteins. In: A. Milford Ward and J. T. Whicher (eds.), *Immunochemistry in Clinical Laboratory Medicine,* pp. 229–236. MTP Press Ltd.
[1126] Thompson, E. J. (1979). Oligoclonal gammaglobulin in relation to clinical classification in multiple sclerosis. In: F. C. Rose (ed.), *Clinical Neuroimmunology,* pp. 338–343. Blackwell Scientific Publications.
[1127] Thompson, E. J. (1979). Oligoclonal immunoglobulins and multiple sclerosis. *Br. Med. J.,* **1**, 1489.
[1128] Thompson, E. J. (1984). Practical problems in spinal fluid diagnosis. In: N. Callaghan and R. Galvin (eds.), *Recent Research in Neurology,* pp. 157–158. Pitman.
[1129] Thompson, E. J. (1985). Nitrocellulose blotting of CSF and urine protides. In: H. Peeters (ed.), *Protides of the Biological Fluids,* pp. 991–992. Pergamon.
[1130] Thompson, E. J. (1987). Advances in CSF protein research and diagnosis: a centennial celebration. pp. 1–149. Lancaster: MTP Press Ltd.
[1131] Thompson, E. J. (1987). Immunoblotting: a rapid and sensitive method for disease diagnosis. *Lab. Manage.,* **25**, 39–46.

[1132] Thompson, E. J. (1987). The use and significance of oligoclonal bands in multiple sclerosis. In: F. C. Rose and R. Jones (eds.), *Multiple Sclerosis: Immunological, Diagnostic and Therapeutic Aspects*, pp. 111–114. John Libbey & Co.
[1133] Thompson, E. J. (1988). Cerebrospinal fluid proteins. *Curr. Opin. Neurol. Neurosurg.*, **1**, 1087–1090.
[1134] Thompson, E. J. (1988). Fitting the data to curves. In: E. J. Thompson (ed.), *The CSF Proteins: A Biochemical Approach*, pp. 67–85. Elsevier.
[1135] Thompson, E. J. (1988). *The CSF Proteins: A Biochemical Approach*. 1 ed. pp. 1–228. Elsevier, Amsterdam.
[1136] Thompson, E. J. (1992). The importance of molecular size in the pathophysiology of CSF proteins. *Clin. Invest.*, **70**, 877–879.
[1137] Thompson, E. J. (1993). Diagnostic significance of CSF analysis. In: K. Felgenhauer, M. Holzgraefe and H. W. Prange (eds.), *CNS Barriers and Modern Diagnosis*, pp. 364–372. Springer-Verlag.
[1138] Thompson, E. J. (1994). Nervous system. In: D. A. Noe and R. C. Rock (eds.), *Laboratory Medicine*, pp. 462–475. Williams & Wilkins.
[1139] Thompson, E. J. (1995). Cerebrospinal fluid. *J. Neurol. Neurosurg. Psychiatry*, **59**, 349–357.
[1140] Thompson, E. J. (1995). Cerebrospinal fluid – an overview of methods. In: A. Townshend, P. J. Worsfold and S. J. Haswell *et al.* (eds.), *Encyclopedia of Analytical Science*, pp. 592–598. Academic Press.
[1141] Thompson, E. J. (1996). Qualitative versus quantitative analysis in the detection of intrathecal synthesis of immunolobulins. *J. Lab. Med.*, **20**, 309–310.
[1142] Thompson, E. J. (1997). Cerebrospinal fluid. In: R. A. C. Hughes (ed.), *Neurological Investigations*, pp. 443–466. BMA Publishing Group.
[1143] Thompson, E. J. (1999). Neurochemistry. In: F. Baynes and M. H. Dominiczak (ed.), *Medical Biochemistry*, pp. 487–495. Mosby.
[1144] Thompson, E. J. (2004). Quality versus quantity: which is better for cerebrospinal fluid IgG? *Clin. Chem.*, **50**, 1721–1722.
[1145] Thompson, E. J., Daroga, B. M., Quick, J. M. S. *et al.* (1974). Quantitative determination of fluorescent labelled proteins by using dansyl chloride or fluorescamine with application to polyacrylamide gels. *Biochem. Soc. Trans.*, **2**, 989–990.
[1146] Thompson, E. J., Goodwin, H. and Cumings, J. N. (1967). Caesium chloride in the preparation of membrane fractions from human cerebral tissue. *Nature*, **215**, 168–169.
[1147] Thompson, E. J. and Green, A. J. (1998). Protein markers of brain damage. *Mult. Scler.*, **4**, 5–6.
[1148] Thompson, E. J. and Johnson, M. H. (1982). Electrophoresis of CSF proteins. *Br. J. Hosp. Med.*, **28**, 600–608.
[1149] Thompson, E. J., Kaufmann, P. and Rudge, P. (1983). Sequential changes in oligoclonal patterns during the course of multiple sclerosis. *J. Neurol. Neurosurg. Psychiatry*, **46**, 547–550.
[1150] Thompson, E. J., Kaufmann, P., Shortman, R. C. *et al.* (1979). Oligoclonal immunoglobulins and plasma cells in spinal fluid of patients with multiple sclerosis. *Br. Med. J.*, **1**, 16–17.
[1151] Thompson, E. J. and Keir, G. (1984). Improved detection of oligoclonal and Bence-Jones proteins by kappa/lambda immunoblotting. *Clin. Chim. Acta*, **143**, 329–335.

[1152] Thompson, E. J. and Keir, G. (1990). Laboratory investigation of cerebrospinal fluid proteins. *Ann. Clin. Biochem.*, **27**, 425–435.
[1153] Thompson, E. J., Learmouth, R. and Kaim, G. L. (2001). Clearer images. *Lancet*, **357**, 642.
[1154] Thompson, E. J., Luxon, L. M., Jethwa, J. *et al.* (1982). On the relationship of CSF pleocytosis to immunoglobulin levels as estimated by different techniques. *J. Neuroimmunol.*, **2**, 321–330.
[1155] Thompson, E. J., Norman, P. M. and MacDermot, J. (1975). The analysis of cerebrospinal fluid. *Br. J. Hosp. Med.*, **14**, 645–652.
[1156] Thompson, E. J., Riches, P. G. and Kohn, J. (1983). Antibody synthesis within the central nervous system: comparisons of CSF IgG indices and electrophoresis. *J. Clin. Pathol.*, **36**, 312–315.
[1157] Thompson, E. J. and Vakaet, A. (1985). Free light chains in multiple sclerosis. In: H. Peeters (ed.), *Protides of the Biological Fluids*, pp. 215–216. Pergamon.
[1158] Thompson, E. J. and Zeman, A. (1992). Fluids of the brain and the pathogenesis of MS. *Neurochem. Res.*, **17**, 901–905.
[1159] Thompson, R. J. (1981). Human nervous system specific proteins. In: K. G. M. M. Alberti and C. P. Price (eds.), *Recent Advances in Clinical Biochemistry*, pp. 295–307. Edinburgh, Churchill Livingstone.
[1160] Thompson, R. J., Doran, J. F., Jackson, P. *et al.* (1983). PGP 9.5 - a new marker for vertebrate neurons and neuroendocrine cells. *Brain Res.*, **278**, 224–228.
[1161] Thompson, R. J., Kynoch, P. A. M. and Willson, V. J. C. (1982). Cellular localization of aldolase C subunits in human brain. *Brain Res.*, **232**, 489–493.
[1162] Thomson, R. B., Jr. and Bertram, H. (2001). Laboratory diagnosis of central nervous system infections. *Infect. Dis. Clin. North. Am.*, **15**, 1047–1071.
[1163] Tibbling, G., Link, H. and Ohman, S. (1977). Principles of albumin and IgG analyses in neurological diseases I. Establishment of reference values. *Scand. J. Clin. Lab. Invest.*, **37**, 385–390.
[1164] Tillyer, C. R. (1992). The estimation of free light chains of immunoglobulins in biological fluids. *Int. J. Clin. Lab. Res.*, **22**, 152–158.
[1165] Tourtellotte, W. W. (1966). Multiple sclerosis: correlation between immunoglobulin-G in cerebrospinal fluid and brain. *Science*, **154**, 1044–1045.
[1166] Tourtellotte, W. W. (1970). Cerebrospinal fluid in multiple sclerosis. In: P. J. Vinken and G. W. Bruyn (eds.), *Handbook of Clin. Neurol.*, pp. 324–382. North-Holland Pub. Co.
[1167] Tourtellotte, W. W. (1970). On cerebrospinal fluid immunoglobulin-G (IgG) quotients in multiple sclerosis and other diseases. *J. Neurol. Sci.*, **10**, 279–304.
[1168] Tourtellotte, W. W. (1975). What is multiple sclerosis? Laboratory criteria for diagnosis. In: A. N. Davison, J. H. Humphrey, A. L. Liversedge *et al.* (eds.), *Multiple Sclerosis Research*, pp. 9–26. HMSO.
[1169] Tourtellotte, W. W., Baumhefner, R. W., Potvin, A. R. *et al.* (1980). Multiple sclerosis de novo CNS IgG synthesis: Effect of ACTH and corticosteroids. *Neurology*, **30**, 1155–1162.
[1170] Tourtellotte, W. W., Haerer, A. F., Heller, G. L. and Somers, J. E. (1964). Post-lumbar puncture headaches. Springfield: Charles C Thomas.
[1171] Tourtellotte, W. W., Henderson, W. G., Tucker, R. P. *et al.* (1972). A randomized, double-blind clinical trial comparing the 22 versus 26 gauge needle in the production

of the post-lumbar puncture syndrome in normal individuals. *Headache*, **12**, 73–78.

[1172] Tourtellotte, W. W. and Ma, B. I. (1978). Multiple sclerosis: The blood-brain-barrier and the measurements of de novo central nervous system IgG systhesis. *Neurology*, **28**, 76–83.

[1173] Tourtellotte, W. W., Ma, B. I., Brandes, D. B. *et al.* (1981). Quantification of de novo central nervous system IgG measles antibody synthesis in SSPE. *Ann. Neurol.*, **9**, 551–556.

[1174] Tourtellotte, W. W., Ma, B. I. and Potvin, A. R. (1981). Positive correlation between multiple sclerosis IgG measles antibody concentration and central nervous system IgG synthesis rate. 12th World Congress of Neurology, Kyoto, Japan. p. 130. Exerpta Medica.

[1175] Tourtellotte, W. W. and Parker, J. A. (1966). Multiple sclerosis: correlation between immunoglobulin G in cerebrospinal fluid and brain. *Science*, **154**, 1044–1046.

[1176] Tourtellotte, W. W. and Parker, J. A. (1967). Multiple sclerosis: Brain immunoglobulin-G and albumin. *Nature*, **214**, 683–686.

[1177] Tourtellotte, W. W. and Potvin, A. R. (1981). Cytomorphology of multiple sclerosis cerebrospinal fluid cells, clinical activity, and de novo central nervous system IgG synthesis. Abstracts of the 12th World Congress of Neurol Internat Cong Series 548. Excerpta Medica.

[1178] Tourtellotte, W. W., Potvin, A. R., Baumhefner, R. W. *et al.* (1980). Multiple sclerosis de novo CNS IgG synthesis. *Arch. Neurol.*, **37**, 620–624.

[1179] Tourtellotte, W. W., Potvin, A. R., Booe, I. *et al.* (1982). Isotachophoresis quantitation of subfractions of multiple sclerosis intra-blood-brain barrier IgG synthesis modulated by ACTH and/or steroids. *Neurology*, **32**, 261–266.

[1180] Tourtellotte, W. W., Potvin, A. R., Fleming, J. O. *et al.* (1980). Multiple sclerosis: Measurement and validation of central nervous system IgG synthesis rate. *Neurology*, **30**, 240–244.

[1181] Tourtellotte, W. W., Potvin, A. R., Mendez, M. *et al.* (1980). Failure of intravenous and intrathecal cytarabine to modify central nervous system IgG synthesis in multiple sclerosis. *Ann. Neurol.*, **8**, 402–408.

[1182] Tourtellotte, W. W., Travolato, B., Parker, J. A. and Comiso, P. (1971). Cerebrospinal fluid electroimmunodiffusion. *Arch. Neurol.*, **25**, 345–350.

[1183] Tourtellotte, W. W., Turpin, R. A. and Cawley, L. W. (1977). A sensitive technic for identifying immunoglobulin-G (IgG) oligoclonals by immunofixation in cerebrospinal fluid (CSF). *Am. J. Clin. Pathol.*, **67**, 210.

[1184] Tovi, D., Nilsson, I. M. and Thulin, C. A. (1972). Fibrinolysis and subarachnoid haemorrhage inhibitory effect of tranexamic acid. *Acta Neurol. Scand.*, **48**, 393–402.

[1185] Tovi, D., Nilsson, I. M. and Thulin, C. A. (1973). Fibrinolytic activity of the cerebrospinal fluid after subarachnoid haemorrhage. *Acta Neurol. Scand.*, **49**, 1–9.

[1186] Traviesa, D. C., Prystowsky, S. D., Nelso, B. J. *et al.* (1978). Cerebrospinal fluid findings in asymptomatic patients with reactive serum fluorescent treponemal antibody absorption tests. *Ann. Neurol.*, **4**, 524–530.

[1187] Triger, D. R., MacCallum, F. O., Kurtz, J. B. and Wright, R. (1972). Raised antibody titres to measles and rubella viruses in chronic active hepatitis. *Lancet*, **i**, 665–667.

[1188] Tripathi, R. C., Millard, C. B., Tripathi, B. J. and Noronha. (2004). Tau fraction of transferrin is present in human aqueous humor and is not unique to cerebrospinal fluid. *Exp. Eye Res.*, **50**, 541–547.

[1189] Trojaborg, W., Bottcher, J. and Saxtrup, O. (1981). Evoked protentials and immunoglobulin abnormalities in multiple sclerosis. *Neurology*, **31**, 866–871.

[1190] Trotter, J. L., Banks, G. and Wang, P. (1977). Isoelectric focusing of gamma globulins in cerebrospinal fluid from patients with multiple sclerosis. *Clin. Chem.*, **23**, 2213–2215.

[1191] Trotter, J. L. and Brooks, B. R. (1980). Pathophysiology of cerebrospinal fluid immunoglobulins. In: J. H. Wood (ed.), *Neurobiology of Cerebrospinal Fluid*, pp. 465–485. Plenum Pub. Corp.

[1192] Trotter, J. L., Wegescheide, C. L. and Garvey, W. F. (1983). Immunoreactive myelin proteolipid fluid and serum of neurological impaired patients. *Ann. Neurol.*, **14**, 554–558.

[1193] Trysberg, E., Nylen, K., Rosengren, L. E. and Tarkowski, A. (2003). Neuronal and astrocytic damage in systemic lupus erythematosus patients with central nervous system involvement. *Arthritis Rheum.*, **48**, 2881–2887.

[1194] Tshabalala, M. A. and Latz, H. W. (1981). Fluorometric identification and microdetermination of proteins labeled with 4-(3'-phenyl-2'-pyrazolin-1'-yl) benzenesulfonyl chloride in polyacrylamide gel matrices. *Anal. Biochem.*, **111**, 343–356.

[1195] Tullberg, M., Mansson, J. E., Fredman, P. *et al.* (2000). CSF sulfatide distinguishes between normal pressure hydrocephalus and subcortical arteriosclerotic encephalopathy. *J. Neurol. Neurosurg. Psychiatry*, **69**, 74–81.

[1196] Tumani, H., Tourtellotte, W. W., Peter, J. B. and Felgenhauer, K., The Optic Neuritis Study Group. (1998). Acute optic neuritis: combined immunological markers and magnetic resonance imaging predict subsequent development of multiple sclerosis. *J. Neurol. Sci.*, **155**, 44–49.

[1197] Tune, L., Gucker, S., Folstein, M. *et al.* (1985). Cerebrospinal fluid acetylcholinesterase activity in senile dementia of the Alzheimer type. *Ann. Neurol.*, **17**, 46–48.

[1198] Tveten, L. (1965). Cerebrospinal-fluid proteins in obstructive lesions of the central nervous system. *Acta Neurol. Scand.*, **41**, 80–91.

[1199] Tyrrell, D. A. J., Crow, T. J., Parry, R. P. and Johnstone, E. (1979). Possible virus in schizophrenia and some neurological disorders. *Lancet*, **i**, 839–841.

[1200] Udeozo, I. O. K., Bezer, A. E., Osunkoya, B. O. *et al.* (1968). Cerebrospinal fluid immunoglobulins in Burkitt lymphoma. *J. Lab. Clin. Med.*, **71**, 912–918.

[1201] Ukkonen, P., Granstrom, M. L., Rasanen, J. *et al.* (1981). Local production of mumps IgG and IgM antibodies in the cerebrospinal fluid of meningitis patients. *J. Med. Virol.*, **8**, 257–265.

[1202] Urade, Y. and Hayaishi, O. (2000). Prostaglandin D synthase: structure and function. *Vitam. Horm.*, **58**, 89–120.

[1203] Vakaet, A. and Thompson, E. J. (1985). Free light chains in the cerebrospinal fluid: an indicator of recent immunological stimulation. *J. Neurol. Neurosurg. Psychiatry*, **48**, 995–998.

[1204] Nilsson, C., Stahlberg, F., Thomsen, C. *et al.* (1992). Circadian variation in human cerebrospinal fluid production measured by magnetic resonance imaging. *Am. J. Physiol.*, **262**, R20–R24.

[1205] Valenzuela, R., Mandler, R. and Goren, H. (1982). Immunonephelometric quantitation of central nervous system IgG daily synthesis in multiple sclerosis. *Am. J. Clin. Pathol.*, **78**, 22–28.

[1206] Valette, I., Pointis, J., Rondeau, Y. *et al.* (1979). Is immunochemical determination of haptoglobin phenotype dependent. *Clin. Chim. Acta*, **99**, 1–6.

[1207] Van Den Bergh, F. A. J. T. and Roos, R. A. C. (1981). Gel isoelectric focusing and IgG index for demonstration of intrathecal IgG synthesis in neurological disorders. *Ann. Clin. Biochem.*, **18**, 153–157.

[1208] Van Der Helm, H. J. and Hische, E. A. (1979). Application of Bayes theorem to results of quantitative clinical chemical determinations. *Clin. Chem.*, **25**, 985–988.

[1209] Van Eijk, H. G., Van Noort, W. L., Dubelaar, M. L. and Van der H. C. (1983). The microheterogeneity of human transferrins in biological fluids. *Clin. Chim. Acta*, **132**, 167–171.

[1210] Van Eijk, H. G., Van Noort, W. L., Kroos, M. J. and Van der H. C. (1982). The heterogeneity of human serum transferrin and human transferrin preparations on isoelectric focusing gels: no functional difference of the fractions in vitro. *Clin. Chim. Acta*, **121**, 209–216.

[1211] Van Eijk, H. G., Van Noort, W. L. and Van der H. C. (1982). Microheterogeneity of human serum transferrins: a consequence for immunochemical determinations? *Clin. Chim. Acta*, **126**, 193–195.

[1212] Van Kammen, D. P., Mann, L. S., Sternberg, D. E. *et al.* (1983). Dopamine-B-hydroxylase activity and homovanillic acid in spinal fluid of schizophrenics with brain atrophy. *Science*, **220**, 974–977.

[1213] Van Kamp, H. J., Mulder, K., Kuiper, M. and Wolters, E. C. (1995). Changed transferrin sialylation in Parkinson's disease. *Clin. Chim. Acta*, **235**, 159–167.

[1214] Van Loon, A. M., Van Der Logt, J. T. M. and Van, D., V. (1981). Diagnosis of herpes encephalitis by ELISA. *Lancet*, **ii**, 1228–1229.

[1215] Van Oss, C. J., Absolom, D. R. and Bronson, P. M. (1982). Affinity diffusion II. Comparison between thermodynamic data obtained by affinity diffusion and precipitation in tubes. *Immunol. Commun.*, **2**, 139–148.

[1216] Van Zanten, A. P., Twijnstra, A., Van, B. V. *et al.* (1985). Cerebrospinal fluid B-glucuronidase activities in patients with central nervous system metastases. *Clin. Chim. Acta*, **147**, 127–134.

[1217] Vandvik, B. (1977). Oligoclonal IgG and free light chains in the cerebrospinal fluid of patients with multiple sclerosis and infectious diseases of the central nervous system. *Scand. J. Immunol.*, **6**, 913–922.

[1218] Vandvik, B. (1977). Oligoclonal measles virus-specific IgG antibodies isolated from sera of patients with subacute sclerosing panencephalitis. *Scand. J. Immunol.*, **6**, 641–649.

[1219] Vandvik, B., Froland, S. S., Hoyeraal, H. M. *et al.* (1973). Immunological features in a case of subacute sclerosing panencephalitis treated with transfer factor. *Scand. J. Immunol.*, **2**, 367–374.

[1220] Vandvik, B., Mellbye, O. J. and Norrby, E. (1977). Local synovial synthesis of oligoclonal measles virus antibodies and of smooth muscle antibodies in a case of atypical rheumatoid arthritis. *Ann. Rheum. Dis.*, **36**, 302–310.

[1221] Vandvik, B., Natvig, J. B. and Norrby, E. (1977). IgG 1 subclass restriction of oligoclonal measles virus-specific IgG antibodies in patients with subacute sclerosing panencephalitis and in a patient with multiple sclerosis. *Scand. J. Immunol.*, **6**, 651–657.

[1222] Vandvik, B., Natvig, J. B. and Wiger, D. (1976). IgG1 subclass restriction of oligoclonal IgG from cerebrospinal fluids and brain extracts in patients with multiple sclerosis and subacute encephalitides. *Scand. J. Immunol.*, **5**, 427–436.

[1223] Vandvik, B., Nilsen, R. E., Vartdal, F. and Norrby, E. (1982). Mumps meningitis: specific and non-specific antibody responses in the central nervous system. *Acta Neurol. Scand.*, **65**, 468–487.

[1224] Vandvik, B. and Nordal, H. (1978). Local synthesis in the central nervous system of diclonal IgM-kappa and homogeneous free kappa light chain proteins in a case of chronic meningoencephalitis. *Eur. Neurol.*, **17**, 23–31.

[1225] Vandvik, B. and Norrby, E. (1973). Oligoclonal IgG antibody response in the central nervous system to different measles virus antigens in subacute sclerosing panencephalitis. *Proc. Natl. Acad. Sci. USA*, **70**, 1060–1063.

[1226] Vandvik, B. and Norrby, E. (1980). Viral antibody responses in the central nervous system of patients with multiple sclerosis. *Prog. MS Res.*, 256–262.

[1227] Vandvik, B., Norrby, E. and Nordal, H. J. (1979). Optic neuritis: local synthesis in the central nervous system of oligoclonal antibodies to measles, mumps, rubella, and herpes simplex viruses. *Acta Neurol. Scand.*, **60**, 204–213.

[1228] Vandvik, B., Norrby, E., Nordal, H. J. and Degre, M. (1976). Oligoclonal measles virus-specific IgG antibodies isolated from cerebrospinal fluids, brain extracts, and sera from patients with subacute sclerosing panencephalitis and multiple sclerosis. *Scand. J. Immunol.*, **5**, 979–992.

[1229] Vandvik, B., Norrby, E., Steen-Johnsen, J. and Stensvold, K. (1978). Mumps meningitis: Prolonged pleocytosis and occurrence of mumps virus-specific oligoclonal 1gG in the cerebrospinal fluid. *Eur. Neurol.*, **13**, 13–22.

[1230] Vandvik, B., Skolldenberg, B., Forsgren, M. et al. (1985). Long-term persistence of intrathecal virus-specific antibody responses after herpes simplex virus encephalitis. *J. Neurol.*, **231**, 307–312.

[1231] Vandvik, B. and Skrede, S. (1973). Electrophoretic examination of cerebrospinal fluid proteins in multiple sclerosis and other neurological diseases. *Eur. Neurol.*, **9**, 224–241.

[1232] Vandvik, B., Vartdal, F. and Norrby, E. (1982). Herpes simplex virus encephalitis: Intrathecal synthesis of oligoclonal virus-specific IgG IgA and IgM antibodies. *J. Neurol.*, **228**, 25–38.

[1233] Vandvik, B., Weil, M. L., Grandien, M. and Norrby, E. (1978). Progressive rubella virus panencephalitis: Synthesis of oligoclonal virus-specific IgG antibodies and homogeneous free light chains in the central nervous system. *Acta Neurol. Scand.*, **57**, 53–64.

[1234] Varitek, V. A. and Day, E. D. (1979). Relative affinity of antisera for myelin basic protein (MBP) and degree of affinity heterogeneity. *Mol. Immunol.*, **16**, 163–172.

[1235] Vartdal, F. and Vandvik, B. (1982). Multiple sclerosis: Electrofocused 'bands' of oligoclonal CSF IgG do not carry antibody activity against measles, varicella-zoster or rotaviruses. *J. Neurol. Sci.*, **54**, 99–107.

[1236] Vartdal, F., Vandvik, B., Michaelsen, T. E. et al. (1982). Neurosyphilis: intrathecal synthesis of oligoclonal antibodies to treponema pallidum. *Ann. Neurol.*, **11**, 35–40.
[1237] Vartdal, F., Vandvik, B. and Norrby, E. (1980). Viral and bacterial antibody responses in multiple sclerosis. *Ann. Neurol.*, **8**, 248–255.
[1238] Vartdal, F., Vandvik, B. and Norrby, E. (1982). Intrathecal synthesis of virus-specific oligoclonal IgG, IgA and IgM antibodies in a case of varicella-zoster meningoencephalitis. *J. Neurol. Sci.*, **57**, 121–132.
[1239] Vecchio, T. J. (1966). Predictive value of a single diagnostic test in unselected populations. *N. Engl. Med. J.*, **274**, 1171–1173.
[1240] Vedeler, C. A., Nyland, H., Fagius, J. et al. (1982). The clinical effect and the effect on serum IgG antibodies to peripheral nerve tissue of plasma exchange in patients with Guillain-Barre syndrome. *J. Neurol.*, **229**, 59–64.
[1241] Verbeek, M. M., de Reus, H. P. and Weykamp, C. W. (2002). Comparison of methods for the detection of oligoclonal IgG bands in cerebrospinal fluid and serum: results of the Dutch Quality Control survey. *Clin. Chem.*, **48**, 1578–1580.
[1242] Verheecke, P. (1974). Agar gel electrophoresis of unconcentrated cerebrospinal fluid: the degenerative type. *Acta Neurol. Belg.*, **74**, 376–382.
[1243] Verheecke, P. (1975). Agar gel electrophoresis of unconcentrated cerebrospinal fluid. *J. Neurol.*, **209**, 59–63.
[1244] Verheecke, P. (1975). On the tau-protein in cerebrospinal fluid. *J. Neurol. Sci.*, **26**, 277–281.
[1245] Verjans, E., Theys, P., Delmotte, P. and Carton, H. (1983). Clinical parameters and intrathecal IgG synthesis as prognosic features in multiple sclerosis. *J. Neurol.*, **229**, 155–165.
[1246] Vesterberg, O. (1980). Quantification of antibodies by reversed zone immunoelectrophoresis. *J. Immunol. Methods*, **37**, 311–314.
[1247] Vesterberg, O. (1980). Quantification of proteins with a new sensitive method - zone immunoelectrophoresis assay. *Hoppe Seylers, Z. Physiol. Chem.*, **361**, 617–624.
[1248] Vesterberg, O. (1981). Quantification of albumin in urine by a new method: zone immuno-electrophoresis assay. *Clin. Chim. Acta*, **113**, 305–310.
[1249] Vesterberg, O. and Breig, U. (1981). Quantitative analysis of multiple molecular forms of transferrin using isoelectric focusing and zone immunoelectrophoresis assay (ZIA). *J. Immunol. Methods*, **46**, 53–62.
[1250] Vesterberg, O. and Hansen, L. (1978). New procedure for concentration and analytical isoelectric focusing of proteins. *Biochim. Biophys. Acta*, **534**, 369–373.
[1251] Vesterberg, O., Hansen, L. and Sjosten, A. (1977). Staining of proteins after isoelectric focusing in gel by new procedures. *Biochim. Biophys. Acta*, **491**, 160–166.
[1252] Vianello, M., Vitaliani, R., Pezzani, R. et al. (2004). The spectrum of antineuronal autoantibodies in a series of neurological patients. *J. Neurol. Sci.*, **220**, 29–36.
[1253] Vincent, A., Buckley, C., Schott, J. M. et al. (2004). Potassium channel antibody-associated encephalopathy: a potentially immunotherapy-responsive form of limbic encephalitis. *Brain*, **127**, 701–712.
[1254] Virella, G. and Fudenberg, H. H. (1977). Comparison of immunoglobulin determinations in pathological sera by radial immunodiffusion and laser nephelometry. *Clin. Chem.*, **23**, 1925–1928.
[1255] Visscher, B. R., Myers, L. W., Ellison, G. W. et al. (1979). HLA types and immunity in multiple sclerosis. *Neurology*, **29**, 1561–1565.

[1256] Vivekanandan, S., Kamalakara, A. P., Selvam, R. and Kanaka, T. S. (1982). Sequential determinations of cerebrospinal fluid lactate dehydrogenase isoenzyme in human brain tumors on treatment. *Acta Neurol. Scand.*, **66**, 347–354.
[1257] Voltz, R. (2004). Autoantibodies in paraneoplastic neurological syndrome. *Lancet Neurology*, **1**, 294–305.
[1258] Wald, A., Hochwald, G. M. and Gandhi, M. (1978). Evidence for the movement of fluid, macromolecules and ions from the brain extracellular space to the CSF. *Brain Res.*, **151**, 283–290.
[1259] Walker, R. W. H., Keir, G., Johnson, M. H. and Thompson, E. J. (1983). A rapid method for detecting oligoclonal IgG in unconcentrated CSF, by agarose isoelectric focusing, transfer to cellulose nitrate and immunoperoxidase staining. *J. Neuroimmunol.*, **4**, 141–148.
[1260] Walker, R. W. H., Keir, G. and Thompson, E. J. (1983). Assessment of cerebrospinal fluid immunoglobulin patterns after isoelectric focusing. *J. Neurol. Sci.*, **58**, 123–134.
[1261] Walker, R. W. H. and Thompson, E. J. (1983). The cerebrospinal fluid in subacute sclerosing panencephalitis and multiple sclerosis. In: P. O. Behan, M. V. ter Mevlen and F. C. Rose (eds.), *Immunology of Nervous System Infections, Progress in Brain Research*, pp. 375–390. Elsevier.
[1262] Walker, R. W. H., Thompson, E. J. and McDonald, W. I. (1985). Cerebrospinal fluid in multiple sclerosis: relationships between immunoglobulins, leucocytes and clinical features. *J. Neurol.*, **232**, 250–259.
[1263] Wallen, W. C., Biggar, R. J., Levine, P. H. and Iivanainen, M. V. (1983). Oligoclonal IgG in CSF of patients with African Burkitt's lymphoma. *Arch. Neurol.*, **40**, 11–13.
[1264] Walsh, M. J., Limos, L. and Tourtellotte, W. W. (1984). Two-dimensional electrophoresis of cerebrospinal fluid and ventricular fluid proteins, identification of enriched and unique proteins, and comparison with serum. *J. Neurochem.*, **43**, 1277–1285.
[1265] Walsh, R. L. and Coles, M. E. (1980). Binding of IgG and other proteins to microfilters. *Clin. Chem.*, **26**, 496–498.
[1266] Wasserstrom, W. R., Schwartz, M. K., Fleisher, M. and Posner, J. B. (1981). Cerebrospinal fluid biochemical markers in central nervous system tumors: A review. *Ann. Clin. Lab. Sci.*, **II**, 239–251.
[1267] Weber, T., Jurgens, S. and Luer, W. (1987). Cerebrospinal fluid immunoglobulins and virus-specific antibodies in disorders affecting the facial nerve. *J. Neurol.*, **234**, 308–314.
[1268] Weil, M. L., Norrby, E., Itabashi, H. H. *et al.* (1979). Immunoglobulin G from subacute sclerosing panencephalitis brain as an immunological reagent. *Infect. Immun.*, **24**, 202–210.
[1269] Weiner, H. L. and Dawson, D. M. (1980). Plasmapheresis in multiple sclerosis: Preliminary study. *Neurology*, **30**, 1029–1033.
[1270] Weisel, J. W., Phillips, G. N. and Cohen, C. (1981). A model from electron microscopy for the molecular structure of fibrinogen and fibrin. *Nature*, **289**, 263.
[1271] Weisner, B. and Bernhardt, W. (1978). Protein fractions of lumbar, cisternal, and ventricular cerebrospinal fluid. *J. Neurol. Sci.*, **37**, 205–214.
[1272] Weisner, B. and Bernhardt, W. (1978). Zusammenhang der immunoglobulinkonzentrationen in liquor und serum. *Fortschr. Med.*, **96**, 1865–1869.

[1273] Weisner, B. and Kauerz, U. (1983). The influence of the choroid plexus on the concentration of prealbumin in CSF. *J. Neurol. Sci.*, **61**, 27–35.
[1274] Weisner, B. and Roethig, H. J. (1983). The concentration of prealbumin in cerebrospinal fluid (CSF), indicator of CSF circulation disorders. *Eur. Neurol.*, **22**, 96–105.
[1275] Welsum, R. A. and Van Der Helm, H. J. (1973). Elevated gamma-globulin concentration of the cerebrospinal fluid in a case of uveomeningoencephalitis. *Eur. Neurol.*, **9**, 315–318.
[1276] Wen, H. L., Lo, C. W. and Ho, W. K. K. (1983). Met-enkephalin level in the cerebrospinal fluid of schizophrenic patients. *Clin. Chim. Acta*, **128**, 367–371.
[1277] Wenzel, D. and Felgenhauer, K. (1976). The development of the blood-CSF barrier after birth. *Neuropadiatrie*, **7**, 175–181.
[1278] Werner, M. and Brooks, S. H., Cohnen, G. (1972). Diagnostic effectiveness of electrophoresis and specific protein assays, evaluated by discriminate analysis. *Clin. Chem.*, **18**, 116–123.
[1279] Whaley, K., Lappin, D. and Barkas, T. (1981). C2 synthesis by human monocytes is modulated by a nicotinic cholinergic receptor. *Nature*, **293**, 580–583.
[1280] Whicher, J. T. (1978). The value of complement assays in clinical chemistry. *Clin. Chem.*, **24**, 7–22.
[1281] Whicher, J. T., Higginson, J., Riches, P. G. and Radford, S. (1980). Clinical applications of immunofixation: detection and quantitation. *J. Clin. Pathol.*, **33**, 781–785.
[1282] Whitacre, C. C., Mattson, D. H., Day, E. D. et al. (1982). Oligoclonal IgG in rabbits with experimental allergic encephalomyelitis. *Neurochem. Res.*, **7**, 1209–1221.
[1283] Whitaker, J. N. (1977). Myelin encephalitogenic protein fragments in cerebrospinal fluid of persons with multiple sclerosis. *Neurology*, **27**, 911–920.
[1284] Whitaker, J. N. (1982). The appearance of a new antigenic determinant during the degradation of myelin basic protein. *J. Neuroimmunol.*, **2**, 201–207.
[1285] Whitaker, J. N., Lisak, R. P., Bashir, R. M. et al. (1980). Immunoreactive myelin basic protein in the cerebrospinal fluid in neurological disorders. *Ann. Neurol.*, **7**, 58–64.
[1286] Whitley, R. (1981). Diagnosis and treatment of herpes simplex encephalitis. *Ann. Rev. Med.*, **32**, 335–340.
[1287] Whittle, H. C. and Greenwood, B. M. (1977). Cerebrospinal fluid immunoglobulins and complement in meningococcal meningitis. *J. Clin. Pathol.*, **30**, 720–722.
[1288] Wiederkehr, F., Ogilvie, A. and Vonderschmitt, D. J. (1985). Two-dimensional gel electrophoresis of cerebrospinal fluid proteins from patients with various neurological diseases. *Clin. Chem.*, **31**, 1537–1542.
[1289] Wierzbicki, A. S., Luxton, R., McLean, B. N. et al. (1993). Hairy leukaemic cell influx into the cerebrospinal fluid secondary to encephalomyelitis. *Postgrad. Med. J.*, **69**, 651–653.
[1290] Wikkelso, C. and Blomstrand, C. (1982). Cerebrospinal fluid 'specific' protein in various degenerative neurological diseases. *Acta Neurol. Scand.*, **66**, 199–208.
[1291] Wikkelso, C., Blomstrand, C. and Ronnback, L. (1980). Separation of cerebrospinal fluid specific proteins - A methodological study. *J. Neurol. Sci.*, **44**, 247–257.
[1292] Wikkelso, C., Blomstrand, C. and Ronnback, L. (1981). Cerebrospinal fluid specific proteins in multiinfarct and senile dementia. *J. Neurol. Sci.*, **49**, 293–303.

[1293] Williams, A., Eldridge, R., McFarland, H. et al. (1980). Multiple sclerosis in twins. *Neurology*, **30**, 1139.
[1294] Williams, A., Papadopoulos, N. and Chase, T. N. (1980). Demonstration of CSF gamma-globulin banding in presenile dementia. *Neurology*, **30**, 882–884.
[1295] Williams, A. C., Mingioli, E. S., McFarland, H. F. et al. (1978). Increased CSF IgM in multiple sclerosis. *Neurology*, **28**, 996–998.
[1296] Willson, V. J. C., Graham, J. G., McQueen, I. N. F. and Thompson, R. J. (1980). Immunoreactive aldolase C in cerebrospinal fluid of patients with neurological disorders. *Ann. Clin. Biochem.*, **17**, 110–113.
[1297] Wilske, B. (2003). Diagnosis of Lyme borreliosis in Europe. *Vector Borne Zoonotic Dis.*, **3**, 215–227.
[1298] Wimperis, J. Z., Brenner, M. K., Prentice, H. G. et al. (1987). B cell development and regulation after T cell-depleted marrow transplantation. *J. Immunol.*, **138**, 2445–2450.
[1299] Winchester, J. S. and Hambling, M. H. (1972). Antibodies to measles, mumps, and herpes simplex virus in cerebrospinal fluid in acute infections and post-infectious diseases of the central nervous system. *J. Med. Microbiol.*, **5**, 137–143.
[1300] Wood, J. H. (1980). Neurochemical analysis of cerebrospinal fluid. *Neurology*, **30**, 645–651.
[1301] Wood, J. H. (1983). Neurobiology of cerebrospinal fluid. 2 ed. Plenum.
[1302] Woyciechowska, J. L., Dambarozia, J., Leinikki, P. et al. (1985). Viral antibodies in twins with multiple sclerosis. *Neurology*, **35**, 1176–1180.
[1303] Wright, G. L. (1971). Separation of proteins in concentrated cerebrospinal fluid by two-dimensional electrophoresis in acrylamide gel. *Clin. Chem.*, **17**, 430–432.
[1304] Wright, G. L., Farrell, K. B. and Roberts, D. B. (1971). Gradient polyacrylamide gel electrophoresis of human serum proteins: improved discontinuous gel electrophoretic technique and identification of individual serum components. *Clin. Chim. Acta*, **32**, 285–296.
[1305] Wylie, D. A., Sherman, L. A. and Klinman, N. R. (1982). Participation of the major histocompatibility complex in antibody recognition of viral antigens expressed on infected cells. *J. Exp. Med.*, **155**, 403–414.
[1306] Xu, X. H., McFarlin, D. E. and Twins with M. S. (1984). Oligoclonal bands in CS. *Neurology*, **34**, 769–774.
[1307] Yahr, M. D., Goldensohn, S. S. and Kabat, E. A. (1954). Further studies of the gamma globulin content of cerebrospinal fluid in multiple sclerosis and other neurological diseases. *Ann. N Y Acad. Sci.* **58**, 613–626.
[1308] Yam, P., Petz, L. D., Tourtellotte, W. W. and Ma, B. I. (1980). Measurement of complement components in cerebral spinal fluid by radioimmunoassay in patients with multiple sclerosis. *Clin. Immunol. Immunopathol.*, **17**, 492–505.
[1309] Yamada, T., Takami, M. S. and Gerner, R. H. (1981). Bombesin-like immunoreactivity in human cerebrospinal fluid. *Brain Res.*, **223**, 214–2.7.
[1310] Yamauchi, K., Tozuka, M., Hidaka, H. et al. (1999). Characterization of apolipoprotein E-containing lipoproteins in cerebrospinal fluid: effect of phenotype on the distribution of apolipoprotein E. *Clin. Chem.*, **45**, 1431–1438.
[1311] Yanagisawa, K., Quarles, R. H., Johnson, D. et al. (1985). A derivative of myelin-associated-glycoprotein in cerebrospinal fluid of normal subjects and patients with neurological disease. *Ann. Neurol.*, **18**, 464–469.

[1312] Yates, C. M., Urquhart, A., Wilson, H. *et al.* (1976). Lysosomal enzymes in cerebral atrophy. *Clin. Chim. Acta*, **71**, 215–219.
[1313] Young, I. R., Hall, A. S., Pallis, C. A. *et al.* (1981). Nuclear magnetic resonance imaging of the brain in multiple sclerosis. *Lancet*, **ii**, 1063–1066.
[1314] Zachrisson, O. C., Balldin, J., Ekman, R. *et al.* (2000). No evident neuronal damage after electroconvulsive therapy. *Psychiatry Res.*, **96**, 157–165.
[1315] Zajicek, J. P., Scolding, N. J., Foster, O. *et al.* (1999). Central nervous system sarcoidosis – diagnosis and management. *Q. J. Med.*, **92**, 103–117.
[1316] Zeman, A., McLean, B. N., Keir, G. *et al.* (1993). The significance of serum oligoclonal bands in neurological diseases. *J. Neurol. Neurosurg. Psychiatry*, **56**, 32–35.
[1317] Zeman, A. Z., Keir, G., Luxton, R. and Thompson, E. J. (1996). Serum oligoclonal IgG is a common and persistent finding in multiple sclerosis, and has a systemic source. *Q. J. Med.*, **89**, 187–193.
[1318] Zeman, A. Z., Kidd, D., McLean, B. N. *et al.* (1996). A study of oligoclonal band negative multiple sclerosis. *J. Neurol. Neurosurg. Psychiatry*, **60**, 27–30.
[1319] Zeman, A. Z., Maurice-Williams, R. S., Luxton, R. *et al.* (1993). Lymphocytic meningitis following insertion of a porcine dermis dural graft. *Surg. Neurol.*, **40**, 75–80.
[1320] Zemlan, F. P., Rosenberg, W. S., Luebbe, P. A. *et al.* (1999). Quantification of axonal damage in traumatic brain injury: affinity purification and characterization of cerebrospinal fluid tau proteins. *J. Neurochem.*, **72**, 741–750.
[1321] Zettervall, O. and Link, H. (1970). Electrophoretic distribution of kappa and lambda immunoglobulin light chain determinants in serum and cerebrospinal fluid in multiple sclerosis. *Clin. Exp. Immunol.*, **7**, 365–372.
[1322] Ziola, B., Reunanen, M. and Salmi, A. (1981). IgM-class rheumatoid factor in serum and cerebrospinal fluid of multiple sclerosis and matched neurological control patients. *J. Neurol. Sci.*, **51**, 101–109.

Index

Abnormal protein patterns, 51
Adrenoleukodystrophy, 145
Affinity, 97, 98, 99
Agar gel electrophoresis, 5
Agar Gel Electrophoresis in Neurology, 3
Agarose, 212
Agarose electrophoresis, 222–3
AIDS, 145
Albumin, 14, 16, 35
   biological scatter, 129
     as indicator of barrier function, 129
   normal value, 128
   percentage transfer, 136
   physiologic functions, 17, 18, 19
   percentage transfer, 15
   Rf value, 224
   size, 22
Albumin quotient, 52
Aldolase, 20
   physiologic functions, 17
   size, 22
Aldolase C4, 16
   physiologic functions, 18
Alpha albumin *see* glial fibrillary acidic protein
Alpha fetoprotein, 19
Alpha lipoprotein, 14
   physiologic functions, 17, 18
Alpha-1-antichymotrypsin, 20
   physiologic functions, 17, 18
Alpha-1-antitrypsin, 14, 16, 19–20, 35, 170
   biological scatter, 129

CSF versus serum forms, 39, 41
   normal ratios, 214
   physiologic functions, 17, 18, 19
   size, 22
Alpha-1-microglobulin, 16
Alpha-2 Hermann Schultz, 16, 35
   physiologic functions, 18
   size, 22
Alpha-2-macroglobulin, 14, 16, 19, 35, 115, 116, 117, 120, 172
   biological scatter, 129
   CSF versus serum forms, 39, 41
   percentage transfer, 136
   physiologic functions, 17, 19
   Rf value, 224
   size, 22
Amniotic fluid, 46
Ampholytes, 212
Anamnestic response, 93
Anatomical classification, 53
Antibody index, 162
Antichymotrypsin, 16
   size, 22
Antigen-specific immunoblotting, 155–7
Antigens, 146–50
   binding, 149
Antithrombin III, 19
Antitrypsin, 19
Apo-A-lipoprotein, 16, 19
Apo-B-lipoprotein, 19
   biological scatter, 129
Apo-C-lipoprotein, 22

Apo-E-lipoprotein, 14, 16
  percent transfer, 38
  size, 22
Apoferritin, 22
Arachnoid villi, 44
Aspergillus, 158
Astroprotein *see* glial fibrillary acidic
    protein
Ataxia telangiectasia, 145

B2 microglobulin, 18
Behçet's disease, 145, 173
Behring antibodies, 25
Bence-Jones proteins, 51, 138
Benign intracranial hypertension, 120
Benign paraproteins, 60
Beta 2 tau, 100
Beta endorphin, 16
Beta lipoprotein, 14, 16, 35
  physiologic functions, 17
  percentage transfer, 15
  Rf value, 224
  size, 22
Beta mercaptoethanol, 210
Beta trace, 14, 16, 34
  percent transfer, 38
  physiologic functions, 18
  pI range, 21
  size, 21, 22
Beta-2 transferrin *see* tau protein
Beta-2-glycoprotein I, 16
Beta-2-microglobulin, 16
  percent transfer, 38
  percentage transfer, 15
  size, 22
Blood-brain barrier, 29, 43, 45, 46
Blood-CSF barrier, 9, 43–63
  damage to, 116
  measures of function, 128
  normal values, 127–38
Blood-urine barrier, 45
Bock, Elizabeth, 4
Borrelia burgdorferi, 158
Brain-specific proteins, 173–6, 202

C-reactive protein, 172
  physiologic functions, 18, 19
  pI range, 30
Calcitonin, 16
Carcinoembryonic antigen, 19
Ceruloplasmin, 16, 20
  physiologic functions, 17, 18
  size, 22
Chemical classification, 55
Chemiluminescence, 86, 204
Choroid *see* prealbumin
Choroidal fluid, 48
Chymotrypsin, 20
  physiologic functions, 19
Circumventricular area, 58
CNS leukemia, 190
Complement, 172–3
  physiologic functions, 19
Complement C′3, 16, 20, 35
  biological scatter, 129
  physiologic functions, 17, 18, 19
  size, 22
Complement C′4, 16, 20
  biological scatter, 129
  physiologic functions, 17, 19
Complement C′9, 16
  percent transfer, 38
  percentage transfer, 37
Complement C′lq, 19
Coomassie blue, 67, 86, 150
Correction factors, 118
Creatine kinase, 20, 180
  physiologic functions, 17, 18
Creatine kinase-BB, 16
CSF antibodies, 161–6
  differential diagnosis of serum
    bands, 163–4
  multiple sclerosis, 164–5
  paraneoplastic serum, 165–6
CSF proteins
  amounts of, 14–15
  charge, 28–31, 39
  classification
    anatomical, 53

chemical, 55
  pathological, 54
functions of, 15–20
half-life, 104
molecular properties, 20–31
normal levels, 9
physiologic functions, 17
rate of transfer, 104
size, 21–8, 39
sources of, 43–63
time to equilibrium, 103
versus serum proteins, 33–41
Curve fitting, 105–25
Cystatin C *see* gamma trace
Cytomegalovirus, 69, 158

D-2 antigen, 16, 178
  percent transfer, 38
Dako antibodies, 25
Delpech and Lichtblau plot, 110, 111
Densitometry, 74
Differential diagnosis, 144–6, 199–200
Diffuse gamma 1 *see* gamma'acidic'
Dilution factor, 162
Discrepancy problem, 87–99
Dissociation-enhanced lanthanide
    fluoro-immuno-assay (DELFIA), 86
Dithiothreitol, 210
Dot-blots, 157–8

Eastern blots, 156
  relative specific antibody, 158–61
ECHO virus, 158
Efficiency, 67
Elastase, 20
Electrophoresis, 84
  agar gel, 5
  agarose, 222–3
  gamma region, 10, 11
  immunoglobulins, 10, 11
  Tiselius apparatus, 3
Emden-Meyerhoff pathway, 4, 20
Encephalitis, 145
  see also herpes encephalitis

Enolase, 4, 16, 20, 177
  neuron-specific, 52
  percent transfer, 38
  physiologic functions, 17, 18
Eosinophil cationic protein, 16, 36
  percent transfer, 38
  percentage transfer, 37
  physiologic functions, 19
Epitopes, 75
Epstein-Barr virus, 158
Equal molar transfer, 111, 114
Essex Wynter, W, 1, 2
Estimation, 214–15
Evans Blue, 97
Experimental allergic encephalomyelitis, 97

Factor X, 19
Felgenhauer, Klaus, 4
  CSF protein size, 23–8
Ferritin, 16, 179–80
  percent transfer, 38
  percentage transfer, 37
  physiologic functions, 18
Fibrinogen, 14, 16, 20
  CSF versus serum forms, 39, 41
  physiologic functions, 17, 18
  pI range, 30
  percentage transfer, 15
  size, 22
Fibrinogen degradation products, 16, 173
  percent transfer, 38
Fibronectin, 18
Fragments, 75
Free immunoglobulin light chains, 173
Froin's syndrome, 23, 61, 116, 117

Gamma 5, normal ratios, 214
Gamma 'acidic', 35, 203
Gamma region, 10, 11
Gamma trace, 14, 16, 19, 34, 203
  percent transfer, 38
  physiologic functions, 18
  pI range, 21, 30
  size, 22

Ganrot and Laurell plot, 110, 111, 131
Gel rod immunoblotting, 216
Glial fibrillary acidic protein, 16, 20, 178
   percent transfer, 38
   physiologic functions, 17, 18
Group components, 16, 19, 35, 169–70
   normal ratios, 214
   physiologic functions, 17
   Rf value, 224
   size, 22
Guillain-Barré syndrome, 49, 62, 116, 117, 121, 164, 172, 197, 202
   prognostic proteins, 198

Haemophilus influenzae, 158
Half-life of CSF proteins, 104
Haptoglobin, 16, 19, 35, 92
   biological scatter, 129
   CSF versus serum forms, 39
   normal value, 128
   physiologic functions, 17, 18, 19
   Rf value, 224
   size, 22
Haptoglobin oligomers, 37
Haptoglobin polymers, 169
Haptoglobins, 56
Harada's disease, 145
Head injury, 198
Hemoglobin
   pI range, 30
   Rf value, 224
   size, 22
Hemopexin, 16, 19, 35
   physiologic functions, 17
   size, 22
Herpes antibody, 78
Herpes encephalitis, 20, 69, 146, 159–60, 190–2
Herpes simplex, 158
Herpes zoster, 69
Horseradish peroxidase, 53, 152
Hot spots, 212
Human lymphocyte antigen protein, 35
Hydrocephalus, 120, 190
Hydrogen bonds, 94

I-CAM protein, 178
IgA, 16
   biological scatter, 129
   normal value, 128
   physiologic functions, 19
   pI range, 21
   size, 21, 22
IgG, 14, 16, 35, 107
   antigen-specific, 158
   biological scatter, 129
   CSF versus serum forms, 39
   local synthesis, 128
   percentage transfer, 136
   physiologic functions, 19
   pI range, 21, 30
   population variance, 130
   qualitative versus quantitative tests, 68
   radiolabeled, 101–2
   size, 21, 22
IgG index, 107, 134
IgG/albumin ratio, 114
IgM, 16
   biological scatter, 129
   CSF versus serum forms, 39
   normal levels, 138
   normal value, 128
   percentage transfer, 136
   physiologic functions, 18, 19
   pI range, 30
   percentage transfer, 15
   size, 22
Immunofixation, 172–3
Immunoglobulins, 137–8
   clonal distribution, 10, 11
   intravenous, 187
   local synthesis, 10, 141–66
      diagnosis, 142–4
      differential diagnosis, 144–6
      qualitative versus quantitative tests, 143–4
      specific antigens, 146–50
   physiologic functions, 19
   see also IgG; IgM

Index, 11
  problems of, 181–2
  see also IgG index
Infection, 164
Inflammation, 164
Inter-cellular adhesion molecules
    (ICAM), 57
Interferon, 187
Interstitial fluid, 49, 60
Intracellular iron binding protein
    see ferritin
Isoelectric focusing, 84, 151–5, 210, 211,
    224–39
  analytical principle, 225–6
  chemicals required, 229–30
  common bands, 152–3
  equipment required, 228–9
  materials required, 229
  pseudo-bands, 153
  purpose of, 226
  reagent preparation, 230
  sample requirements, 226–8
  technique, 230–4
  troubleshooting, 234–8
Isotachophoresis, 85

JC papova virus, 158

Kabat, Elvin, 3
Kappa light chains, 212
  size, 22
Karger, Denise, 3
Kurtzke score, 145, 194
Kveim, 158

Lactoferrin, 16, 36
  percent transfer, 38
  percentage transfer, 37
  physiologic functions, 19
Lambda light chains, 212
  size, 22
Laterre, Christian, 5
Laurell, Carl-Bertil, 5
Laurell rocket technique, 4

Link, Hans, 4
Lipoprotein A
  pI range, 21
  size, 21
Local synthesis, 68
Longston, Maurice, 98
Lowenthal, Armand, 3
Lupus, diagnosis, 145
Lyme disease, 145, 146
Lymphocytes, 57
Lysine, polymerized, 187
Lysozyme, 16, 20
  percent transfer, 38
  percentage transfer, 15
  physiologic functions, 17, 19
  pI range, 30
  size, 22
Lysozyme index, 36

Macrophages, 36
Magnetic resonance imaging, 161
Major oligodendrocyte
    glycoprotein, 179
Malaria, 158
Mancini plate, 87
Measles, 69, 158
Measles antibody, 148
Meninges, 50
Meningitis, 116, 117
  diagnosis, 145
  prognostic proteins, 198
  tuberculous, 188–90
Methodology, 83–104
  discrepancy problem, 87–99
  quantification, 86
  radiolabeled proteins, 99–104
  separation, 84–5
  visualization, 85
Michaelis-Menton equation, 149
Mitoxantrone, 187
Molecular aggregates, 75
Monitoring of therapy, 185–92
Monoclonal, 10, 76
Moore, Blake, 4
MRZ test, 69

Multiple sclerosis, 69, 106, 202
   clinical-pathological correlations, 146
   diagnosis, 145, 164
   magnetic resonance imaging, 161
   monitoring therapy, 185–7
   prognostic proteins, 193–6
   serum oligoclonal bands, 164–5
Mumps, 158
Mycobacterium tuberculosis, 158
Myelin associated glycoprotein, 179
Myelin basic protein, 16, 20, 176–7
   percent transfer, 38
   physiologic functions, 17, 18

N-CAM protein *see* D-2 antigen
NANA, 52, 92
Neopterin, 181
Neurofilament protein, 180–1
Neuron-specific enolase, 52
Neurosyphilis
   diagnosis, 145
   monitoring therapy, 187–8
Nisseria meningitidis, 158
Non-immunoglobulin proteins, 167–83
Normal ratios, 214
Normal values, 127–38
   biologic variability, 128–37
Northern blots, 156

Ohman formula, 72
Oligoclonal, 10
Oligodendrocytes, 176
Optic neuritis, 196
Optical density, 162
Orosomucoid, 16, 35, 170
   biological scatter, 129
   normal ratios, 214
   normal value, 128
   physiologic functions, 18, 19
   Rf value, 224
   size, 22

Paraneoplastic antibodies, 40, 165–6
Paraproteins, 89, 170, 213
   isoelectric focusing, 152

Pathological classification, 54
Pauling, Linus, 93
Peptides, 16
Percent transfer, 37, 38
Percentage of total IgG, 9
Percentage transfer, 9, 10, 15, 105, 107, 136
   see also individual proteins
Pharmalytes, 154, 212, 225
Plasmin, 20
   physiologic functions, 19
Plasminogen, 16, 20
   physiologic functions, 17
   pI range, 30
   size, 22
Point of care testing, 205–6
Polyacrylamide disc electrophoresis, 215–22
   apparatus, 217–18
   gel preparation, 218–20
   PAGE immunoblotting, 221–2
   reagents, 217
   sample preparation, 221
   solutions, 218
Polyacrylamide gel electrophoresis, 213–14, 215
   abnormal protein patterns, 51
   differential diagnosis, 199–200
   immunoblotting, 221–2
   normal ratios, 214
Polyclonal, 10, 76
Polymerase chain reaction, 150
Polymorphs, 36, 57
Post-lumbar Puncture Headaches, 3
Prealbumin, 14, 16, 19, 35, 36, 47, 170–1, 203
   CSF versus serum forms, 39
   elevated, 51
   normal ratios, 214
   percent transfer, 38
   percentage transfer, 15
   physiologic functions, 17, 18
   Rf value, 224
   size, 22
Proenzymes, 20
Prognostic protein levels, 193–9
Proportional composition, 107
Prostaglandin synthase *see* beta trace

INDEX

14-3-3 Protein, 181
Protein no. 23, 35
Proteolipid protein, 179
Proteomics, 203–4

Qualitative analysis, 65–79
Quantification, 86
Quantitative analysis, 65–79
Queckensted maneuver, 122
Quotient, 10
    problems of, 181–2
Quotient data, 27–30
Quotient Limit, 70

Radial immunodiffusion, 88, 90
Radiolabeled proteins, 99–104
Range of variability, 131
Rate of transfer, 104
Ratio, 10
Readily visualized proteins, 168–72
Reiber formula, 72, 112, 120, 131
Relative specific antibody, 11, 147
    Eastern blots, 158–61
Retinol-binding protein, 19
    CSF versus serum forms, 39, 41
    physiologic functions, 18
    size, 22
Rett's syndrome, 169
Rf values, 224
Rubella, 69, 158

S-100 protein, 16, 178–9
    percent transfer, 38
    physiologic functions, 18
Sandberg-Wollheim, Magnhild, 4
Sarcoid, diagnosis, 145
Schizophrenia, 61
Schuller formula, 111, 113
SDS gel electrophoresis, 211
Selectivity, 11
Sensitivity, 67
Separation, 84–5, 150–61, 209–14
Serum amyloid protein, 18, 19
Serum hypergammaglobulinemia, 124

Serum leak, 119
Serum proteins, 33–41
Seward antibodies, 25
Sialic acid, 29
Silver stain, 73, 74
Sips plot, 149
Sodium dodecyl sulfate, 209
Solubility, 96
Southern blots, 156
Specific activity, 99, 103
Specificity, 67
Spinal roots, 49
SSPE, 145, 148–9
Steroids, 187
Stiff man syndrome, 166
Stokes-Einstein radius, 51, 55
Streptococcus pneumoniae, 158
Subacute sclerosing panencephalitis
    see SSPE
Sydenham's chorea, 4, 166

Tau protein, 34, 35, 169, 180
    CSF versus serum forms, 39
    elevated, 51
    normal ratios, 214
    percent transfer, 38
    physiologic functions, 18
    Rf value, 224
Three-dimensional configuration, 75
Thrombin, 19, 20
    physiologic functions, 17
Time to equilibrium, 103
Tiselius electrophoresis apparatus, 3
Total protein, 106, 223–4
Tourtellotte formula, 111, 112, 114, 134, 135
Tourtellotte, Wallace, 3
Toxoplasma, 69, 158
Transferrin, 14, 16, 19, 35, 51, 100
    biological scatter, 129
    cellular cycle, 52
    normal value, 128
    physiologic functions, 17, 18
    Rf value, 224
    size, 22

Transferrin index, 92
Transthyretin *see* prealbumin
Treponema pallidum, 158
Trypsin, 19
   physiologic functions, 19
Tuberculous meningitis, 188–90
Tumor, 145, 164

Vandvik, Bodvar, 4
Varicella zoster, 69, 158
Vasopressin, 16

Virchow-Robin spaces, 56
Visualization, 85, 150–61

Western blots, 155, 156

Xanthine oxidase, 20

Zinc alpha-2-glycoprotein, 16
Zymogens, 20